# Calculations for A-level
# Chemistry

E.N.Ramsden

**FOURTH EDITION**

First published in 1982 by Stanley Thornes (Publishers) Ltd
Second edition 1987
Third edition 1993

Fourth edition published in 2001 by:

Nelson Thornes
Delta Place
27 Bath Road
Cheltenham
GL53 7TH

    02  03  04  05 / 10  9  8  7  6  5  4  3  2

A catalogue record of this book is available from the British Library.

ISBN 0 7487 5839 9

Typeset by Mathematical Composition Setters Ltd, Salisbury, Wiltshire
Printed and bound in Great Britain by Redwood Books, Trowbridge, Wiltshire

# Contents

# List of Exercises

## Abbreviations of Awarding Bodies

OCR = Oxford, Cambridge and RSA Examinations

L = London Examinations, Edexcel Foundation

L(N) = London Examinations (Nuffield specification), Edexcel Foundation

NEAB = Northern Examinations and Assessment Board, Assessment and Qualifications Alliance

NI = Northern Ireland Council for the Curriculum, Examinations and Assessment

WJEC = Welsh Joint Education Committee

# Preface

Many topics in Chemistry involve numerical problems. Textbooks are not long enough to include sufficient examples and problems to give students the practice which they need to become thoroughly confident in tackling calculations. This book aims to fill that need.

Chapter 1 is a revision of mathematical techniques, with special reference to the use of the calculator, and some hints on how to tackle chemical calculations. In later chapters of the book, reminders are given to direct students back to Chapter 1 for the relevant mathematics which they need for a particular type of problem.

With each topic a theoretical background is given, leading to worked examples followed by a large number of problems and a selection of questions from past AS- and A-level examination papers. The theoretical sections do not aim to give the full treatment which text books give. They are intended to make it easier for a student to use the book for individual study as well as class work. The inclusion of answers is also an aid to private study.

Students coming to an AS-level or A-level course from GCSE Science Double-Award have done some quantitative work and will find parts of the earlier chapters familiar. I have, however, made a fresh start in this book. I have started from the beginning in case any of the GCSE work has been forgotten. A few of the topics covered are not in the A-level specifications of all the Awarding Bodies these are marked with an asterisk. Students should be sufficiently familiar with the specification they are following to omit topics outside their course if they wish.

In the fourth edition of the book I have dropped a number of topics which no longer feature in GCE specifications. I have given more help with some problems. In some problems I point out the *Steps to take* to arrive at the answer, in others I give a simple *Hint*; in others there is no help. The selection of questions from past examination papers has been updated.

E N Ramsden

Oxford 2000

# Acknowledgements

In the first edition of this book I was fortunate to receive excellent advice from Professor R R Baldwin, Professor R P Bell, FRS, Mr G H Davies, Professor W C E Higginson, Dr K A Holbrook, Dr R B Moyes and Dr J R Shorter. I thank these chemists for their help. I thank Mr C Parker for his advice on the Basic Mathematics chapter of this edition. Mr J P D Taylor checked the answers to the problems in the first edition.

Many numerical values are taken from the *Chemistry Data Book* by J G Stark and H G Wallace (published by John Murray). For help with definitions and physical constants, reference has been made to *Physico-chemical Quantities and Units* by M L McGlashan (published by The Royal Society of Chemistry).

I thank the following Awarding Bodies for permission to print questions from recent papers.

The Assessment and Qualifications Alliance, AQA, for the Northern Examinations and Assessment Board (NEAB)
The Edexcel Foundation, EDEXCEL, for London Examinations (L and L(N))
Northern Ireland Council for the Curriculum Examinations and Assessment (NI)
Oxford, Cambridge and RSA Examinations, OCR for the University of Cambridge Local Examinations Syndicate (OCR)
Welsh Joint Education Committee (WJEC)

The Awarding Bodies take no responsibility for the answers to the problems.

I thank Stanley Thornes (Publishers) for the care which they have given to this new edition, Tony Wayte, Science Commissioning Editor, for his interest and encouragement, John Hepburn for the meticulous thought and care which he has given to the book and Janet Oswell for her careful editing. I thank my family for their encouragement.

E N Ramsden
Oxford 2000

# Basic mathematics

To succeed in solving numerical problems you need two things. The first is an understanding of the chemistry involved. The second is the ability to tackle simple mathematics.

First the chemistry: A numerical problem gives you some data and asks you to obtain some other numerical values. The connection between the data you are given and the information you are asked for is a chemical relationship. You will need to know your chemistry to spot what that relationship is.

Secondly the mathematics: This chapter is a reminder of some of the mathematics which you studied earlier in your school career. A few problems are included to help you to brush up your mathematical skills. As you tackle the calculations in the book, you will see *reminders* that direct you back to this chapter for the mathematical techniques which you need for a particular type of problem.

## 1.1 USING EQUATIONS

Scientists measure quantities such as pressure, volume and temperature. Sometimes they find that when one quantity changes another changes as a result. The related quantities are described as **variables**. The relationship between the variables can be written in the form of a mathematical equation.

### Proportionality

For example the mass of a sample of a substance is proportional to its volume:

$$\text{Mass} \propto \text{Volume} \quad (\propto \text{ means 'is proportional to'})$$

This means that as the volume increases the mass increases in proportion. The relationship can be written as

$$\text{Mass} = k \times \text{Volume, where } k \text{ is a constant}$$

The constant $k$ is the density of the substance:

$$\text{Mass} = \text{Density} \times \text{Volume}$$

### Inverse proportionality

The volume of a fixed mass of gas is inversely proportional to its pressure, that is, as the pressure increases the volume decreases in proportion:

$$\text{Volume} \propto 1/\text{Pressure}$$

$$\text{Volume} = k/\text{Pressure} \text{ (where } k \text{ is a constant)}$$

### Rearranging equations

The aim of rearranging equations is to put the quantity you want

- by itself on one side of the equation
- positive.

**EXAMPLE 1**　You might want to use the equation

$$\text{Mass} = \text{Density} \times \text{Volume}$$

to find the density of an object from its mass and volume. Then you rearrange the equation to put density by itself on one side of the equation, that is, in the form

$$\text{Density} = ?$$

To obtain density by itself you must divide the right-hand side of the equation by volume. Naturally you must do the same to the left-hand side.

$$\frac{\text{Mass}}{\text{Volume}} = \frac{\text{Density} \times \text{Volume}}{\text{Volume}}$$

$$\text{Density} = \text{Mass}/\text{Volume}$$

**To sum up:** If you add to or subtract from or multiply or divide one side of the equation, you must do the same to the other side.

### A triangle for rearranging equations

A short cut for rearranging equations is to put the quantities into a triangle. For the equation

$$X = Y \times Z$$

the triangle is

Cover up the letter you want; then what you see is the equation you need to use. Cover up Y; then you see $X/Z$ so you know that $Y = X/ Z$. Cover up Z; then you read $X/Y$ so you know that $Z = X/Y$. Draw a triangle for Mass = Density × Volume to try out the method.

### Cross-multiplying

Once you have understood the idea behind rearranging equations you can try the method of cross-multiplying.

**EXAMPLE 2**

For one mole of a gas the relationship between pressure, temperature and volume is

$$\frac{P}{T} = \frac{R}{V}$$

If you cross-multiply:

$$\frac{P}{T} \diagdown \diagup \frac{R}{V}$$

you obtain the equation

$$PV = RT$$

Cross-multiplying is putting into practice the method of multiplying or dividing both sides of the equation ($P/T = R/V$) by the same quantity. Check it:

- first multiply both sides of the equation by $T \Rightarrow P = RT/V$
- then multiply both sides of the equation by $V \Rightarrow PV = RT$

which is the equation you obtained by cross-multiplying.

**EXAMPLE 3**

A zinc electrode and a copper electrode can be connected to form a chemical cell. The voltage of the cell under standard conditions is $E^{\ominus}$. It is given by

$$E^{\ominus}_{cell} = E^{\ominus}_{Zn} - E^{\ominus}_{Cu}$$

If $E^{\ominus}_{cell} = -1.10$ V and $E^{\ominus}_{Zn} = -0.76$ V, what is $E^{\ominus}_{Cu}$?

You need to get $E^{\ominus}_{Cu}$ by itself and positive. You must do the same to both sides of the equation. First, add $E^{\ominus}_{Cu}$ to both sides of the equation.

$$E^{\ominus}_{Cu} + E^{\ominus}_{cell} = E^{\ominus}_{Zn}$$

Now subtract $E^{\ominus}_{cell}$ from both sides:

$$E^{\ominus}_{Cu} = E^{\ominus}_{Zn} - E^{\ominus}_{cell}$$
$$= -0.76 \text{ V} - (-1.10 \text{ V}) = -0.76 \text{ V} + 1.10 \text{ V} = +0.34 \text{ V}$$

### Change the side; change the sign

You will have noticed that when a quantity changes from one side of the equation to the other side, it changes sign, from (+) to (−) or from (−) to (+).

**EXERCISE 1.1**  **Equations**

**1**  Rewrite each of the following equations:
   **a**  $a = b + c$ in the form $c = ?$   **b**  $x - y = z$ in the form   **(i)** $x = ?$   **(ii)** $y = ?$

   **c**  $p - q = r + s$ in the form   **(i)** $p = ?$   **(ii)** $q = ?$

**2**  Rewrite each of the following equations:
   **a**  Speed = Distance/Time in the form   **(i)** distance = ?   **(ii)** time = ?

   **b**  $K_c = \dfrac{[PCl_5]}{[PCl_3][Cl_2]}$ in the form   **(i)** $[PCl_5] = ?$   **(ii)** $[Cl_2] = ?$

**c** $K_c = \dfrac{[HI]^2}{[H_2][I_2]}$ in the form **(i)** $[H_2]$ = ? **(ii)** $[HI]$ = ?

**3** The pressure, volume and temperature of one mole of a gas are related to the gas constant $R$ by the equation

$$\frac{P}{T} = \frac{R}{V}$$

Rearrange the equation to obtain equations for **(i)** $T$ and **(ii)** $V$.

**4** The concentration of a solution can be expressed

$$\text{Concentration} = \frac{\text{Amount of solute}}{\text{Volume of solution}}$$

**a** Rearrange the equation into the form **(i)** Amount of solute = ? and **(ii)** Volume of solution = ?

**b** Draw a triangle for rearranging this equation.

**5** The expression for the dissociation constant of a weak acid is

$$K_a = \frac{[H^+]^2}{[HA]}$$

Rearrange this expression to give an equation for $[H^+]$.

**6** Draw a triangle to represent the equation

$$\text{Mass} = \text{Amount} \times \text{Molar mass}$$

Give equations for **a** Amount = ? and **b** Molar mass = ?

## 1.2 RATIO CALCULATIONS

Many of the calculations you meet involve ratios. You met this type of problem in your maths course. Do not forget how to solve them when you meet them in chemistry!

### Multiplier method

**EXAMPLE 1**

A cook makes a pie for 6 people. She uses 300 g flour, 150 g margarine and 450 g of fruit. What quantities should she use to make a pie for 8 people?

**METHOD**

You know that the quantities needed are proportional to the number of people. The method is to multiply by a ratio called the **multiplier** (or **scale factor**).

$$\text{Multiplier} = \frac{\text{Size you want}}{\text{Size you have}}$$

Size cook wants = 8 people. Size she has = 6 people; therefore multiplier = $\frac{8}{6} = \frac{4}{3}$.

Flour needed = 300 g $\times \frac{4}{3}$ = 400 g

Margarine needed = 150 g $\times \frac{4}{3}$ = 200 g

Fruit needed = 450 g $\times \frac{4}{3}$ = 600 g

**ANSWER** Flour 400 g, margarine 200 g, fruit 600 g.

**EXAMPLE 2** If 24 g of magnesium give 40 g of magnesium oxide, what mass of magnesium is needed to give 60 g of magnesium oxide?

**METHOD**
$$\text{Multiplier} = \frac{\text{Mass of MgO you want}}{\text{Mass of MgO you have}} = \frac{60 \text{ g}}{40 \text{ g}} = \frac{3}{2}$$

Multiply 24 g magnesium by $\frac{3}{2}$ to get 36 g magnesium.

**ANSWER** The mass of magnesium needed = 36 g.

### Unitary method

There is an alternative to the multiplier method. It is a bit longer but completely safe. In the unitary method we put in an extra step.

**EXAMPLE 1 (ABOVE)** This time using the unitary method

**METHOD** A pie for 6 people needs 300 g flour + 150 g margarine + 450 g fruit.

A pie for 1 person needs $\frac{1}{6} \times$ 300 g flour + $\frac{1}{6} \times$ 150 g margarine + $\frac{1}{6} \times$ 450 g fruit

$$= 50 \text{ g flour} + 25 \text{ g margarine} + 75 \text{ g fruit}$$

A pie for 8 people needs 8 $\times$ 50 g flour + 8 $\times$ 25 g margarine + 8 $\times$ 75 g fruit

$$= 400 \text{ g flour} + 200 \text{ g margarine} + 600 \text{ g fruit}$$

**ANSWER** Quantities needed are flour 400 g, margarine 200 g, fruit 600 g.

The answer is the same as when we used the multiplier method – naturally!

**EXAMPLE 2 (ABOVE)** This time using the unitary method

**METHOD** 40 g of magnesium oxide are formed by 24 g of magnesium

therefore 1 g of magnesium oxide is formed by $\frac{24}{40}$ g of magnesium

and 60 g of magnesium oxide are formed by 60 $\times \frac{24}{40}$ g of magnesium = 36 g

**ANSWER** Mass of magnesium = 36 g (as before).

## EXERCISE 1.2 Ratios

**1** The mass of copper deposited by electroplating is proportional to the time for which a constant current is passed. If the mass of copper deposited in 12.5 minutes is 27.5 mg, what mass of copper is deposited in 75 minutes?

**2** Zinc reacts with dilute acids to give hydrogen. If 0.0400 g of hydrogen is formed when 1.30 g of zinc reacts with an excess of acid, what mass of zinc is needed to produce 6.00 g of hydrogen?

**3** If 0.030 g of a gas has a volume of 150 cm$^3$ what is the volume of 45 g of the gas (at the same temperature and pressure)?

**4** 88 g of iron(II) sulphide is the maximum quantity that can be obtained from the reaction of an excess of sulphur with 56 g of iron. What is the maximum quantity of iron(II) sulphide that can be obtained from 7.00 g of iron?

**5** A lime kiln obtains 56 tonnes of calcium oxide from 100 tonnes of limestone. What mass of limestone must be decomposed to yield 280 tonnes of calcium oxide?

## 1.3 PERCENTAGES

'Per cent' means 'per hundred'. 15% means 15 per hundred or $\frac{15}{100}$ or 0.15. The symbol for percent is %.

To convert a fraction into a percentage, multiply by 100. For example,

$$\frac{1}{2} = \frac{1}{2} \times 100\% = 50\%$$

$$\frac{3}{4} = \frac{3}{4} \times 100\% = 75\%$$

$$\frac{14}{254} = \frac{14}{254} \times 100\% = 5.5\%$$

**EXAMPLE 1**

A jar contains 250 g of jam. Of this, 75 g are fruit, 150 g are sugar and the rest is water. Find the percentages of **a** fruit **b** sugar **c** water.

**METHOD**

Mass of fruit = 75 g and total mass = 250 g

**a**  % fruit = $\dfrac{\text{Mass of fruit}}{\text{Total mass}} \times 100\% = \dfrac{75 \text{ g}}{250 \text{ g}} \times 100\% = 30\%$

**b**  % sugar = $\dfrac{\text{Mass of sugar}}{\text{Total mass}} \times 100\% = \dfrac{150 \text{ g}}{250 \text{ g}} \times 100\% = 60\%$

**c**  % water = $\dfrac{\text{Mass of water}}{\text{Total mass}} \times 100\% = \dfrac{25 \text{ g}}{250 \text{ g}} \times 100\% = 10\%$

Check: Do the percentages add up to 100%? The sum 30% + 60% + 10% = 100% ; that's reassuring!

**EXAMPLE 2**

A compound consists of phosphorus P, oxygen O and chlorine Cl. In a sample of the compound of mass 1.536 g the masses of the elements are P: 0.310 g, O: 0.160 g, Cl: 1.066 g. What percentage of each element is present?

**METHOD**

Percentage of P = $\dfrac{\text{Mass of P}}{\text{Total mass}} \times 100\% = \dfrac{0.310 \text{ g}}{1.536 \text{ g}} \times 100\% = 20.2\%$

Percentage of O = $\dfrac{\text{Mass of O}}{\text{Total mass}} \times 100\% = \dfrac{0.160 \text{ g}}{1.536 \text{ g}} \times 100\% = 10.4\%$

$$\text{Percentage of Cl} = \frac{\text{Mass of Cl}}{\text{Total mass}} \times 100\% = \frac{1.066\text{ g}}{1.536\text{ g}} \times 100\% = 69.4\%$$

**ANSWER**    Percentages are P 20.2 %, O 10.4 %, Cl 69.4%.

Check: Do the percentages add up to 100%?

**EXAMPLE 3**    A chemist prepared a compound **Y**. From the equation for the reaction she calculated that she should obtain 79 g of **Y**. In fact the yield was only 61 g. What percentage yield did she obtain?

**METHOD**    Mass obtained = 61 g out of a possible total of 79 g

$$\text{Percentage yield} = \frac{61\text{ g}}{79\text{ g}} \times 100\% = 77\%$$

**ANSWER**    Percentage yield = 77%.

**EXAMPLE 4**    Calcium makes up 40.0% by mass of pure marble. What is the mass of calcium in a 12.75 kg slab of pure marble?

**METHOD**    Since 40 parts by mass out of one hundred parts by mass of marble are calcium,

$$\text{Mass of calcium} = \frac{40.0}{100} \times 12.75\text{ kg} = 5.10\text{ kg}$$

**ANSWER**    Mass of calcium = 5.10 kg.

**EXERCISE 1.3    Percentages**

1    Convert into percentages (to 2 significant figures): $\frac{2.0}{5.0}, \frac{2.0}{3.0}, \frac{56}{84}, \frac{24}{120}$

2    At a line-dance, 75% of the dancers are women. What percentage are men?
What fraction of the group are men?

3    17.5% of the mass of a crystalline compound is water of crystallisation. What mass of water is present in 15.6 g of crystals?

4    A lump of impure titanium(IV) oxide contains 52.65 kg of titanium. From this sample, 49.75 kg of titanium are extracted. What percentage yield is this?

5    When 250 g of a crystalline compound were heated to drive off water of crystallisation, 160 g of solid remained. Calculate the percentage of water of crystallisation in the crystals.

## 1.4 NEGATIVE NUMBERS

You will often meet physical quantities with negative values. Electrode potentials can be positive or negative. Enthalpy changes can be positive (in endothermic reactions ) or negative (in exothermic reactions).

Remember how to deal with negative numbers:

- adding a negative number, e.g. $7 + (-6) = 7 - 6 = 1$
- subtracting a negative number, e.g. $7 - (-6) = 7 + 6 = 13$
- multiplying by a negative number, e.g. $3 \times (-6) = -18$
- dividing by a negative number, e.g. $18 \div (-3) = -6$

**EXAMPLE**

In the reaction shown below the energy content of each species is shown beneath it in kJ mol$^{-1}$.

$$\begin{array}{ccccccc} \mathbf{A} & + & \mathbf{B} & \longrightarrow & \mathbf{C} & + & \mathbf{D} \\ (-100) & & (+50) & & (-200) & & (+60) \end{array}$$

**a** What is the total energy of the products?
**b** What is the total energy of the reactants?
**c** What is the energy change: (energy of products) – (energy of reactants)?

**METHOD**

**a** The total energy of $\mathbf{C} + \mathbf{D} = (-200) + (+60) = -140$ kJ mol$^{-1}$

**b** The total energy of $\mathbf{A} + \mathbf{B} = (-100) + (+50) = -50$ kJ mol$^{-1}$

**c** The difference $(-140) - (-50) = -140 + 50 = -90$ kJ mol$^{-1}$

## EXERCISE 1.4    Negative numbers

**1** A process consists of four steps. The heat changes in the four steps are +29.5 kJ mol$^{-1}$, –41.6 kJ mol$^{-1}$, –61.0 kJ mol$^{-1}$, +42.5 kJ mol$^{-1}$. Calculate the heat change for the complete process.

**2** The voltage of a cell is given by (if RHS = right-hand side)

$$E_{cell} = E_{RHS\ electrode} - E_{LHS\ electrode}$$

For the following combinations of electrode, state $E_{cell}$.

|   | LHS electrode | | RHS electrode | |
|---|---|---|---|---|
| **a** | Nickel | –0.25 V | Zinc | –0.76 V |
| **b** | Iron(II) | –0.44 V | Iron(III) | +0.77 V |
| **c** | Zinc | –0.76 V | Nickel | –0.25 V |
| **d** | Lead | –0.13 V | Silver | +0.80 V |
| **e** | Tin | –0.14 V | Zinc | –0.76 V |

## 1.5 STANDARD FORM

For working with large numbers and small numbers, it is convenient to write them in the form known as **scientific notation** or **standard form** or **standard index form**. Instead of writing two million as 2 000 000, we can write:

$$2\ 000\ 000 = 2 \times 10 \times 10 \times 10 \times 10 \times 10 \times 10 = 2 \times 10^6$$

In standard form, $2\,000\,000 = 2 \times 10^6$

Note

$$1 \times 10^2 = 100 = 1 \text{ hundred}$$
$$1 \times 10^3 = 1000 = 1 \text{ thousand}$$
$$1 \times 10^6 = 1\,000\,000 = 1 \text{ million}$$

Numbers less than one can also be written in standard form:

$$0.0002 = \frac{2}{10\,000} = \frac{2}{10^4} = 2 \times 10^{-4}$$

In standard form, $0.0002 = 2 \times 10^{-4}$

Note

$$1 \times 10^{-1} = 0.1 = \text{one tenth}$$

$$1 \times 10^{-2} = 0.01 = \text{one hundredth}$$

$$1 \times 10^{-3} = 0.001 = \text{one thousandth}$$

$$1 \times 10^{-6} = 0.000\,001 = \text{one millionth}$$

You can see that in standard form, the number is written as a product of two factors. In the first factor the decimal point comes after the first digit. The second factor is a multiple of ten. For example,

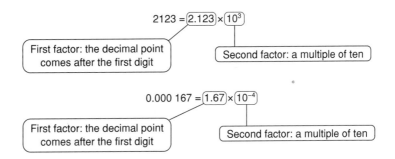

The number 3 or −4 is called the **power** or the **index** or the **exponent**, and the number 10 is the **base**. $10^3$ is referred to as '10 to the power 3'.

If the power (exponent) is

- increased by 1, the decimal point must be moved one place to the left,

- decreased by 1, the decimal point must be moved one place to the right:

$$2.5 \times 10^3 = 0.25 \times 10^4 = 25 \times 10^2 = 250 \times 10^1 = 2500 \times 10^0.$$

Since $10^0 = 1$, this factor is omitted and $2500 \times 10^0$ is written as $2500$.

### To change a number into standard form

- put a decimal point after the first digit

- count how many places the decimal point has moved. This number gives the power of ten.

- If the decimal place has moved to the left the power is positive; if the decimal place has moved to the right the power is negative.

**EXAMPLE 1**    $1234 = 1.234 \times 10^3$

The decimal point has moved 3 places to the left so the new number is multiplied by $10^3$.

**EXAMPLE 2**    $0.0037 = 3.7 \times 10^{-3}$

The decimal point has moved 3 places to the right so the new number is multiplied by $10^{-3}$.

### How to enter powers (indices) in a calculator

- To enter $1.44 \times 10^6$, you enter 1.44; then press the EXP key, then the 6 key. The display reads 1.44 06 or 1.44 E 06.

- To enter $4.50 \times 10^{-2}$, you enter 4.5; then press the EXP key, then the 2 key and lastly the +/– key. The display will read 4.50 –02 or 4.50 E –02

- To enter $10^{-3}$, you enter 1; then press the EXP key, then the 3 key and finally the +/–key. The display will show 1. –03 or 1. E –03

When you are writing down an answer you have read from your calculator, write it as e.g. $1.44 \times 10^6$ and not as 1.44 06.

In multiplication, powers are added, e.g.

$$(1.40 \times 10^6) \times (1.20 \times 10^{-3}) = (1.40 \times 1.20) \times (10^6 \times 10^{-3}) = 1.68 \times 10^3$$

In division, powers are subtracted, e.g.

$$(2.8 \times 10^6) \div (1.4 \times 10^3) = (2.8 \div 1.4) \times (10^6 \div 10^3) = 2.0 \times 10^3$$

**EXERCISE 1.5**    **Standard form**

**1**    Express the following numbers in standard form:

**a** 600 000    **b** 4000    **c** 45 000    **d** 0.0123    **e** 0.006    **f** 0.000 49

## 1.6 DECIMAL PLACES

Sometimes you are asked to give an answer to a certain number of decimal places. For example you are asked to quote your answer to four decimal places, and your calculator reads 0.0867318. You count four digits from the decimal point to obtain 0.0867.

However, if the display reads 0.0867618, the answer to four decimal places is 0.0868 because it has been rounded up. The first digit dropped after the 7 is a 6 so the fourth digit is rounded up from 7 to 8. If the first digit dropped is 5 or greater than 5, the last digit is rounded up. Note that 4.602 to 2 decimal places is 4.60.

## 1.7 SIGNIFICANT FIGURES

Often your calculator will display an answer containing more digits than the numbers you fed into it. Suppose you are calculating the concentration of a sodium hydroxide solution. You have found that 18.6 cm$^3$ of aqueous sodium hydroxide neutralise 25.0 cm$^3$ of hydrochloric acid of concentration 0.100 mol dm$^{-3}$. On putting the numbers into your calculator you obtain a value of 0.134 408 6 mol dm$^{-3}$. If you take this as your answer, you are claiming an accuracy of one part in a million! Since you read the burette to three figures, you quote your answer to three figures, 0.134 mol dm$^{-3}$. The figures you are sure of are called the **significant figures**. The insignificant figures are dropped. This operation is called **rounding off**. If the answer had been 0.134 708 6, it would have been rounded off to 0.135. When the first of the figures to be dropped is 5 or greater, the last of the significant figures is rounded up to the next digit.

The number 123 has 3 significant figures. The number $1.23 \times 10^4$ has 3 significant figures, but 12 300 has 5 significant figures because the final zeros mean that each of these digits is known to be zero and not some other digit.

**TO SUM UP**

The number of significant figures is the number you are sure of.

Rounding off is dropping the insignificant figures.

If the first figure to be dropped is 5 or greater, round up the last significant figure.

**EXERCISE 1.6**  **Decimal places and significant figures**

1 State the number of significant figures in:
   **a** $1.0 \times 10^{-2}$  **b** $1.00 \times 10^{-2}$  **c** 0.135  **d** 1.2345
   **e** 2500  **f** 25  **g** $2.5 \times 10^{-3}$  **h** $2.500 \times 10^{-3}$

2 Write the following quantities to three significant figures:
   **a** a volume of 15.814 cm$^3$  **f** a concentration of 0.14535 mol dm$^{-3}$

   **b** a mass of 205.5 tonnes  **g** an enthalpy change of $-108.6$ kJ mol$^{-1}$

   **c** a volume of 1120 cm$^3$  **h** a rate constant of 2.048 dm$^3$ mol$^{-1}$ s$^{-1}$

   **d** a percentage of 49.517%  **i** the ideal gas constant 0.082 0578 dm$^3$ atm K$^{-1}$ mol$^{-1}$

   **e** an amount of 3.3333 mol

3 Write the results of these calculations to the appropriate number of significant figures:
   **a** $2.00 \times 41.72$  **b** $3.00 \times 20.02 \times 40.1$  **c** $20.5/32.0$

4 Give each number correct to two decimal places:
   **a** 47.123  **b** 8.334  **c** 8.345  **d** 0.0643  **e** 0.567

## 1.8 ESTIMATING YOUR ANSWER

One advantage of standard index form is that very large and very small numbers can easily be entered on a calculator. Another advantage is that you can easily estimate the answer to a calculation to the correct order of magnitude (the correct power of ten).

For example, $\dfrac{2456 \times 0.0123}{5223 \times 60.7}$

Putting the numbers into standard form gives

$$\frac{2.456 \times 10^3 \times 1.23 \times 10^{-2}}{5.223 \times 10^3 \times 6.07 \times 10}$$

This is approximately

$$\frac{2 \times 1}{5 \times 6} \times \frac{10^3 \times 10^{-2}}{10^3 \times 10} = \frac{2 \times 10^{-3}}{30} = 0.07 \times 10^{-3} = 7 \times 10^{-5}$$

By putting the numbers into standard form you can estimate the answer very quickly. A complete calculation gives the answer $9.53 \times 10^{-5}$. The rough estimate is sufficiently close to this to reassure you that you have not made any slips with powers of ten.

## 1.9 LOGARITHMS

The **logarithm** of a number (written as log or lg) is the power to which 10 must be raised to give the number. For example, the number $100 = 10^2$ therefore the logarithm of 100 to the base 10 is 2: lg 100 = 2.

There is also a set of logarithms to the base e. They are called **natural logarithms** because e is a significant quantity in mathematics. Natural logarithms are written as ln $N$ and are related to logs to the base ten by

$$\ln N = 2.303 \lg N$$

### To obtain the logarithm of a number

● enter the number on your calculator

● press the log key.

Some calculators, however, require you to press the log key first, then the number. The value of the log will appear in the display. This will happen whether you enter the number in standard form or another form. For example, lg 12 345 = 4.0915, whether you enter the number as 12 345 or as $1.2345 \times 10^4$.

### To obtain the natural logarithm of a number

● enter the number in your calculator

● press the ln key

(unless your calculator requires the ln key first followed by the number).

Natural logarithms arise in problems in physical chemistry (e.g. §13.7).

Logarithms to the base 10 are used when a graph must be plotted of measurements over a wide range of values. You would find it difficult to fit the numbers 10, 1000 and 1 000 000 onto the same graph, but their logs, 1, 3 and 6 can easily fit onto the same scale.

## 1.10 GRAPHS

Here are some hints for drawing graphs:

**1** If the quantities $x$ and $y$ are related by the equation $y = ax + b$, then a plot of experimental values of $y$ against corresponding values of $x$ will give a straight line [see Figure A]. Values of $y$ are plotted on the vertical axis, the $y$-axis, and values of $x$ are plotted on the horizontal axis, the $x$-axis. From the graph,

$$\text{gradient} = (\text{increase in } y)/(\text{increase in } x) = a$$

$$\text{intercept on the } y\text{-axis} = b$$

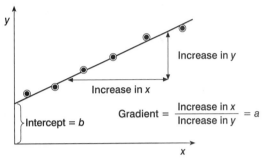

Fig. A  *Plotting a graph*

**2** Choose a scale which will allow the graph to cover as much of the piece of paper as possible. There is no need to start at zero. If the points lie between 80 and 100 to start at zero would cramp your graph into a small section at the top of the page [see Figure B].

**3** Label the axes with the quantities and the units. Make the scale units as simple as possible. Instead of plotting $1 \times 10^{-3}$ mol dm$^{-3}$, $2 \times 10^{-3}$ mol dm$^{-3}$, $3 \times 10^{-3}$ mol dm$^{-3}$ etc, plot 1, 2, 3 etc and label the axis as (Concentration/ mol dm$^{-3}$) $\times 10^3$ [see Figure C]. The sign (/) means 'divided by'. The numbers 1, 2, 3 etc. are the values of the concentration divided by the unit, mol dm$^{-3}$ and multiplied by $10^3$.

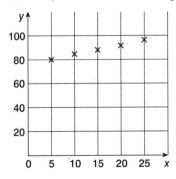

Fig. B  *Don't cramp your graph!*

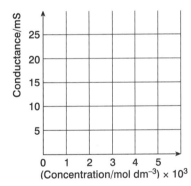

Fig. C  *Label the axes*

**4** When you draw a straight line through the points, draw the best straight line you can, to pass through or close to as many points as possible. In Figure D, a line can be drawn close to four of the points, but one point does not fit in. It is better to assume that some experimental error was made in this point and draw the best fit to the other four points.

**5** If you are drawing a curve, draw a smooth curve [Figure E]. Do not join up the points with straight lines.

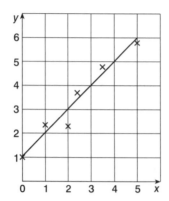

Fig. D   *Drawing the best line*

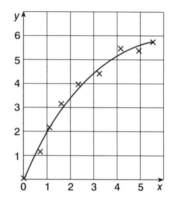

Fig. E   *Drawing a smooth curve*

**EXERCISE 1.7   Graphs**

**1**   Plot the values of $y$ against the corresponding values of $x$:

| $x$ | 3 | 6 | 9 | 12 |
|-----|----|----|----|----|
| $y$ | 13 | 22 | 31 | 40 |

Deduce the equation of the straight line plot.

**2**   Plot a graph of values of $y$ against $x$:

| $x$ | 56 | 28 | 14 | 7 |
|-----|----|----|----|----|
| $y$ | 1 | 2 | 4 | 8 |

Deduce the equation of the straight line.

**3**   Hydrogen peroxide decomposes to give oxygen.

$$2H_2O_2(aq) \longrightarrow O_2(g) + 2H_2O(l)$$

The rate of decomposition is given by (change in concentration)/time and has the units mol dm$^{-3}$ s$^{-1}$. The rate depends on the hydrogen peroxide concentration at the start of the reaction.

| rate/mol dm$^{-3}$ s$^{-1}$ | $[H_2O_2]$/mol dm$^{-3}$ |
|-----------------------------|--------------------------|
| $3.64 \times 10^{-5}$ | 0.05 |
| $7.41 \times 10^{-5}$ | 0.10 |
| $1.51 \times 10^{-4}$ | 0.20 |
| $2.21 \times 10^{-4}$ | 0.30 |

Plot the values of rate (on the $y$-axis) against the concentration of hydrogen peroxide (on the $x$-axis). Write an equation relating rate and concentration of hydrogen peroxide.

**4** Krypton is a radioactive element. The amount of radioactivity is measured by the count per second registered on a detector. Use the results shown to plot a graph of radioactivity (on the *y*-axis) against time (on the *x*-axis). From the graph, read off the time taken for the radioactivity to drop to half its original value.

| Time/minute | 0 | 20 | 40 | 60 | 80 | 100 | 120 | 140 | 160 | 180 |
|---|---|---|---|---|---|---|---|---|---|---|
| Radioactivity/count per second | 100 | 92 | 85 | 78 | 72 | 66 | 61 | 56 | 52 | 48 |

## 1.11 INCLUDE THE UNITS

When you are doing a calculation, include the units of each physical quantity as well as the numerical values. If the answer you obtain has the correct units, it is likely that you have the correct answer.

**EXAMPLE**  A student is calculating the concentration of a solution which contains 1.25 moles of solute in 250 cm$^3$ of solution. She obtains the answer 0.2 dm$^3$ mol$^{-1}$. Immediately she realises that since concentration has the unit mol dm$^{-3}$ her answer is wrong. This time she puts in the units.

$$\text{Concentration} = \frac{\text{Amount of solute}}{\text{Volume of solution}}$$

$$\text{Concentration} = 1.25 \text{ mol}/250 \text{ cm}^3 = 1.25 \text{ mol}/250 \times 10^{-3} \text{ dm}^3$$

$$= 5.0 \text{ mol dm}^{-3}$$

Now the unit of concentration is correct, and the student can have more confidence in her answer.

## 1.12 CHECK YOUR ANSWER

Always look critically at your answer. Ask yourself the questions:

● Is it a reasonable answer?

● Does it have the right power of ten for the data? Be sure it is not a hundred times or a thousand times the correct answer.

● Is it in the right units?

● Have you given the correct number of significant figures?

Many errors can be detected by a check of this kind.

Note: You can find the answers to all of these exercises on p 232.

# Formulae and equations

**2**

Calculations are based on formulae and on equations. In order to tackle the calculations in this book you will have to be quite sure you can work out the formulae of compounds correctly, and that you can balance equations. This section is a revision of work on formulae and equations.

## 2.1 FORMULAE

Electrovalent compounds consist of oppositely charged ions. The compound formed is neutral because the charge on the positive ion (or ions) is equal to the charge on the negative ion (or ions). In sodium chloride, NaCl, one sodium ion, $Na^+$, is balanced in charge by one chloride ion, $Cl^-$.

*This is how the formulae of electrovalent compounds can be worked out:*

| | |
|---|---|
| *Compound* | *Zinc chloride* |
| Ions present are | $Zn^{2+}$ and $Cl^-$ |
| Now balance the charges | One $Zn^{2+}$ ion needs two $Cl^-$ ions |
| Ions needed are | $Zn^{2+}$ and $2Cl^-$ |
| The formula is | $ZnCl_2$ |

| | |
|---|---|
| *Compound* | *Sodium sulphate* |
| Ions present are | $Na^+$ and $SO_4^{2-}$ |
| Now balance the charges | Two $Na^+$ balance one $SO_4^{2-}$ |
| Ions needed are | $2Na^+$ and $SO_4^{2-}$ |
| The formula is | $Na_2SO_4$ |

| | |
|---|---|
| *Compound* | *Aluminium sulphate* |
| Ions present are | $Al^{3+}$ and $SO_4^{2-}$ |
| Now balance the charges | Two $Al^{3+}$ balance three $SO_4^{2-}$ |
| Ions needed are | $2Al^{3+}$ and $3SO_4^{2-}$ |
| The formula is | $Al_2(SO_4)_3$ |

| | |
|---|---|
| *Compound* | *Iron(II) sulphate* |
| Ions present are | $Fe^{2+}$ and $SO_4^{2-}$ |
| Now balance the charges | One $Fe^{2+}$ balances one $SO_4^{2-}$ |
| Ions needed are | $Fe^{2+}$ and $SO_4^{2-}$ |
| The formula is | $FeSO_4$ |

*Compound*          *Iron(III) sulphate*
Ions present are       $Fe^{3+}$ and $SO_4^{2-}$
Now balance the charges   Two $Fe^{3+}$ balance three $SO_4^{2-}$
Ions needed are      $2Fe^{3+}$ and $3SO_4^{2-}$
The formula is       $Fe_2(SO_4)_3$

You need to know the charges of the ions in the table below. Then you can work out the formula of any electrovalent compound.

You will notice that the compounds of iron are named iron(II) sulphate and iron(III) sulphate to show which of its valencies iron is using in the compound. This is always done with the compounds of elements of variable oxidation number (see § 8.5).

| Name | Symbol | Charge | Name | Symbol | Charge |
|---|---|---|---|---|---|
| Hydrogen | $H^+$ | +1 | Hydroxide | $OH^-$ | −1 |
| Ammonium | $NH_4^+$ | +1 | Nitrate | $NO_3^-$ | −1 |
| Potassium | $K^+$ | +1 | Chloride | $Cl^-$ | −1 |
| Sodium | $Na^+$ | +1 | Bromide | $Br^-$ | −1 |
| Silver | $Ag^+$ | +1 | Iodide | $I^-$ | −1 |
| Copper(I) | $Cu^+$ | +1 | Hydrogen-carbonate | $HCO_3^-$ | −1 |
| Barium | $Ba^{2+}$ | +2 | | | |
| Calcium | $Ca^{2+}$ | +2 | Oxide | $O^{2-}$ | −2 |
| Copper(II) | $Cu^{2+}$ | +2 | Sulphide | $S^{2-}$ | −2 |
| Iron(II) | $Fe^{2+}$ | +2 | Sulphite | $SO_3^{2-}$ | −2 |
| Lead | $Pb^{2+}$ | +2 | Sulphate | $SO_4^{2-}$ | −2 |
| Magnesium | $Mg^{2+}$ | +2 | Carbonate | $CO_3^{2-}$ | −2 |
| Zinc | $Zn^{2+}$ | +2 | | | |
| Aluminium | $Al^{3+}$ | +3 | Phosphate | $PO_4^{3-}$ | −3 |
| Iron(III) | $Fe^{3+}$ | +3 | | | |

## 2.2 EQUATIONS

Having symbols for elements and formulae for compounds gives us a way of representing chemical reactions.

**EXAMPLE 1** Copper(II) carbonate decomposes to form copper(II) oxide and carbon dioxide. We can write a **word equation** for the reaction:

Copper(II) carbonate $\longrightarrow$ Copper(II) oxide + Carbon dioxide

In a **chemical equation** we write symbols for elements and formulae for compounds:

$$CuCO_3 \longrightarrow CuO + CO_2$$

The atoms we finish with are the same as the atoms we start with. We start with one atom of copper, one atom of carbon and three atoms of oxygen on the left-hand side of the equation and we finish with the same on the right-hand side. The two sides of the expression are equal; this is why it is called an equation. A simple way of adding more information is to include state symbols. These are (s) = solid, (l) = liquid, (g) = gas and (aq) = in solution in water. The equation

$$CuCO_3(s) \longrightarrow CuO(s) + CO_2(g)$$

tells you that solid copper(II) carbonate dissociates to form solid copper(II) oxide and the gas carbon dioxide.

**EXAMPLE 2**

The equation

$$Zn(s) + H_2SO_4(aq) \longrightarrow ZnSO_4(aq) + H_2(g)$$

tells you that solid zinc reacts with a solution of sulphuric acid to give a solution of zinc sulphate and hydrogen gas. Hydrogen is written as $H_2$, since each molecule of hydrogen gas contains two atoms.

**EXAMPLE 3**

Sodium carbonate reacts with dilute hydrochloric acid to give carbon dioxide and a solution of sodium chloride. The equation could be

$$Na_2CO_3(s) + HCl(aq) \longrightarrow CO_2(g) + NaCl(aq) + H_2O(l)$$

but, when you add up the atoms on the right, you find that they are not equal to the atoms on the left. The equation is not '*balanced*', so, the next step is to **balance** it. Multiplying NaCl by two gives

$$Na_2CO_3(s) + HCl(aq) \longrightarrow CO_2(g) + 2NaCl(aq) + H_2O(l)$$

This makes the number of sodium atoms on the right-hand side equal to the number on the left-hand side. But there are two chlorine atoms on the right-hand side, therefore the HCl must be multiplied by two:

$$Na_2CO_3(s) + 2HCl(aq) \longrightarrow CO_2(g) + 2NaCl(aq) + H_2O(l)$$

The equation is now *balanced*.

When you are balancing a chemical equation, the only way to do it is to multiply the number of atoms or molecules. You never try to alter a formula. In the above example, you got two chlorine atoms by multiplying HCl by two, not by altering the formula to $HCl_2$, which does not exist.

**STEPS TO TAKE 1** Write a word equation.

**2** Put in the symbols and formulae (symbols for elements, formulae for compounds) and state symbols.

**3** Balance the equation.

**EXAMPLE 4**

When methane burns,

$$Methane + Oxygen \longrightarrow Carbon\ dioxide + Water$$

$$CH_4(g) + O_2(g) \longrightarrow CO_2(g) + H_2O(g)$$

There is one carbon atom on the left-hand side and one carbon atom on the right-hand side. There are four hydrogen atoms on the left-hand side, and therefore we need to put four hydrogen atoms on the right-hand side. Putting $2H_2O$ on the right-hand side will do this:

$$CH_4(g) + O_2(g) \longrightarrow CO_2(g) + 2H_2O(g)$$

There is one molecule of $O_2$ on the left-hand side and four O atoms on the right-hand side. We can make the two sides equal by putting $2O_2$ on the left-hand side:

$$CH_4(g) + 2O_2(g) \longrightarrow CO_2(g) + 2H_2O(g)$$

This is a balanced equation. The numbers of atoms of carbon, hydrogen and oxygen on the left-hand side are equal to the numbers of atoms of carbon, hydrogen and oxygen on the right-hand side.

## EXERCISE 2.1    Practice with equations

1    For practice, try writing the equations for the reactions:

   **a**  Hydrogen + Copper oxide $\longrightarrow$ Copper + Water

   **b**  Carbon + Carbon dioxide $\longrightarrow$ Carbon monoxide

   **c**  Carbon + Oxygen $\longrightarrow$ Carbon dioxide

   **d**  Magnesium + Sulphuric acid $\longrightarrow$ Hydrogen + Magnesium sulphate

   **e**  Copper + Chlorine $\longrightarrow$ Copper(II) chloride

2    Now try writing balanced equations for the reactions:

   **a**  Calcium + Water $\longrightarrow$ Hydrogen + Calcium hydroxide solution

   **b**  Copper + Oxygen $\longrightarrow$ Copper(II) oxide

   **c**  Sodium + Oxygen $\longrightarrow$ Sodium oxide

   **d**  Iron + Hydrochloric acid $\longrightarrow$ Iron(II) chloride solution

   **e**  Iron + Chlorine $\longrightarrow$ Iron(III) chloride

3    Balance these equations:

   **a**  $Na_2O(s) + H_2O(l) \longrightarrow NaOH(aq)$

   **b**  $KClO_3(s) \longrightarrow KCl(s) + O_2(g)$

   **c**  $H_2O_2(aq) \longrightarrow H_2O(l) + O_2(g)$

   **d**  $Fe(s) + O_2(g) \longrightarrow Fe_3O_4(s)$

   **e**  $Mg(s) + N_2(g) \longrightarrow Mg_3N_2(s)$

   **f**  $NH_3(g) + O_2(g) \longrightarrow N_2(g) + H_2O(g)$

   **g**  $Fe(s) + H_2O(g) \longrightarrow Fe_3O_4(s) + H_2(g)$

   **h**  $H_2S(g) + O_2(g) \longrightarrow H_2O(g) + SO_2(g)$

   **i**  $H_2S(g) + SO_2(g) \longrightarrow H_2O(l) + S(s)$

# Relative atomic mass

## 3.1 RELATIVE ATOMIC MASS

Atoms are tiny: one atom of hydrogen has a mass of $1.66 \times 10^{-24}$ g; one atom of carbon has a mass of $1.99 \times 10^{-23}$ g. Numbers as small as this are awkward to handle, and instead of atomic masses we use **relative atomic masses**. The standard with which other atoms are compared is the mass of one atom of carbon-12. Carbon-12 is the most plentiful isotope of carbon. **Isotopes** are forms of an element whose atoms possess the same number of protons and electrons but different numbers of neutrons and therefore have different masses. The term **relative atomic mass** is defined by:

$$\text{Relative atomic mass} = \frac{\text{Mass of one atom of the element}}{\frac{1}{12} \text{ Mass of one atom of carbon-12}}$$

By definition, carbon-12 has a relative atomic mass of 12.000 000.

Since relative atomic masses are ratios of two masses they have no unit. A table of relative atomic masses is given on p 243.

The calculation of relative atomic masses from mass spectrometer measurements is covered in § 9.1.

## 3.2 RELATIVE MOLECULAR MASS

We can find the mass of a molecule by adding the masses of the atoms in it. We can find the relative molecular mass of a compound by adding the relative atomic masses of all the atoms in a molecule of the compound. For example, we can work out the relative molecular mass of carbon dioxide as follows:

The formula is $CO_2$.

$$1 \text{ atom of C, relative atomic mass } 12 = 12$$
$$2 \text{ atoms of O, relative atomic mass } 16 = 32$$
$$\text{Total} = 44$$
$$\text{Relative molecular mass of } CO_2 = 44$$

The symbol for relative molecular mass is $M_r$.

### Relative formula mass

A vast number of compounds consist of ions, not molecules. The compound sodium chloride, for example, consists of sodium ions and chloride ions. You cannot correctly refer to a 'molecule of sodium chloride'. For ionic compounds, the term **formula unit** is used to describe the ions which make up the compound. A formula unit of sodium chloride is $NaCl$. A formula unit of copper(II) sulphate-5-water is $CuSO_4 . 5H_2O$.

$$\text{Relative formula mass} = \frac{\text{Mass of one formula unit}}{(\frac{1}{12})\,\text{Mass of one atom of carbon-12}}$$

The symbol for relative formula mass is $M_r$.

**EXAMPLE 1**

We work out the relative formula mass of calcium chloride as follows:

The formula is $CaCl_2$.

$$1 \text{ atom of Ca, relative atomic mass } 40 = 40$$
$$2 \text{ atoms of Cl, relative atomic mass } 35.5 = 71$$
$$\text{Total} = 111$$
$$\text{Relative formula mass of } CaCl_2 = 111$$

**EXAMPLE 2**

We work out the relative formula mass of aluminium sulphate as follows:

The formula is $Al_2(SO_4)_3$.

$$2 \text{ atoms of Al, relative atomic mass } 27 = 54$$
$$3 \text{ atoms of S, relative atomic mass } 32 = 96$$
$$12 \text{ atoms of O, relative atomic mass } 16 = 192$$
$$\text{Total} = 342$$
$$\text{Relative formula mass of } Al_2(SO_4)_3 = 342$$

**EXERCISE 3.1** **Problems on relative formula mass**

**1** Work out the relative formula masses of these compounds:

| | | |
|---|---|---|
| $SO_2$ | $NaOH$ | $KNO_3$ |
| $MgCO_3$ | $PbCl_2$ | $MgCl_2$ |
| $Mg(NO_3)_2$ | $Zn(OH)_2$ | $ZnSO_4$ |
| $H_2SO_4$ | $HNO_3$ | $MgSO_4 . 7H_2O$ |
| $CaSO_4$ | $Pb_3O_4$ | $P_2O_5$ |
| $Na_2CO_3$ | $Ca(OH)_2$ | $CuCO_3$ |
| $CuSO_4$ | $Ca(HCO_3)_2$ | $CuSO_4 . 5H_2O$ |
| $Fe_2(SO_4)_3$ | $Na_2CO_3 . 10H_2O$ | $FeSO_4 . 7H_2O$ |

## 3.3 PERCENTAGE COMPOSITION

From the formula of a compound, we can work out the percentage by mass of each element present in the compound.

**EXAMPLE 1**     Calculate the percentage of silicon and oxygen in silicon(IV) oxide (silica).

**METHOD**     First, work out the relative formula mass. The formula is $SiO_2$.

$$1 \text{ atom of silicon, relative atomic mass } 28 = 28$$

$$2 \text{ atoms of oxygen, relative atomic mass } 16 = 32$$

$$\text{Total} = \text{Relative formula mass} = 60$$

$$\text{Percentage of silicon} = \frac{28}{60} \times 100 = \frac{7}{15} \times 100\%$$

$$= \frac{7 \times 20}{3} = 47\%$$

$$\text{Percentage of oxygen} = \frac{32}{60} \times 100 = \frac{8}{15} \times 100\%$$

$$= \frac{8 \times 20}{3} = 53\%$$

**ANSWER**     Silicon(IV) oxide contains 47% silicon and 53% oxygen by mass.

In general,

Percentage of element **A** =

$$\frac{\text{Relative atomic mass of } \mathbf{A} \times \text{No. of atoms of } \mathbf{A} \text{ in formula}}{\text{Relative formula mass of compound}} \times 100\%$$

**EXAMPLE 2**     Find the percentage by mass of magnesium, oxygen and sulphur in magnesium sulphate.

**METHOD**     First calculate the relative formula mass. The formula is $MgSO_4$.

$$1 \text{ atom of magnesium, relative atomic mass } 24 = 24$$

$$1 \text{ atom of sulphur, relative atomic mass } 32 = 32$$

$$4 \text{ atoms of oxygen, relative atomic mass } 16 = 64$$

$$\text{Total} = \text{Relative formula mass, } M_r = 120$$

$$\text{Percentage of magnesium} = \frac{A_r(\text{Mg}) \times \text{No. of Mg atoms}}{M_r(\text{MgSO}_4)} \times 100\%$$

$$= \frac{24}{120} \times 100\%$$

$$= 20\%$$

$$\text{Percentage of sulphur} = \frac{A_r(\text{S}) \times \text{No. of S atoms}}{M_r(\text{MgSO}_4)} \times 100\%$$

$$= \frac{32}{120} \times 100\%$$

$$= 27\%$$

$$\text{Percentage of oxygen} = \frac{A_r(O) \times \text{No. of O atoms}}{M_r(MgSO_4)} \times 100\%$$

$$= \frac{16 \times 4}{120} \times 100\%$$

$$= 53\%$$

**ANSWER**  Magnesium 20%; sulphur 27%; oxygen 53%. You can check on the calculation by adding up the percentages to see whether they add up to 100. In this case $20 + 27 + 53 = 100$.

**EXAMPLE 3**  Calculate the percentage of water in copper sulphate crystals.

**METHOD**  Find the relative formula mass. The formula is $CuSO_4 \cdot 5H_2O$.

1 atom of copper, relative atomic mass $64 = 64$ (approx.)

1 atom of sulphur, relative atomic mass $32 = 32$

4 atoms of oxygen, relative atomic mass $16 = 64$

5 molecules of water, $5 \times [(2 \times 1) + 16] = 5 \times 18 = 90$

Total = Relative formula mass = 250

Mass of water = 90

$$\text{Percentage of water} = \frac{\text{Mass of water in formula}}{\text{Relative formula mass}} \times 100\%$$

$$= \frac{90}{250} \times 100\%$$

$$= 36\%$$

**ANSWER**  The percentage of water in copper sulphate crystals is 36%.

**EXERCISE 3.2**  **Problems on percentage composition**

For relative atomic masses refer to the table on p 243.

Section 1  **Reminder :** For percentages see § 1.3.

**1**  Calculate the percentages by mass of:
**a** carbon and hydrogen in ethane, $C_2H_6$

**STEPS TO TAKE**  $A_r$ of carbon =

No of atoms of carbon per molecule =

$A_r$ of H =

Number of atoms of H per molecule =
$M_r$ of $C_2H_6$ =

$$\text{Percentage of carbon} = \frac{A_r(C) \times \text{No. of atoms of C}}{M_r(C_2H_6)} \times 100\%$$

$$\text{Percentage of hydrogen} = \frac{A_r(\text{H}) \times \text{No. of atoms of H}}{M_r(\text{C}_2\text{H}_6)} \times 100\%$$

**Check:** Do the percentages add up to 100%?

**b**  sodium, oxygen and hydrogen in sodium hydroxide, NaOH

**STEPS TO TAKE**

$$\text{Percentage of Na} = \frac{A_r(\text{Na}) \times 1 \text{ atom of Na in formula}}{M_r(\text{NaOH})} \times 100\%$$

$$\text{Percentage of O} = \frac{A_r(\text{O}) \times 1 \text{ atom of O in formula}}{M_r(\text{NaOH})} \times 100\%$$

Now find the percentage of H in a similar way.

**Check:** Do the percentages add up to 100%?

**c**  sulphur and oxygen in sulphur trioxide, $SO_3$

**STEPS TO TAKE**

$$\text{Percentage of S} = \frac{A_r(\text{S}) \times 1 \text{ atom of S in formula}}{M_r(\text{SO}_3)} \times 100\%$$

$$\text{Percentage of O} = \frac{A_r(\text{O}) \times 3 \text{ atoms of O in formula}}{M_r(\text{SO}_3)} \times 100\%$$

**Check:** Do the percentages add up to 100%?

**d**  carbon and hydrogen in propyne, $C_3H_4$

**STEPS TO TAKE**

$$\text{Percentage of C} = \frac{A_r(\text{C}) \times 3}{M_r(\text{C}_3\text{H}_4)} \times 100\%$$

$$\text{Percentage of H} = \frac{A_r(\text{H}) \times 4}{M_r(\text{C}_3\text{H}_4)} \times 100\%$$

Check the sum of the percentages as usual.

**2**  Calculate the percentages by mass of

**a**  carbon and hydrogen in heptane, $C_7H_{16}$

**b**  magnesium and nitrogen in magnesium nitride, $Mg_3N_2$

**c**  sodium and iodine in sodium iodide, NaI

**d**  calcium and bromine in calcium bromide, $CaBr_2$

Section 2

**1**  Calculate the percentage by mass of

**a**  carbon and hydrogen in pentene, $C_5H_{10}$

**b**  nitrogen, hydrogen and oxygen in ammonium nitrate, $NH_4NO_3$

**c**  iron, oxygen and hydrogen in iron(II) hydroxide, $Fe(OH)_2$

**d**  carbon, hydrogen and oxygen in ethanedioic acid, $C_2O_4H_2$

**2**  Calculate the percentages of

  **a**  carbon, hydrogen and oxygen in propanol, $C_3H_7OH$

  **b**  carbon, hydrogen and oxygen in ethanoic acid, $CH_3CO_2H$

  **c**  carbon, hydrogen and oxygen in methyl methanoate, $HCO_2CH_3$

  **d**  aluminium and sulphur in aluminium sulphide, $Al_2S_3$

**3**  Haemoglobin contains 0.33% by mass of iron. There are 2 Fe atoms in 1 molecule of haemoglobin. What is the relative molecular mass of haemoglobin?

---

**Hint :** Remember from § 3.3

$$\% \; \mathbf{A} = \frac{A_r(\mathbf{A}) \times \text{No. of atoms of } \mathbf{A} \text{ in formula}}{M_r(\text{compound})} \times 100\%$$

---

**4**  An adult's bones weigh about 11 kg, and 50% of this mass is calcium phosphate, $Ca_3(PO_4)_2$. What is the mass of phosphorus in the bones of an average adult?

---

**Hint :** The steps to take are: mass of calcium phosphate in bones ... % of P in $Ca_3(PO_4)_2$ ... mass of phosphorus.

---

# The mole

4

## 4.1 THE MOLE

Looking at equations tells us a great deal about chemical reactions. For example,

$$Fe(s) + S(s) \longrightarrow FeS(s)$$

tells us that iron and sulphur combine to form iron(II) sulphide, and that one atom of iron combines with one atom of sulphur. Chemists are interested in the exact quantities of substances which react together in chemical reactions. For example, in the reaction between iron and sulphur, if you want to measure out just enough iron to combine with, say, 10 g of sulphur, how do you go about it? What you need to do is to count out equal numbers of atoms of iron and sulphur. This sounds a formidable task, and it puzzled a chemist called Avogadro, working in Italy early in the nineteenth century. He managed to solve this problem with a piece of clear thinking which makes the problem look very simple once you have followed his argument.

### Avogadro's reasoning

We know from their relative atomic masses that an atom of carbon is 12 times as heavy as an atom of hydrogen. Therefore, we can say:

| If | 1 atom of carbon is | 12 times as heavy as 1 atom of hydrogen, |
| --- | --- | --- |
| then | 1 dozen C atoms are | 12 times as heavy as 1 dozen H atoms, |
| and | 1 hundred C atoms are | 12 times as heavy as 1 hundred H atoms, |
| and | 1 million C atoms are | 12 times as heavy as 1 million H atoms, |

and it follows that when we see a mass of carbon which is 12 times as heavy as a mass of hydrogen, the two masses must contain equal numbers of atoms. If we have 12 g of carbon and 1 g of hydrogen, we know that we have the same number of atoms of carbon and hydrogen. The same argument applies to any element. Take the relative atomic mass of an element in grams:

| 40 g Calcium | 24 g Magnesium | 32 g Sulphur | 12 g Carbon | 1 g Hydrogen |
| --- | --- | --- | --- | --- |

All these masses contain the same number of atoms. This number is $6.022 \times 10^{23}$. The amount of an element that contains $6.022 \times 10^{23}$ atoms is called **one mole** of the element. (The symbol for mole is mol.)

The ratio $6.022 \times 10^{23}$ mol$^{-1}$ is called the **Avogadro constant**.

The mole is defined as the amount of a substance that contains as many elementary units as there are atoms in 12 grams of carbon-12. The units may be atoms, ions, molecules or formula units.

### How the mole concept helps

We can count out $6 \times 10^{23}$ atoms of any element by weighing out its relative atomic mass in grams. If we want to react iron and sulphur so that there is an atom of sulphur for every atom of iron, we can count out $6 \times 10^{23}$ atoms of sulphur by weighing out 32 g of sulphur and we can count out $6 \times 10^{23}$ atoms of iron by weighing out 56 g of iron. Since one atom of iron reacts with one atom of sulphur to form one formula unit of iron(II) sulphide, one mole of iron reacts with one mole of sulphur to form one mole of iron(II) sulphide:

$$Fe(s) + S(s) \longrightarrow FeS(s)$$

and 56 g of iron react with 32 g of sulphur to form 88 g of iron(II) sulphide.

### One mole of a compound

Just as one mole of an element is the relative atomic mass in grams, one mole of a compound is the relative formula mass in grams. If you want to weigh out one mole of sodium hydroxide, you first work out its relative formula mass.

The formula is NaOH.

$$\text{Relative formula mass} = 23 + 16 + 1$$
$$= 40$$

If you weigh out 40 g of sodium hydroxide, you have one mole of sodium hydroxide. The quantity $40 \text{ g mol}^{-1}$ is the **molar mass** of sodium hydroxide. The molar mass of a compound is the relative formula mass in grams per mole. The molar mass of an element is the relative atomic mass in grams per mole. The molar mass of sodium hydroxide is $40 \text{ g mol}^{-1}$, and the molar mass of sodium is $23 \text{ g mol}^{-1}$.

Remember that most gaseous elements consist of molecules, not atoms. Chlorine exists as $Cl_2$ molecules, oxygen as $O_2$ molecules, hydrogen as $H_2$ molecules, and so on. To work out the mass of one mole of chlorine molecules, you must use the relative molecular mass of $Cl_2$.

$$\text{Relative atomic mass of chlorine} = 35.5$$
$$\text{Relative molecular mass of } Cl_2 = 2 \times 35.5 = 71$$
$$\text{Mass of 1 mole of chlorine, } Cl_2 = 71 \text{ grams}$$

The noble gases, helium, neon, argon, krypton and xenon, exist as atoms. Since the relative atomic mass of helium is 4, the mass of 1 mole of helium is 4 g.

## 4.2 RELATIVE FORMULA MASS

The **relative formula mass** of a compound is defined by the expression:

$$\text{Relative formula mass} = \frac{\text{Mass of one formula unit of the compound}}{\frac{1}{12} \text{ Mass of one atom of carbon-12}}$$

The relative formula mass of a compound expressed in grams is one mole of the compound. For example 44 g of carbon dioxide is one mole of carbon dioxide and contains $6.022 \times 10^{23}$ molecules. A formula unit of sodium sulphate is $2Na^+SO_4^{2-}$. The relative formula mass in grams, 142 g, is one mole of sodium sulphate.

## 4.3 MOLAR MASS

The mass of one mole of an element or compound is called its **molar mass** (symbol **M**). The molar mass of copper is 63.5 g $mol^{-1}$. The molar mass of sodium hydroxide is 40 g $mol^{-1}$. There is a quantity that relates to the number of particles in a mass of substance. This quantity is called the **amount** of substance.

$$\text{Amount of substance} = \frac{\text{Mass}}{\text{Molar mass}}$$

If mass is stated in g and molar mass in g $mol^{-1}$, then the unit of amount is mol.

$$\text{Amount of substance/mol} = \frac{\text{Mass/g}}{\text{Molar mass/g mol}^{-1}}$$

### Calculation of molar mass

The molar mass of a compound is the sum of the relative atomic masses of the atoms in a molecule or formula unit of the compound expressed in g $mol^{-1}$.

**EXAMPLE**

What is the molar mass of glucose, $C_6H_{12}O_6$?

**METHOD**

Sum of relative atomic masses $= 6A_r(C) + 12A_r(H) + 6A_r(O)$

$$= (6 \times 12) + (12 \times 1) + (6 \times 16) = 180$$

Molar mass $= 180$ g $mol^{-1}$

## EXERCISE 4.1     Problems on the mole

For relative atomic masses see the table on p 243.

**Reminder :** Amount = Mass/Molar mass § 4.1.

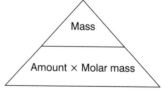

Section 1

1    State the mass of each element in:

    **a**  0.5 mol chromium    **b**  $\frac{1}{7}$ mol iron    **c**  $\frac{1}{3}$ mol carbon

    **d**  $\frac{1}{4}$ mol magnesium    **e**  $\frac{1}{7}$ mol nitrogen molecules    **f**  $\frac{1}{4}$ mol oxygen molecules

    Remember that nitrogen and oxygen exist as diatomic molecules, $N_2$ and $O_2$.

2    Calculate the amount of each element in:

    **a**  46 g sodium    **b**  130 g zinc    **c**  10 g calcium

    **d**  2.4 g magnesium    **e**  13 g chromium

3    Find the mass of each element in:

    **a**  10 mol lead    **b**  $\frac{1}{6}$ mol copper    **c**  0.1 mol iodine molecules

    **d**  10 mol hydrogen molecules    **e**  0.25 mol calcium

    **f**  0.25 mol bromine molecules    **g**  $\frac{3}{4}$ mol iron    **h**  0.20 mol zinc

    **i**  $\frac{1}{2}$ mol chlorine molecules    **j**  0.1 mol neon

4    State the amount of substance (mol) in:

    **a**  58.5 g sodium chloride

    **b**  26.5 g anhydrous sodium carbonate

    **c**  50.0 g calcium carbonate

    **d**  15.9 g copper(II) oxide

    **e**  8.00 g sodium hydroxide

    **f**  303 g potassium nitrate

    **g**  9.8 g sulphuric acid

    **h**  499 g copper(II) sulphate-5-water

5    Given Avogadro's constant is $6 \times 10^{23}$ $mol^{-1}$, calculate the number of atoms in:

    **a**  35.5 g chlorine    **b**  27 g aluminium    **c**  3.1 g phosphorus

    **d**  336 g iron    **e**  48 g magnesium    **f**  1.6 g oxygen

    **g**  0.4 g oxygen    **h**  216 g silver

**Hint :** Mass ... amount in moles ... multiply by $6 \times 10^{23}$ $mol^{-1}$ ... number of atoms.

6    How many grams of zinc contain:

    **a**  $6 \times 10^{23}$ atoms    **b**  $6 \times 10^{20}$ atoms?

7   How many grams of aluminium contain:

a   $2 \times 10^{23}$ atoms   b   $6 \times 10^{20}$ atoms?

8   What mass of carbon contains:

a   $6 \times 10^{23}$ atoms   b   $2 \times 10^{21}$ atoms?

9   Write down:

a   the mass of calcium which has the same number of atoms as 12 g of magnesium

b   the mass of silver which has the same number of atoms as 3 g of aluminium

c   the mass of zinc with the same number of atoms as 1 g of helium

d   the mass of sodium which has 5 times the number of atoms as 39 g of potassium.

## Section 2

Use Avogadro constant = $6 \times 10^{23}$ mol$^{-1}$.

1   Imagine a hardware store is having a sale. The knock-down price of titanium is one billion ($10^{9}$) atoms for 1p. How much would you have to pay for 1 milligram ($1 \times 10^{-3}$ g) of titanium?

**Hint :** 1 mg of Ti contains how many moles? ... how many atoms? ... and the price?

2   Ethanol, $C_2H_6O$, is the alcohol in alcoholic drinks. If you have 9.2 g of ethanol, how many moles do you have of the following?

a   ethanol molecules   b   carbon atoms

c   hydrogen atoms   d   oxygen atoms

3   A car releases about 5 g of nitrogen oxide into the air for each mile driven. How many molecules of nitrogen oxide are emitted per mile?

**Hint :** Trace the connection mass ... moles ... number of molecules.

4   How many moles of $H_2O$ are there in 1.00 dm$^3$ of water?

**Hint :** Trace the connection volume of water ... mass of water ... amount (moles) of water.

5   What is the amount (mol) of sucrose, $C_{12}H_{22}O_{11}$ in a one kilogram bag of sugar?

# Equations and the mole

5

You will find that the mole concept, which you studied in Chapter 4, helps with all your chemical calculations. In chemistry, calculations are related to the equations for chemical reactions. The quantities of substances that react together are expressed in moles.

## 5.1 CALCULATIONS BASED ON CHEMICAL EQUATIONS

Equations tell us not only what substances react together but also what amounts of substances react together. The equation for the action of heat on sodium hydrogencarbonate

$$2NaHCO_3(s) \longrightarrow Na_2CO_3(s) + CO_2(g) + H_2O(g)$$

tells us that 2 moles of $NaHCO_3$ give 1 mole of $Na_2CO_3$. Since the molar masses are $NaHCO_3 = 84$ g mol$^{-1}$ and $Na_2CO_3 = 106$ g mol$^{-1}$, it follows that 168 g of $NaHCO_3$ give 106 g of $Na_2CO_3$.

The amounts of substances that react, as given by the balanced chemical equation, are called the **stoichiometric** amounts. **Stoichiometry** is the relationship between the amounts of reactants and products in a chemical reaction.

If one reactant is present in excess of the stoichiometric amount required for reaction with another of the reactants, then the excess of one reactant will be left unused at the end of the reaction.

**EXAMPLE 1**

How many moles of iodine can be obtained from $\frac{1}{6}$ mole of potassium iodate(V)?

**METHOD**

The equation

$$\text{Potassium} + \text{Potassium} + \text{Hydrogen} \longrightarrow \text{Iodine} + \text{Potassium} + \text{Water}$$
$$\text{iodate(V)} \quad \text{iodide} \quad \text{ion} \qquad\qquad\qquad \text{ion}$$

$$KIO_3(aq) + 5KI(aq) + 6H^+(aq) \longrightarrow 3I_2(aq) + 6K^+(aq) + 3H_2O(l)$$

tells us that 1 mol of $KIO_3$ gives 3 mol of $I_2$. Therefore:

**ANSWER**

$\frac{1}{6}$ mol of $KIO_3$ gives $\frac{1}{6} \times 3$ mol of $I_2 = \frac{1}{2}$ mol of $I_2$.

**EXAMPLE 2**

What is the maximum mass of ethyl ethanoate that can be obtained from 0.1 mol of ethanol?

**METHOD**

Write the equation:

$$\text{Ethanol} + \text{Ethanoic acid} \longrightarrow \text{Ethyl ethanoate} + \text{Water}$$

$$C_2H_5OH(l) + CH_3CO_2H(l) \longrightarrow CH_3CO_2C_2H_5(l) + H_2O(l)$$

1 mol of $C_2H_5OH$ gives 1 mol $CH_3CO_2C_2H_5$

0.1 mol of $C_2H_5OH$ gives 0.1 mol $CH_3CO_2C_2H_5$

The molar mass of $CH_3CO_2C_2H_5$ is 88 g $mol^{-1}$. Therefore:

**ANSWER**   0.1 mol of ethanol gives 8.8 g of ethyl ethanoate.

**EXAMPLE 3**   What mass of iron(III) oxide must be reduced in the blast furnace to give 100 tonnes of iron? (1 tonne = $10^3$ kg = $10^6$ g)

**METHOD**   The reduction is

$$Fe_2O_3(s) + 3CO(g) \longrightarrow 2Fe(s) + 3CO_2(g)$$

Iron(III) oxide + Carbon monoxide $\longrightarrow$ Iron + Carbon dioxide

1 mol $Fe_2O_3$ gives 2 mol Fe

160 g $Fe_2O_3$ gives 112 g Fe

2 mol Fe are produced from 1 mol $Fe_2O_3$

$\Rightarrow$ 112 g Fe are produced from 160 g $Fe_2O_3$

1 tonne ($1 \times 10^6$ g) Fe is produced from $\dfrac{1 \times 10^6}{112} \times 160$ g $Fe_2O_3$

$= 1.43 \times 10^6$ g = 1.43 tonnes $Fe_2O_3$

**ANSWER**   Mass of $Fe_2O_3$ = 1.43 tonnes

**EXAMPLE 4**   A mixture of 5.00 g of sodium carbonate and sodium hydrogencarbonate is heated. The loss in mass is 0.31 g. Calculate the percentage by mass of sodium carbonate in the mixture.

**METHOD**   On heating the mixture, the reaction

$$NaHCO_3(s) \longrightarrow Na_2CO_3(s) + CO_2(g) + H_2O(g)$$

Sodium hydrogencarbonate $\longrightarrow$ Sodium carbonate + Carbon dioxide + Water

takes place. The loss in mass is due to the decomposition of $NaHCO_3$.

Since 2 mol $NaHCO_3$ form 1 mol $CO_2$ + 1 mol $H_2O$

$2 \times 84$ g $NaHCO_3$ form 44 g $CO_2$ and 18 g $H_2O$

168 g $NaHCO_3$ lose 62 g in mass.

The observed loss in mass of 0.31 g is due to the decomposition of

$$\frac{0.31}{62} \times 168 \text{ g } NaHCO_3 = 0.84 \text{ g}$$

The mixture contains 0.84 g $NaHCO_3$

The difference, 5.00 – 0.84 = 4.16 g $Na_2CO_3$.

**ANSWER**   Percentage of $Na_2CO_3 = \dfrac{4.16}{5.00} \times 100 = 83.2\%$.

## EXERCISE 5.1    Problems on reacting masses of solids

Section 1

**Reminder :** Ratio calculations § 1.2. Balancing equations § 2.2.
Amount = Mass/Molar mass § 4.1.

**1** A sulphuric acid plant uses 2500 tonnes of sulphur dioxide each day. What mass of sulphur must be burned to produce this quantity of sulphur dioxide?

STEPS TO TAKE

**a** Write the equation connecting $SO_2$ and S.

**b** How many moles of S produce 1 mole of $SO_2$?

**c** How many moles of $SO_2$ does the plant produce daily?

**d** How many moles of S must be burned?

**e** What mass of S is required?

**2** An antacid tablet contains 0.1 g of magnesium hydrogencarbonate, $Mg(HCO_3)_2$. What mass of stomach acid, HCl, will it neutralise? The reaction between $Mg(HCO_3)_2$ and HCl is

$$Mg(HCO_3)_2(s) + 2HCl(aq) \longrightarrow MgCl_2(aq) + 2CO_2(g) + H_2O(l)$$

STEPS TO TAKE

**a** Work out, from the $M_r$, the amount of $Mg(HCO_3)_2$ in 0.1 g.

**b** Find **(i)** the amount of HCl **(ii)** the mass of HCl which it can neutralise.

**3** Aspirin, $C_9H_8O_4$, is made by the reaction:

$$Salicylic\ acid + Ethanoic\ anhydride \longrightarrow Aspirin + Ethanoic\ acid$$
$$C_7H_6O_3 + C_4H_6O_3 \longrightarrow C_9H_8O_4 + C_2H_4O_2$$

How many grams of salicylic acid, $C_7H_6O_3$, are needed to make one aspirin tablet, which contains 0.33 g of aspirin?

STEPS TO TAKE

**a** What amount (moles) of aspirin is contained in 0.33 g?

**b** What amount of salicylic acid is needed to make it?

**c** What mass of salicylic acid is this?

**Note:** If you became worried when you saw the formulae for aspirin and salicylic acid, you will see now that you can tackle the question and get the correct answer without knowing the chemistry of aspirin!

**4** Aluminium sulphate is used to treat sewage. It can be made by the reaction

$$Aluminium\ hydroxide + Sulphuric\ acid \longrightarrow Aluminium\ sulphate + Water$$

Find the masses of aluminium hydroxide and sulphuric acid needed to make 1.00 kg of aluminium sulphate.

STEPS TO TAKE

**a** Balance the unbalanced equation for this reaction:

$$Al(OH)_3(s) + H_2SO_4(aq) \longrightarrow Al_2(SO_4)_3(aq) + H_2O(l)$$

**b** What amount of $Al_2(SO_4)_3$ is present in 1.00 kg?

**c** What amounts of **(i)** $Al(OH)_3$ and **(ii)** $H_2SO_4$ are needed to make it?

**d** What masses of **(i)** $Al(OH)_3$ and **(ii)** $H_2SO_4$ are needed?

**5** Washing soda, $Na_2CO_3 . 10H_2O$, loses some of its water of crystallisation if it is not kept in an air-tight container to form $Na_2CO_3 . H_2O$. A shopkeeper buys a 10 kg bag of washing soda at 30p/kg. While it is standing in his store room, the bag punctures, and the crystals turn into a powder, $Na_2CO_3 . H_2O$. He sells this powder at 50p/kg. Does he make a profit or a loss?

**STEPS TO TAKE**

**a** Write the equation for the reaction.

**b** From the values of $M_r(Na_2CO_3 . 10H_2O)$ and $M_r(Na_2CO_3 . H_2O)$ deduce the ratio,

$$\frac{\text{Mass of } Na_2CO_3 . H_2O}{\text{Mass of } Na_2CO_3 . 10H_2O}$$

**c** Work out the mass of $Na_2CO_3 . H_2O$ formed from 10 kg of $Na_2CO_3 . 10H_2O$.

**d** What does it sell for? How does this compare with the purchase price?

**6** When you take a warm bath, the power station has to burn 1.2 kg of coal to provide enough energy to heat the water. If the coal contains 3% of sulphur, what mass of sulphur dioxide does the power station emit as a result? Multiply this by the number of warm baths you take in a year. This is only a part of your contribution to pollution. What can be done to reduce this source of pollution – apart from taking cold baths?

**STEPS TO TAKE**

**a** Find the mass of sulphur in 1.2 kg of coal.

**b** Write the equation for the burning of sulphur.

**c** Use values of $A_r(S)$ and $M_r(SO_2)$ to find what mass of sulphur dioxide is produced by the mass of sulphur which you found in **a**.

**d** Multiply the answer to **c** by your choice of a number between 52 and 365.

**7** Nitrogen monoxide, NO, is a pollutant gas which comes out of vehicle exhausts. One technique for reducing the quantity of nitrogen monoxide in vehicle exhausts is to inject a stream of ammonia, $NH_3$, into the exhaust. Nitrogen monoxide is converted into the harmless products nitrogen and water:

$$4NH_3(g) + 6NO(g) \longrightarrow 5N_2(g) + 6H_2O(l)$$

An average vehicle emits 5 g of nitrogen monoxide per mile. Assuming a mileage of 10 000 miles a year, what mass of ammonia would be needed to clean up the exhaust?

**STEPS TO TAKE**

**a** Find the mass of NO emitted per year.

**b** What amount (mol) of NO is this?

**c** What amount of $NH_3$ does it require?

**d** What mass of $NH_3$ is this?

**8** Some industrial plants, for example aluminium smelters, emit fluorides. In the past, there have been cases of fluoride pollution affecting the teeth and joints of cattle. The Union Carbide Corporation has invented a process for removing fluorides from waste gases. It involves the reaction:

$$\text{Fluoride ion } (F^-) + \text{Charcoal (C)} \xrightarrow{\text{high temperature}} \text{Carbon tetrafluoride } (CF_4)$$

The product, $CF_4$, is harmless. The firm claims that 1 kg of charcoal will remove 6.3 kg of fluoride ion. Do you believe this claim? Explain your answer.

**STEPS TO TAKE**

**a** What amount of $F^-$ is present in 6.3 kg of fluoride ion?

**b** Find **(i)** the amount of C **(ii)** the mass of C that will combine with it.

## Section 2

**1**  The element **X** has a relative atomic mass of 35.5. It reacts with a solution of the sodium salt of **Y** according to the equation

$$X_2 + 2NaY \longrightarrow Y_2 + 2NaX$$

If 14.2 g of $X_2$ displace 50.8 g of $Y_2$, what is the relative atomic mass of **Y**?

**2**  TNT is an explosive. The name stands for trinitrotoluene. The compound is made by the reaction

$$\text{Toluene} + \text{Nitric acid} \longrightarrow \text{TNT} + \text{Water}$$

$$C_7H_8(l) + 3HNO_3(l) \longrightarrow C_7H_5N_3O_6(s) + 3H_2O(l)$$

Calculate the masses of **a** toluene and **b** nitric acid that must be used to make 10.00 tonnes of TNT. (1 tonne = 1000 kg.)

**3**  A large power plant produces about 500 tonnes of sulphur dioxide in a day. One way of removing this pollutant from the waste gases is to inject limestone. This converts sulphur dioxide into calcium sulphate.

**(i)**

$$\text{Limestone} + \text{Sulphur dioxide} + \text{Oxygen} \longrightarrow \text{Calcium sulphate} + \text{Carbon dioxide}$$

$$2CaCO_3(s) + 2SO_2(g) + O_2(g) \longrightarrow 2CaSO_4(s) + 2CO_2(g)$$

Another method of removing sulphur dioxide is to 'scrub' the waste gases with ammonia. The product is ammonium sulphate.

**(ii)**

$$\text{Ammonia} + \text{Sulphur dioxide} + \text{Oxygen} + \text{Water} \longrightarrow \text{Ammonium sulphate}$$

$$4NH_3(g) + 2SO_2(g) + O_2(g) + 2H_2O(l) \longrightarrow 2(NH_4)_2SO_4(aq)$$

**a** Calculate the mass of calcium sulphate produced in a day by method **(i)** and the mass of ammonium sulphate produced in a day by method **(ii)**.

**b** Say whether limestone and ammonia are found naturally occurring or are manufactured.

**c** Which do you think is the more valuable by-product, calcium sulphate or ammonium sulphate? Explain your answer.

**4**  What mass of glucose must be fermented to give 5.00 kg of ethanol?

$$C_6H_{12}O_6(aq) \longrightarrow 2C_2H_5OH(aq) + 2CO_2(g)$$

5    What mass of sulphuric acid can be obtained from 1000 tonnes of an ore which contains 32.0% of $FeS_2$?

6    What mass of silver chloride can be precipitated from a solution which contains $1.000 \times 10^{-3}$ mol of silver ions?

7    When potassium iodate(V) is allowed to react with acidified potassium iodide, iodine is formed.

$$KIO_3(aq) + KI(aq) + 6H^+(aq) \longrightarrow 3I_2(aq) + 6K^+(aq) + 3H_2O(l)$$

What mass of $KIO_3$ is required to give 10.00 g of iodine?

8    How many tonnes of iron can be obtained from 10.00 tonnes of
**a** $Fe_2O_3$ and **b** $Fe_3O_4$?

9    What mass of quicklime can be obtained by heating 75.0 g of limestone, which is 86.8% calcium carbonate?

10    What mass of barium sulphate can be precipitated from a solution which contains 4.000 g of barium chloride?

## 5.2 WORKING OUT THE EQUATION FOR A REACTION

The equation for a reaction can be used to enable you to calculate the masses of chemicals taking part in the reaction. The converse is also true. If you know the mass of each substance taking part in a reaction, you can calculate the number of moles of each substance taking part in the reaction, and this will tell you the equation.

**EXAMPLE 1**

Iron burns in chlorine to form iron chloride. An experiment showed that 5.60 g of iron combined with 10.65 g of chlorine. Deduce the equation for the reaction.

**METHOD**

5.60 g of iron combine with 10.65 g of chlorine

Relative atomic masses are: Fe = 56, Cl = 35.5

Amount (mol) of iron $= \dfrac{5.60}{56} = 0.10$

Amount (mol) of chlorine molecules $= \dfrac{10.65}{71.0} = 0.15$

The equation must be:        $Fe + 1.5Cl_2 \longrightarrow$

or                     $2Fe + 3Cl_2 \longrightarrow$

To balance the equation, the right-hand side must read $2FeCl_3$. Therefore,

**ANSWER**

$$2Fe(s) + 3Cl_2(g) \longrightarrow 2FeCl_3(s)$$

**EXAMPLE 2**

17.0 g of sodium nitrate react with 19.6 g of sulphuric acid to give 12.6 g of nitric acid. Deduce the equation for the reaction.

**METHOD**

Molar masses/g $mol^{-1}$ are: $NaNO_3$ = 85, $H_2SO_4$ = 98, $HNO_3$ = 63

Amount of $NaNO_3$ = 17.0 g/85 g $mol^{-1}$ = 0.20 mol

Amount of $H_2SO_4$ = 19.6 g/98 g mol$^{-1}$ = 0.20 mol

Amount of $HNO_3$ = 12.6 g/63 g mol$^{-1}$ = 0.20 mol

0.2 mol $NaNO_3$ reacts with 0.2 mol $H_2SO_4$ to form 0.2 mol of $HNO_3$

1 mol $NaNO_3$ reacts with 1 mol $H_2SO_4$ to form 1 mol of $HNO_3$

$$NaNO_3(s) + H_2SO_4(l) \longrightarrow HNO_3(l)$$

The equation must be balanced by inserting $NaHSO_4$ on the right-hand side:

**ANSWER**

$$NaNO_3(s) + H_2SO_4(l) \longrightarrow HNO_3(l) + NaHSO_4(s)$$

**EXAMPLE 3**

An unsaturated hydrocarbon of molar mass 54 g mol$^{-1}$ reacts with bromine. If 1.00 g bromine adds to 0.169 g of the hydrocarbon, what is the equation for the reaction?

**METHOD**

$$\text{Amount of unsaturated hydrocarbon} = \frac{0.169\ g}{80\ g\,mol^{-1}} = 3.13 \times 10^{-3}\ mol$$

$$\text{Amount of bromine (Br}_2) = \frac{1.00\ g}{160\ g\,mol^{-1}} = 6.25 \times 10^{-3}\ mol$$

Amount of $Br_2$: amount of hydrocarbon = 2 : 1

Before we can write the equation we have to work out the formula of the unsaturated hydrocarbon.

$$\frac{\text{Mass of 1 molecule of hydrocarbon}}{\text{Mass of 1 atom of carbon}} = \frac{54}{12}$$

therefore the maximum number of carbon atoms per molecule = 4

The number of H atoms = 54 − (4 × 12) = 6, and the formula is $C_4H_6$

$$C_4H_6 + 2Br_2 \longrightarrow C_4H_6Br_4$$

(The hydrocarbon is buta-1,4-diene $CH_2{=}CH{-}CH{=}CH_2$.)

---

**Reminder :** Mass = molar mass × amount § 4.1. Ratio calculations § 1.2.

Balancing equations § 2.2.

---

**EXERCISE 5.2**    **Problems on deriving equations**

**1**    A mass of 0.65 g of zinc powder was added to a beaker containing silver nitrate solution. When all the zinc had reacted 2.15 g of silver were obtained. Derive the ionic equation for the reaction.

**STEPS TO TAKE**    Calculate:

**a**  the amount of zinc used (from mass and $A_r$)

**b**  the amount of silver formed

**c**  the amount of silver produced by 1 mol of zinc

**d** write a balanced ionic equation for the reaction.

**2** Titanium(IV) chloride, $TiCl_4$, is reduced by magnesium to titanium. Given that 9.5 g of $TiCl_4$ are reduced by 2.4 g of magnesium, derive the equation for the reaction.

STEPS TO TAKE

**a** Calculate the amount of $TiCl_4$ ($M_r = 190$).

**b** Calculate the amount of Mg ($A_r = 24$).

**c** Work out the ratio $TiCl_4$ : Mg.

Nearly there now! What is the equation for the reaction?

**3** To a solution containing 1.198 g of sodium persulphate, $Na_2S_2O_8$, is added an excess of aqueous potassium iodide. A reaction occurs between persulphate ions and iodide ions to form sulphate ions and 1.27 g of iodine. The unbalanced equation is

$$S_2O_8{}^{2-}(aq) + I^-(aq) \longrightarrow SO_4{}^{2-}(aq) + I_2(aq)$$

Use the data to balance the equation for the reaction.

STEPS TO TAKE

**a** Calculate **(i)** the amount of persulphate **(ii)** the amount of iodine.

**b** Find the ratio moles of persulphate : moles of iodine. Then balance the equation.

**4** Aminosulphuric acid, $H_2NSO_3H$, reacts with warm aqueous sodium hydroxide to give ammonia and a solution which contains sulphate ions. 0.485 g of the acid, when treated

---

**Hint :** Amount of reactant ... amount of product ... ratio ... equation.

---

with an excess of alkali gave 120 cm$^3$ of ammonia (at r.t.p.). Deduce the equation for the reaction. (GMV = 24.0 dm$^3$ at r.t.p.)

**5** An unsaturated hydrocarbon of molar mass 54 g mol$^{-1}$ reacts with bromine. If 0.169 g of hydrocarbon reacts with 1.00 g of bromine, what is **a** the formula of the hydrocarbon **b** the equation for the reaction?

**6** Given that 1.00 g of phenylamine $C_6H_5NH_2$, reacts with 5.16 g of bromine, with the formation of HBr, derive an equation for the reaction.

## 5.3 PERCENTAGE YIELD

There are many reactions which do not go to completion. Reactions between organic

$$\text{Percentage yield} = \frac{\text{Actual mass of product}}{\text{Calculated mass of product}} \times 100\%$$

compounds do not often give a 100% yield of product. The actual yield is compared with

the yield calculated from the molar masses of the reactants. The equation
is used to give the percentage yield.

**EXAMPLE**     From 23 g of ethanol are obtained 36 g of ethyl ethanoate by esterification with ethanoic
acid in the presence of concentrated sulphuric acid. What is the percentage yield of the
reaction?

**METHOD**     Write the equation:

$$CH_3CO_2H(l) + C_2H_5OH(l) \longrightarrow CH_3CO_2C_2H_5(l) + H_2O(l)$$

46 g of $C_2H_5OH$ forms 88 g of $CH_3CO_2C_2H_5$

23 g of $C_2H_5OH$ should give $\dfrac{23}{46} \times 88$ g $= 44$ g of $CH_3CO_2C_2H_5$

Actual mass obtained $= 36$ g

Percentage yield $= \dfrac{\text{Actual mass of product}}{\text{Calculated mass of product}} \times 100\%$

**ANSWER**     Percentage yield $= \dfrac{36}{44} \times 100\% = 82\%$

---

**Reminder :** Percentages § 1.3. Ratio calculations § 1.2.

---

## EXERCISE 5.3     Problems on percentage yield

**1**     When 25.0 g of ethanol, $C_2H_5OH$, were oxidised, 29.5 g of ethanoic acid, $CH_3CO_2H$,
were obtained. What percentage of the theoretical yield was obtained?

**STEPS TO TAKE**     Since

$$C_2H_5OH + 2[O] \longrightarrow CH_3CO_2H + H_2O$$

using the molar masses, 46 g of $C_2H_5OH$ gives 60 g of $CH_3CO_2H$

Therefore 25.0 g $C_2H_5OH$ gives ? g of $CH_3CO_2H$

Actual yield of $CH_3CO_2H = 29.5$ g

so percentage yield $= ?$

**2**     Phenol, $C_6H_5OH$, is converted into trichlorophenol, $C_6H_2Cl_3OH$. If 488 g of product are
obtained from 250 g of phenol, calculate the percentage yield.

**STEPS TO TAKE**     Since

$$C_6H_5OH + 3Cl_2 \longrightarrow C_6H_2Cl_3OH + 3HCl$$

using molar masses, 94 g of $C_6H_5OH$ gives 197.5 g of $C_6H_2Cl_3OH$

therefore 250 g of $C_6H_5OH$ gives ? g of $C_6H_2Cl_3OH$

Actual yield of product $= 488$ g

so percentage yield $= ?$

**3**     When 10.2 g phenylamine $C_6H_5NH_2$ react with an excess of ethanoyl chloride $CH_3COCl$, the yield of purified N-phenylethanamide, $C_6H_5NHCOCH_3$, is 9.20 g. Calculate the percentage yield.

4   A sample of ethanoyl chloride, 7.85 g, reacted with an excess of ammonia to form 4.43 g of ethanamide.

$$CH_3COCl + NH_3 \longrightarrow CH_3CONH_2 + HCl$$

The percentage yield was **A** 37.5% **B** 50% **C** 75% **D** 100%

5   When an excess of tin is refluxed with iodine in an organic solvent tin(IV) chloride is formed.

$$Sn + 2I_2 \longrightarrow SnI_4$$

When 3.18 g of iodine were used with an excess of tin 1.95 g of $SnI_4$ crystals were obtained. Calculate the percentage yield.

6   0.8500 g of hexanone, $C_6H_{12}O$, is converted into its 2,4-dinitrophenylhydrazone. After isolation and purification, 2.1180 g of product, $C_{12}H_{18}N_4O_4$, are obtained. What percentage yield does this represent?

7   Benzaldehyde, $C_7H_6O$, forms a hydrogensulphite compound of formula $C_7H_7SO_4Na$. From 1.210 g of benzaldehyde, a yield of 2.181 g of the product was obtained. Calculate the percentage yield.

8   100 cm$^3$ of barium chloride solution of concentration 0.0500 mol dm$^{-3}$ were treated with an excess of sulphate ions in solution. The precipitate of barium sulphate formed was dried and weighed. A mass of 1.1558 g was recorded. What percentage yield does this represent?

## 5.4 LIMITING REACTANT

In a chemical reaction, the reactants are often added in amounts which are not stoichiometric. One or more of the reactants is in excess and is not completely used up in the reaction. The amount of product is determined by the amount of the reactant that is not in excess and is used up completely in the reaction. This is called the **limiting reactant**. You first have to decide which is the limiting reactant before you can calculate the amount of product formed.

**EXAMPLE**   5.00 g of iron and 5.00 g of sulphur are heated together to form iron(II) sulphide. Which reactant is present in excess? What mass of product is formed?

**METHOD**   Write the equation:

$$Fe(s) + S(s) \longrightarrow FeS(s)$$

1 mole of Fe + 1 mole of S form 1 mole of FeS

∴   56 g Fe and 32 g S form 88 g FeS

5.00 g Fe is $\dfrac{5}{56}$ mol = 0.0893 mol Fe

5.00 g S is $\dfrac{5}{32}$ mol = 0.156 mol S

There is insufficient Fe to react with 0.156 mol S; iron is the limiting reactant.

0.0893 mol Fe forms 0.0893 mol FeS = 0.0893 × 88 g = 7.86 g

**ANSWER**

Mass formed = 7.86 g.

## EXERCISE 5.4    Problems on limiting reactant

**1**    When 9.0 g of aluminium were treated with an excess of chlorine, 20.0 g of $Al_2Cl_6(s)$ were collected.

$$2Al(s) + 3Cl_2(g) \longrightarrow Al_2Cl_6(s)$$

What was the percentage yield?

**STEPS TO TAKE**

**a**  Which is the limiting reactant?

**b**  What amount of this reactant is present?

**c**  What amount of product should be formed?

**d**  What mass of $Al_2Cl_6$ is this?

**e**  What was the percentage yield?

**2**    In the manufacture of calcium carbide

$$CaO(s) + 3C(s) \longrightarrow CaC_2(s) + CO(g)$$

What is the maximum mass of calcium carbide that can be obtained from 40 kg of quicklime and 40 kg of coke?

**STEPS TO TAKE**

Amount of CaO = $40 \times 10^3$ g/56 g mol$^{-1}$ = 714 mol

This amount of CaO needs 3 × 714 mol of coke.

What amount of coke (carbon) is present?

Which reactant is in excess and which is the limiting reactant?

Now focus on the limiting reactant and use the equation to work out the amount and from this the mass of $CaC_2$ formed.

**3**    In the manufacture of the fertiliser ammonium sulphate

$$H_2SO_4(aq) + 2NH_3(g) \longrightarrow (NH_4)_2SO_4(aq)$$

What is the maximum mass of ammonium sulphate that can be obtained from 2.0 kg of sulphuric acid and 1.0 kg of ammonia?

Calculate the amount of **a** $H_2SO_4$ **b** $NH_3$. Decide which is the limiting reactant. Then focus on the limiting reactant to work out the yield of ammonium sulphate.

**Hint :** Look out for the factor 2.

**4**  In the Solvay process, ammonia is recovered by the reaction

$$2NH_4Cl(s) + CaO(s) \longrightarrow CaCl_2(s) + H_2O(g) + 2NH_3(g)$$

What is the maximum mass of ammonia that can be recovered from $2.00 \times 10^3$ kg of ammonium chloride and 500 kg of quicklime?

**5**  In the Thermit reaction

$$2Al(s) + Cr_2O_3(s) \longrightarrow 2Cr(s) + Al_2O_3(s)$$

Calculate the percentage yield when 180 g of chromium are obtained from a reaction between 100 g of aluminium and 400 g of chromium(III) oxide.

**EXERCISE 5.5**  **Questions from examination papers**

**1**  The composition of a mixture of two solid sodium halides was investigated in two separate experiments.

Experiment (1)

> When a large excess of chlorine gas was bubbled through a concentrated solution of the mixture, orange-brown fumes and a black precipitate were produced.

Experiment (2)

> 0.545 g of the solid mixture was dissolved in water and an excess of silver nitrate solution was added. The mass of the mixture of silver halide precipitates formed was 0.902 g. After washing the mixture of precipitates with an excess of concentrated aqueous ammonia the mass of the final precipitate was 0.564 g.

Write equations for each of the reactions occurring in these experiments and explain how these results enable you to identify the halide ions present. Use the information given above to calculate the percentage by mass of each halide ion present in the solid mixture.

*NEAB, 99 (15)*

# Finding formulae

6

## 6.1 EMPIRICAL FORMULAE

The formula of a compound is determined by finding the mass of each element present in a certain mass of the compound. Remember,

$$\text{Amount (in moles) of substance} = \frac{\text{Mass of substance}}{\text{Molar mass of the substance}}$$

**EXAMPLE 1**

Given that 127 g of copper combine with 32 g of oxygen, what is the formula of copper oxide?

| Elements | Copper | | Oxygen |
|---|---|---|---|
| Symbols | Cu | | O |
| Masses | 127 g | | 32 g |
| Relative atomic masses | 63.5 | | 16 |
| Amounts | 127/63.5 | | 32/16 |
| | = 2 mol | | = 2 mol |
| Divide through by 2 | = 1 mol | to | 1 mol |
| Ratio of atoms | = 1 atom | to | 1 atom |
| Formula | | CuO | |

We divide through by two to obtain the simplest formula for copper oxide which will fit the data. **The simplest formula which represents the composition of a compound is called the empirical formula.**

**EXAMPLE 2**

When 127 g of copper combine with oxygen, 143 g of an oxide are formed. What is the empirical formula of the oxide?

**METHOD**

You will notice here that the mass of oxygen is not given to you. You obtain it by subtraction.

Mass of copper = 127 g

Mass of oxide = 143 g

Mass of oxygen = 143 − 128 = 16 g

Now you can carry on as before:

| Elements | Copper | | Oxygen |
|---|---|---|---|
| Symbols | Cu | | O |
| Masses | 127 g | | 16 g |
| Relative atomic masses | 63.5 | | 16 |
| Amounts | 127/63.5 | | 16/16 |
| | = 2 | | = 1 |
| Ratio of atoms | = 2 | to | 1 |

**ANSWER**

Empirical formula $Cu_2O$.

**EXAMPLE 3**     An oxide of iron contains 72.4% iron and 27.6% oxygen. Work out its formula.

Reminder : Percentages § 1.3.

**METHOD**

| Elements | Iron | Oxygen |
|---|---|---|
| Symbols | Fe | O |
| Percentages | 72.4 | 27.6 |
| Relative atomic masses | 56 | 16 |
| Amounts | $\dfrac{72.4}{56}$ | $\dfrac{27.6}{16}$ |
| | = 1.29 | 1.725 |

Divide both numbers by the smaller number ⟹
Ratio of amounts           = 1 to 1.334
In whole numbers of atoms    = 3 to 4

**ANSWER**     Empirical formula $Fe_3O_4$.

**Notice** that in this example we used percentages in just the same way as masses in earlier examples.

**Notice** how we simplified the ratio 1.29 : 1.725 by dividing both numbers by the smaller number to give 1 : 1.334 or $1 : 1\frac{1}{3}$.

We need a formula with whole numbers of atoms so:

the ratio $1 : 1\frac{1}{3} = 1 : \frac{4}{3} = \frac{3}{3} : \frac{4}{3} = 3 : 4$

**EXAMPLE 4**     Find $n$ in the formula $MgSO_4 . nH_2O$. A sample of 7.38 g of magnesium sulphate crystals lost 3.78 g of water on heating.

**METHOD**

| Compounds present | Magnesium sulphate | Water |
|---|---|---|
| Mass | 3.60 g | 3.78 g |
| Molar mass | 120 g mol$^{-1}$ | 18 g mol$^{-1}$ |
| Amount | 3.60/120 | 3.78/18 |
| | = 0.030 mol | = 0.21 mol |
| Ratio of amounts | $\dfrac{0.030}{0.030}$ | $\dfrac{0.21}{0.030}$ |
| | = 1 | = 7 |

**ANSWER**     The empirical formula is $MgSO_4 . 7H_2O$.

**EXERCISE 6.1**     **Problems on empirical formulae**

> **Reminder :** Ratio calculations § 1.2. Percentage § 1.3.
> Refer to the table of relative atomic masses on p 243.

**1**    An oxide of nitrogen, 0.010 mol, is mixed with an excess of hydrogen and passed over a hot catalyst. The reaction which occurs is:

$$N_xO_y \longrightarrow xNH_3(g) + H_2O(l)$$

The mass of water produced is 7.20 g. The ammonia produced is neutralised by 200 cm$^3$ of 1.00 mol dm$^{-3}$ hydrochloric acid. Which is the formula of the oxide?

**A** $N_2O$   **B** $NO$   **C** $NO_2$   **D** $N_2O_4$

**2**    The percentage by mass composition of **X** is: C 54.5%, H 9.10%, O 36.4%.

Find the empirical formula of **X**.

**3**    A compound has the percentage by mass composition Ni 37.9%, S 20.7%, O 41.4%.

Calculate the empirical formula.

**4**    A compound of carbon, hydrogen and chlorine has the percentage by mass composition shown: C 18.24%, H 0.76%, Cl 81.00%.

Calculate the empirical formula.

**5**    Tin reacts with iodine in an organic solvent to form a covalent compound $SnI_x$. When 0.4650 g of tin (an excess) was used all the iodine reacted. A mass 0.1230 g of unreacted tin and 1.8020 g of $SnI_x$ were obtained. Find the value of $x$.

**6**    0.72 g of magnesium combine with 0.28 g of nitrogen.

**STEPS TO TAKE**

   **a**  How many moles of magnesium does this represent?

   **b**  How many moles of nitrogen atoms combine?

   **c**  How many moles of magnesium combine with one mole of nitrogen atoms?

   **d**  What is the formula of magnesium nitride?

**7**    1.68 g of iron combine with 0.64 g of oxygen.

**STEPS TO TAKE**

   **a**  How many moles of iron does this mass represent?

   **b**  How many moles of oxygen atoms combine?

   **c**  How many moles of iron combine with one mole of oxygen atoms?

   **d**  What is the formula of this oxide of iron?

**8**    Calculate the empirical formula of the compound formed when 2.70 g of aluminium form 5.10 g of its oxide.

**STEPS TO TAKE**

   **a**  What is the mass of aluminium?

   **b**  What is the mass of oxygen (not oxide)?

**c** How many moles of aluminium combine?

**d** How many moles of oxygen atoms combine?

**e** What is the ratio of moles of aluminium to moles of oxygen atoms?

**f** What is the formula of aluminium oxide?

**9** Barium chloride forms a hydrate which contains 85.25% barium chloride and 14.75% water of crystallisation. What is the formula of this hydrate?

STEPS TO TAKE      **a** What is the mass of barium chloride in 100 g of the hydrate?

**b** What is the mass of water in 100 g of the hydrate?

**c** What is the relative formula mass of barium chloride?

**d** What is the relative molecular mass of water?

**e** How many moles of barium chloride are present in 100 g of the hydrate?

**f** How many moles of water are present in 100 g of the hydrate?

**g** What is the ratio of moles of barium chloride to moles of water?

**h** What is the formula of barium chloride hydrate?

**10** Calculate the empirical formula of the compound formed when 414 g of lead form 478 g of a lead oxide.

STEPS TO TAKE      **a** What mass of lead is present?

**b** How many moles of lead are present?

**c** What mass of oxygen (not oxide) is present?

**d** How many moles of oxygen atoms are present?

**e** What is the formula of this oxide of lead?

## 6.2 MOLECULAR FORMULAE

The molecular formula is a simple multiple of the empirical formula. If the empirical formula is $CH_2O$, the molecular formula may be $CH_2O$ or $C_2H_4O_2$ or $C_3H_6O_3$ and so on. You can tell which molecular formula is correct by finding out which gives the correct molar mass.

For methods of finding molar masses, see Chapters 9 and 10. The molar mass is a multiple of the empirical formula mass.

EXAMPLE      A compound has the empirical formula $CH_2O$ and molar mass 180 g mol$^{-1}$. What is its molecular formula?

METHOD      Empirical formula mass = 30 g mol$^{-1}$
Molar mass = 180 g mol$^{-1}$

The molar mass is 6 times the empirical formula mass. Therefore the molecular formula is 6 times the empirical formula. Therefore:

ANSWER      The empirical formula is $C_6H_{12}O_6$.

## EXERCISE 6.2    Problems on formulae

**Reminder :** Ratio calculations § 1.2. Percentage § 1.3. Amount = Mass/Molar mass § 4.3. Refer to the table of relative atomic masses on p 243.

**1** Find the molecular formula for each of the following compounds from the empirical formula and the relative molecular mass:

| | Empirical formula $M_r$ | | | Empirical formula $M_r$ | |
|---|---|---|---|---|---|
| a | $CF_2$ | 100 | e | $CH_2$ | 42 |
| b | $C_2H_4O$ | 88 | f | $CH_3O$ | 62 |
| c | $CH_3$ | 30 | g | $CH_2Cl$ | 99 |
| d | $CH$ | 78 | h | $C_2HNO_2$ | 213 |

**2** A metal **M** forms a chloride of formula $MCl_2$ and relative molecular mass 127. The chloride reacts with sodium hydroxide solution to form a precipitate of the metal hydroxide $M(OH)_2$. What is the relative molecular mass of the hydroxide?

**A** 56    **B** 71    **C** 73    **D** 90    **E** 146

**3** A porcelain boat was weighed. After a sample of the oxide of a metal **M**, of $A_r$ = 119, was placed in the boat, the boat was reweighed. Then the boat was placed in a reduction tube, and heated while a stream of hydrogen was passed over it. The oxide was reduced to the metal **M**. The boat was allowed to cool with hydrogen still passing over it, and then reweighed. Then it was reheated in hydrogen, cooled again and reweighed. The following results were obtained.

Mass of boat = 6.10 g

Mass of boat + metal oxide = 10.63 g

Mass of boat + metal (1) = 9.67 g

Mass of boat + metal (2) = 9.67 g

**a** Explain why hydrogen was passed over the boat while it was cooling.

**b** Explain why the boat + metal was reheated.

**c** Find the empirical formula of the metal oxide.

**d** Write an equation for the reaction of the oxide with hydrogen.

**4** Find the empirical formulae of the compounds formed in the reactions described below:

**a** 10.800 g magnesium form 18.000 g of an oxide

**b** 3.400 g calcium form 9.435 g of a chloride

**c** 3.528 g iron form 10.237 g of a chloride

**d** 2.667 g copper form 4.011 g of a sulphide

**e** 4.662 g lithium form 5.328 g of a hydride.

5 Calculate the empirical formulae of the compounds with the following percentage composition:

a 77.7% Fe   22.3% O   b 70.0% Fe   30.0% O   c 72.4% Fe   27.6% O

d 40.2% K   26.9% Cr   32.9% O   e 26.6% K   35.4% Cr   38.0% O

6 Samples of the following hydrates are weighed, heated to drive off the water of crystallisation, cooled and reweighed. From the results obtained, calculate the values of **a–c** in the formulae of the hydrates:

a 0.869 g of $CuSO_4 . aH_2O$ gave a residue of 0.556 g

b 1.173 g of $CoCl_2 . bH_2O$ gave a residue of 0.641 g

c 1.886 g of $CaSO_4 . cH_2O$ gave a residue of 1.492 g

7 A liquid, **Y**, of molar mass 44 g $mol^{-1}$ contains 54.5% carbon, 36.4% oxygen and 9.1% hydrogen.

a Calculate the empirical formula of **Y**

b deduce its molecular formula.

8 An organic compound contains 58.8% carbon, 9.8% hydrogen and 31.4% oxygen. The molar mass is 102 g $mol^{-1}$.

a Calculate the empirical formula

b deduce the molecular formula of the compound.

9 An organic compound has molar mass 150 g $mol^{-1}$ and contains 72.00% carbon, 6.67% hydrogen and 21.33% oxygen. What is its molecular formula?

# Reacting volumes of gases

# 7

## 7.1 COMBINING VOLUMES OF GASES

In a reaction between gases the volumes of gases that react are in simple ratio. For example,

$$1 \text{ dm}^3 \text{ of hydrogen} + 1 \text{ dm}^3 \text{ of chlorine} \longrightarrow 2 \text{ dm}^3 \text{ of hydrogen chloride}$$

The Italian chemist Amadeus Avogadro gave an explanation of this behaviour. In 1811 he made a suggestion which we know as **Avogadro's Hypothesis**. This states:

> Equal volumes of gases (at the same temperature and pressure) contain the same number of molecules.

It follows from Avogadro's Hypothesis that whenever we see an equation for a reaction between gases we can substitute volumes of gases in the same ratio as numbers of molecules. The equation

$$N_2(g) + 3H_2(g) \longrightarrow 2NH_3(g)$$

means that since

1 molecule of nitrogen + 3 molecules of hydrogen form 2 molecules of ammonia, then
1 volume of nitrogen + 3 volumes of hydrogen form 2 volumes of ammonia. For example,
$1 \text{ dm}^3$ of nitrogen + $3 \text{ dm}^3$ of hydrogen form $2 \text{ dm}^3$ of ammonia.

## 7.2 GAS MOLAR VOLUME

Starting from the statement,

> equal volumes of gases (at the same temperature and pressure) contain the same number of molecules,

it follows that

> equal numbers of molecules occupy the same volume

That is, a number $L$ molecules of carbon dioxide, $L$ molecules of hydrogen, $L$ molecules of oxygen and so on, occupy the same volume.

If $L$ is the number of molecules in one mole of gas, the volume is the **gas molar volume**.

In stating the volume of a gas we need to state the temperature and pressure at which the volume is measured. There are two ways of doing this:

- standard temperature and pressure (s.t.p.) = 0 °C and 1 atm (273 K and 101 kPa)

- room temperature and pressure (r.t.p.) = 20 °C and 1 atm (293 K and 101 kPa)

> The gas molar volume (the volume occupied by one mole of gas)
> $$= 22.4 \text{ dm}^3 \text{ mol}^{-1} \text{ at } 0 \text{ °C and 1 atm (s.t.p.)}$$
> $$= 24.0 \text{ dm}^3 \text{ mol}^{-1} \text{ at } 20 \text{ °C and 1 atm (r.t.p.)}$$

## 7.3 EQUATIONS AND VOLUMES

The gas molar volume makes calculations on reacting volumes of gases very simple. An equation which shows how many moles of different gases react together also shows the ratio of the volumes of the different gases that react. For example, the equation

$$2NO(g) + O_2(g) \longrightarrow 2NO_2(g)$$

tells us that 2 moles of NO + 1 mole of $O_2$ form 2 moles of $NO_2$

$\therefore$ 44.8 dm$^3$ of NO + 22.4 dm$^3$ of $O_2$ form 44.8 dm$^3$ of $NO_2$

In general, 2 volumes of NO + 1 volume of $O_2$ form 2 volumes of $NO_2$.

**EXAMPLE 1**

What is the volume of oxygen needed for the complete combustion of 2 dm$^3$ of propane?

**METHOD**

Write the equation:

$$C_3H_8(g) + 5O_2(g) \longrightarrow 3CO_2(g) + 4H_2O(g)$$

1 mole of $C_3H_8$ needs 5 moles of $O_2$

1 volume of $C_3H_8$ needs 5 volumes of $O_2$. Therefore:

**ANSWER**

2 dm$^3$ of propane need 10 dm$^3$ of oxygen.

**EXAMPLE 2**

10 cm$^3$ of a gas were mixed with 100 cm$^3$ of oxygen (an excess) in a sealed tube at 20 °C and 1 atm. The gas was ignited by a spark and burned completely in the oxygen. After the reaction the volume of gas at the same temperature and pressure was 105 cm$^3$. Which one of the following was the gas?

**A** hydrogen $H_2$     **B** carbon monoxide CO     **C** methane $CH_4$     **D** ethene $C_2H_4$

**METHOD**

As usual begin with the equations

**A** $2H_2(g) + O_2(g) \longrightarrow 2H_2O(l)$

10 cm$^3$ $H_2$ react with 5 cm$^3$ $O_2$ to form no gas (only liquid water), therefore the volume decreases by 15 cm$^3$.

**B** $2CO(g) + O_2(g) \longrightarrow 2CO_2(g)$

10 cm$^3$ CO reacts with 5 cm$^3$ $O_2$ to form 10 cm$^3$ $CO_2$, a decrease in volume of 5 cm$^3$

**C** $CH_4(g) + 2O_2(g) \longrightarrow CO_2(g) + 2H_2O(l)$

10 cm$^3$ $CH_4$ reacts with 20 cm$^3$ $O_2$ to form 10 cm$^3$ $CO_2$, a decrease in volume of 20 cm$^3$

**D** $C_2H_4(g) + 3O_2(g) \longrightarrow 2CO_2(g) + 2H_2O(l)$

10 cm$^3$ $C_2H_4$ reacts with 30 cm$^3$ $O_2$ to form 20 cm$^3$ $CO_2$, a decrease in volume of 10 cm$^3$

The decrease in volume observed was 5 cm$^3$ which shows that the gas is carbon monoxide.

**ANSWER**

The gas is **B**, carbon monoxide.

## 7.4 FINDING FORMULAE BY COMBUSTION

The formula of a hydrocarbon can be found from the results of a combustion experiment.

**1**   The hydrocarbon in the vapour phase is burned in an excess of oxygen. The mixture of gases formed consists of carbon dioxide, water vapour and unused oxygen.

**2**   The volume of water vapour is found by cooling the mixture to room temperature to condense the water and noting the decrease in volume.

**3**   The volume of carbon dioxide is found by absorbing it in an alkali and noting the decrease in volume.

**4**   The remaining gas is unused oxygen.

The results are used as in the following examples.

**EXAMPLE 1**   When 100 cm$^3$ of the hydrocarbon **X** burn in 500 cm$^3$ of oxygen, 50 cm$^3$ of oxygen remain unused. 300 cm$^3$ of carbon dioxide and 300 cm$^3$ of water vapour are formed. Deduce the equation for the combustion and the formula of the hydrocarbon.

**METHOD**
$$X(g) \; + \; O_2(g) \longrightarrow CO_2(g) + H_2O(g)$$
$$100 \text{ cm}^3 \quad 450 \text{ cm}^3 \qquad 300 \text{ cm}^3 \quad 300 \text{ cm}^3$$

The volumes of gases reacting tell us that

$$X(g) + 4\tfrac{1}{2}O_2(g) \longrightarrow 3CO_2(g) + 3H_2O(g)$$

With 3 C atoms on the RHS and 6 H atoms on the RHS, to balance the equation **X** must be $C_3H_6$ and

$$C_3H_6(g) + 4\tfrac{1}{2}O_2(g) \longrightarrow 3CO_2(g) + 3H_2O(g)$$

**ANSWER**   **X** is $C_3H_6$.

$$2C_3H_6(g) + 9O_2(g) \longrightarrow 6CO_2(g) + 6H_2O(g)$$

**EXAMPLE 2**   10 cm$^3$ of a hydrocarbon were exploded with an excess of oxygen. The products were 30 cm$^3$ carbon dioxide and 40 cm$^3$ of water vapour (all volumes at 80 °C). Deduce the formula of the hydrocarbon and write the equation for its combustion.

**METHOD**   Let the formula of the hydrocarbon be $C_xH_y$.

$$C_xH_y(g) + O_2(g) \longrightarrow CO_2(g) + H_2O(g)$$
$$10 \text{ cm}^3 \qquad 30 \text{ cm}^3 \quad 40 \text{ cm}^3$$

The ratio of the volumes of the gases is the ratio of amounts of the compounds, therefore

$$1C_xH_y(g) + nO_2(g) \longrightarrow 3CO_2(g) + 4H_2O(g)$$

With 3 C atoms needed on the RHS of the equation, $x = 3$.

With 8 H atoms needed on the RHS of the equation, $y = 8$.

**ANSWER**   The hydrocarbon is $C_3H_8$.

With 10 O atoms needed on the RHS of the equation, $5O_2$ are needed on the LHS.

**ANSWER**          The equation is

$$C_3H_8(g) + 5O_2(g) \longrightarrow 3CO_2(g) + 4H_2O(g)$$

**EXERCISE 7.1**     **Problems on reacting volumes of gases**

Section 1

**Reminder :** Ratio calculations § 1.2. Percentages § 1.3.
GMV = 24.0 dm$^3$ mol$^{-1}$ at r.t.p. and 22.4 dm$^3$ mol$^{-1}$ at s.t.p.

**1**    The problem is to find the percentage by volume composition of a mixture of hydrogen and
ethane. When 75 cm$^3$ of the mixture was burned in an excess of oxygen the volume of
carbon dioxide produced was 60 cm$^3$ (all volumes at r.t.p.). The equation for the complete
combustion of ethane is

$$2C_2H_6(g) + 7O_2(g) \longrightarrow 4CO_2(g) + 6H_2O(g)$$

**a**  Say what volume of ethane would give 60 cm$^3$ of carbon dioxide.

**b**  Calculate the percentage of ethane in the mixture.

**2**    25 cm$^3$ of carbon monoxide are ignited with 25 cm$^3$ of oxygen. All gas volumes are
measured at the same temperature and pressure.

**a**  Write the equation for the combustion of carbon monoxide to form carbon dioxide.

**b**  What volume of CO$_2$ is produced when 25 cm$^3$ of CO burn?

**c  (i)**  Give the volume of O$_2$ that combines with 25 cm$^3$ of CO and
  **(ii)**  give the volume of oxygen that remains unused.

**d  (i)**  What is the total volume of carbon dioxide and oxygen and
  **(ii)**  the reduction in the total volume?
          **A** 2.5 cm$^3$   **B** 10.0 cm$^3$   **C** 12.5 cm$^3$   **D** 15.0 cm$^3$   **E** 25.0 cm$^3$

**3**    Ethene reacts with oxygen according to the equation

$$C_2H_4(g) + 3O_2(g) \longrightarrow 2CO_2(g) + 2H_2O(l)$$

15.0 cm$^3$ of ethene were mixed with 60.0 cm$^3$ of oxygen and the mixture was sparked to
complete the reaction. The volume of the products (all volumes measured at room
temperature and pressure) was

**A** 15 cm$^3$   **B** 30 cm$^3$   **C** 45 cm$^3$   **D** 60 cm$^3$   **E** 75 cm$^3$

**4**    The table gives the formulae and relative molecular masses of some gases.

| Formula | Ne | C$_2$H$_2$ | O$_2$ | Ar | NO$_2$ | SO$_2$ | SO$_3$ |
|---|---|---|---|---|---|---|---|
| $M_r$ | 20 | 26 | 32 | 40 | 46 | 64 | 80 |
| Volume (cm$^3$) occupied by 1 g of gas at s.t.p. | 1120 | 861 | 700 | 560 | 485 | 350 | 280 |

**a** Plot a graph of volume (on the vertical axis) against $M_r$ (on the horizontal axis).

**b** Use the graph to predict the volumes occupied at s.t.p. by
**(i)** 1 g of fluorine, $F_2$ **(ii)** 1 g of $Cl_2$

**c** What is the relative molecular mass of a gas which occupies 508 cm$^3$ per gram at s.t.p.? If the gas contains only carbon and oxygen, what is its formula?

**5 a** The gas molar volume is 24 dm$^3$ at r.t.p. Air is 20% oxygen.

Calculate:

**(i)** the minimum volume of air needed for complete combustion of 24 dm$^3$ of methane,
**(ii)** the minimum air: methane ratio by volume for complete combustion.

**b** Gas fires use a 16 : 1 mixture by volume of air and methane. Why do manufacturers choose to use a different ratio from that calculated in **a**?

**6** 40 cm$^3$ of butene react with 20 cm$^3$ of hydrogen in the presence of a nickel catalyst at 140 °C to give a mixture of butane and butene. What is the ratio of butane to butene in the mixture? (All gas volumes are at the same temperature and pressure.)
**A** 1 : 1 **B** 2 : 1 **C** 1 : 2 **D** 1 : 4

**7** 10 cm$^3$ of a hydrocarbon were burned in 70 cm$^3$ of oxygen (an excess) to give 30 cm$^3$ CO$_2$ and 20 cm$^3$ O$_2$. The formula of the hydrocarbon is

**A** $C_2H_6$ **B** $C_3H_6$ **C** $C_3H_8$ **D** $C_4H_{10}$

**STEPS TO TAKE**

**a** Write the equations for the combustion of **A**, **B**, **C** and **D**.

**b** Which hydrocarbons give 30 cm$^3$ CO$_2$ per 10 cm$^3$ of hydrocarbon?

**c** Of these, which uses 50 cm$^3$ O$_2$ per 10 cm$^3$ of hydrocarbon?

**8** When North Sea gas burns completely, it forms carbon dioxide and water and no other products. When 250 cm$^3$ of North Sea gas burn, they need 500 cm$^3$ of oxygen, and they form 250 cm$^3$ of carbon dioxide and 500 cm$^3$ of water vapour (all volumes being measured under the same conditions). Deduce the equation for the reaction and the formula of North Sea gas.

**9** 10 cm$^3$ of a hydrocarbon $C_aH_b$ were exploded with an excess of oxygen. The products were 30 cm$^3$ of carbon dioxide and 40 cm$^3$ of water vapour. Deduce the values of $a$ and $b$ in the formula of the hydrocarbon.

## Section 2

**1** Ethene, $H_2C{=}CH_2$, and hydrogen react in the presence of a nickel catalyst to form ethane.

**a** Write a balanced equation for the reaction.

**b** If a mixture of 30 cm$^3$ of ethene and 20 cm$^3$ of hydrogen is passed over a nickel catalyst, what is the composition of the final mixture? (Assume that the reaction is complete and that all gas volumes are at s.t.p.)

**2** State the volume of oxygen (all gas volumes at s.t.p.) required to burn exactly:

**a** 1 dm$^3$ of methane, according to the reaction

$$CH_4(g) + 2O_2(g) \longrightarrow CO_2(g) + 2H_2O(g)$$

**b** 500 cm$^3$ of hydrogen sulphide, according to the reaction

$$2H_2S(g) + 3O_2(g) \longrightarrow 2SO_2(g) + 2H_2O(g)$$

**c** 250 cm$^3$ of ethyne, according to the equation

$$2C_2H_2(g) + 5O_2(g) \longrightarrow 4CO_2(g) + 2H_2O(g)$$

**d** 750 cm$^3$ of ammonia, according to the reaction

$$4NH_3(g) + 5O_2(g) \longrightarrow 4NO(g) + 6H_2O(g)$$

**e** 1 dm$^3$ of phosphine, according to the reaction

$$PH_3(g) + 2O_2(g) \longrightarrow H_3PO_4(s)$$

**3** 1 dm$^3$ of H$_2$S and 1 dm$^3$ of SO$_2$ were allowed to react, according to the equation

$$2H_2S(g) + SO_2(g) \longrightarrow 2H_2O(l) + 3S(s)$$

What volume of gas will remain after the reaction?

**4** Hydrogen sulphide burns in oxygen in accordance with the following equation:

$$2H_2S(g) + 3O_2(g) \longrightarrow 2H_2O(g) + 2SO_2(g)$$

If 4 dm$^3$ of H$_2$S are burned in 10 dm$^3$ of oxygen at 1 atmosphere pressure and 120 °C, what is the final volume of the mixture?

**A** 6 dm$^3$　**B** 8 dm$^3$　**C** 10 dm$^3$　**D** 12 dm$^3$　**E** 14 dm$^3$

## 7.5 REACTIONS OF SOLIDS AND GASES

Many reactions involve both solids and gases. Calculations may involve converting

- mass of solid ÷ molar mass = amount of solid
- volume of gas ÷ molar volume = amount of gas

**EXAMPLE 1** Calcium carbonate dissociates to form calcium oxide and carbon dioxide.

$$CaCO_3(s) \longrightarrow CaO(s) + CO_2(g)$$

Calculate the volume of CO$_2$ (at r.t.p.) produced when 60 g of calcium carbonate dissociates completely.

**METHOD** Amount of CaCO$_3$ = $\dfrac{60\ \text{g}}{100\ \text{g mol}^{-1}}$ = 0.60 mol

From the equation, Amount of CO$_2$ = 0.60 mol

Gas molar volume at r.t.p. = 24.0 dm$^3$ mol$^{-1}$

Volume at r.t.p = 0.60 mol × 24.0 dm$^3$ mol$^{-1}$ = 14.4 dm$^3$

**ANSWER** The volume of CO$_2$ (at r.t.p.) produced = 14.4 dm$^3$

**EXAMPLE 2**  What volume of hydrogen is obtained when 3.00 g of zinc react with an excess of dilute sulphuric acid at s.t.p.?

**METHOD**  Write the equation:

$$Zn(s) + H_2SO_4(aq) \longrightarrow H_2(g) + ZnSO_4(aq)$$

1 mole of Zn forms 1 mole of $H_2$

65 g of Zn form 22.4 dm$^3$ of $H_2$ (at s.t.p.)

3.00 g of Zn form $\dfrac{3.00}{65} \times 22.4$ dm$^3$ = 1.03 dm$^3$ $H_2$

**ANSWER**  3.00 g of zinc give 1.03 dm$^3$ of hydrogen at s.t.p.

**EXERCISE 7.2**  **Problems on reactions of solids and gases**

Section 1

**Reminder :** Ratio calculations §1.2.
GMV = 24.0 dm$^3$ mol$^{-1}$ at r.t.p. and 22.4 dm$^3$ mol$^{-1}$ at s.t.p.
For relative atomic masses refer to the table on p 243.

**1**  In the reaction between marble and hydrochloric acid, the equation is

$$CaCO_3(s) + 2HCl(aq) \longrightarrow CaCl_2(aq) + CO_2(g) + H_2O(l)$$

What mass of marble would be needed to give 11.00 g of carbon dioxide?

What volume would this gas occupy at r.t.p.?

**2**  Zinc reacts with aqueous hydrochloric acid to give hydrogen

$$Zn(s) + 2HCl(aq) \longrightarrow H_2(g) + ZnCl_2(aq)$$

What mass of zinc would be needed to give 100 g of hydrogen? What volume would this gas occupy at r.t.p.?

**3**  Sodium hydrogencarbonate decomposes on heating, with evolution of carbon dioxide:

$$2NaHCO_3(s) \longrightarrow Na_2CO_3(s) + CO_2(g) + H_2O(g)$$

What volume of carbon dioxide (at r.t.p.) can be obtained by heating 4.20 g of sodium hydrogencarbonate? If 4.20 g of sodium hydrogencarbonate react with an excess of dilute hydrochloric acid, what volume of carbon dioxide (at r.t.p.) is evolved?

**4**  Many years ago, bicycle lamps used to burn the gas ethyne, $C_2H_2$. The gas was produced by allowing water to drip on to calcium carbide. The *unbalanced* equation for the reaction is:

$$\text{Calcium carbide} + \text{Water} \longrightarrow \text{Ethyne} + \text{Calcium hydroxide}$$
$$CaC_2(s) + H_2O(l) \longrightarrow C_2H_2(g) + Ca(OH)_2(s)$$

**a**  Balance the equation.

**b**  Calculate the mass of calcium carbide which would be needed to produce 500 cm$^3$ of ethyne (at r.t.p.).

**5** Dinitrogen oxide, $N_2O$, is commonly called *laughing gas*. It can be made by heating ammonium nitrate, $NH_4NO_3$. The *unbalanced* equation for this reaction is:

$$NH_4NO_3(s) \longrightarrow N_2O(g) + H_2O(l)$$

**a** Balance the equation.

**b** Calculate the mass of ammonium nitrate that must be heated to give
   **(i)** 8.8 g of laughing gas
   **(ii)** 12.0 dm$^3$ of laughing gas (at r.t.p.).

## Section 2

**1** An oxide of potassium reacts with carbon dioxide:

$$4KO_2(s) + 2CO_2(g) \longrightarrow 2K_2CO_3(s) + 3O_2(g)$$

This reaction is sometimes used to regenerate oxygen in submarines. What volume of oxygen (at r.t.p.) could be obtained from 355 g of this oxide of potassium?

**2** A cook is making a small cake. It needs 500 cm$^3$ of carbon dioxide (at r.t.p.) to make the cake rise. The cook decides to add baking powder, which contains sodium hydrogencarbonate. This compound generates carbon dioxide when heated:

$$2NaHCO_3(s) \longrightarrow CO_2(g) + Na_2CO_3(s) + H_2O(g)$$

What mass of sodium hydrogencarbonate must the cook add to the cake mixture?

**3** 120 cm$^3$ of a mixture of ethene and ethane at r.t.p. were treated with bromine. 0.400 g of bromine added to the ethene in the mixture. Calculate the percentage of ethene in the mixture.

**4** Carbon tetrachloride can be made by the reaction of chlorine and carbon disulphide.

$$CS_2(l) + 3Cl_2(g) \longrightarrow CCl_4(l) + S_2Cl_2(l)$$

Calculate:

**a** the volume of chlorine at r.t.p. which reacts with 57 kg of carbon disulphide

**b** the mass of carbon tetrachloride formed.

**5** Hydrogen can be used as a fuel for motor vehicles. One source of hydrogen is the action of the enzyme glucose dehydrogenase on D-glucose:

$$C_6H_{12}O_6(aq) + H_2O(l) \longrightarrow C_6H_{12}O_7(aq) + H_2(g)$$
$$\text{D-glucose} \qquad\qquad\qquad \text{D-gluconic acid}$$

Calculate the mass of glucose required to fill a car fuel tank of 48.0 dm$^3$ with hydrogen at r.t.p.

**6** Nickel is purified by preparing and then decomposing nickel tetracarbonyl, $Ni(CO)_4$.

Calculate the volume of carbon monoxide at r.t.p. needed to prepare 683 kg of nickel tetracarbonyl.

**7** A certain industrial plant emits 90 tonnes of nitrogen monoxide, NO, daily through its chimneys. The firm decides to remove nitrogen monoxide from its waste gases by means of the reaction

$$\text{Methane} + \text{Nitrogen monoxide} \longrightarrow \text{Nitrogen} + \text{Carbon dioxide} + \text{Water}$$

$$CH_4(g) + 4NO(g) \longrightarrow 2N_2(g) + CO_2(g) + 2H_2O(l)$$

If methane (North Sea gas) costs 0.50 p per cubic metre, what will this clean-up reaction cost the firm to run? (Ignore the cost of installing the process, which will in reality be high.)

---

**Hint :** Tonnes of NO ... moles of NO ... moles of $CH_4$ ... volume of $CH_4$ ... cost of $CH_4$.
$(1 \text{ m}^3 = 1000 \text{ dm}^3)$

---

**8** In the Solvay process

$$NaCl(aq) + NH_3(g) + H_2O(l) + CO_2(g) \longrightarrow NaHCO_3(s) + NH_4Cl(aq)$$

What volume of carbon dioxide (at s.t.p.) is required to produce 1.00 kg of sodium hydrogencarbonate?

**9** Find the volume of ethyne (at s.t.p.) that can be prepared from 10.0 g of calcium carbide by the reaction

$$CaC_2(s) + 2H_2O(l) \longrightarrow Ca(OH)_2(aq) + C_2H_2(g)$$

**10** Find the mass of phosphorus required for the preparation of 200 cm³ of phosphine (at s.t.p.) by the reaction

$$P_4(s) + 3NaOH(aq) + 3H_2O(l) \longrightarrow 3NaH_2PO_4(aq) + PH_3(g)$$

**11** Calculate the mass of ammonium chloride required to produce 1.00 dm³ of ammonia (at s.t.p.) in the reaction

$$2NH_4Cl(s) + Ca(OH)_2(s) \longrightarrow 2NH_3(g) + CaCl_2(s) + 2H_2O(g)$$

**12** What mass of potassium chlorate(V) must be heated to give 1.00 dm³ of oxygen at s.t.p.? The reaction is

$$2KClO_3(s) \longrightarrow 2KCl(s) + 3O_2(g)$$

**13** What volume of chlorine (at s.t.p.) can be obtained from the electrolysis of a solution containing 60.0 g of sodium chloride?

**14** What volume of oxygen (at s.t.p.) is needed for the complete combustion of 1.00 kg of octane? The reaction is

$$2C_8H_{18}(l) + 25O_2(g) \longrightarrow 16CO_2(g) + 18H_2O(g)$$

## EXERCISE 7.3 Questions from examination papers

**1 a** Name all the substances used in the manufacture of iron and give a **brief** account of the chemistry of the process. (4)

**b** The halide $M(hal)_2$, of a metal **M**, was dissolved in water and the colourless solution was treated as follows, with the observations stated.

1. To the solution of $M(hal)_2$ was added aqueous sodium hydroxide. A white precipitate was formed; the precipitate was insoluble in excess of the reagent.

2. To the solution of $M(hal)_2$ was added an equal volume of dilute sulphuric acid. There was no change, the solution remaining colourless even after standing.

3. When chlorine was bubbled through the solution of $M(hal)_2$, an orange-red solution was formed. On the addition of starch solution there was no change.

Identify the salt $M(hal)_2$, explain the observations, and write **ionic** equations for the reactions occurring. (5)

**c (i)** Two chromium(III) salts, **T** and **U**, have the same formula $CrCl_3(H_2O)_6$. **T** and **U** each contain a complex ion having six ligands around the central chromium ion. When dissolved in water, one mole of **T** gives one mole of free chloride ions and one mole of **U** gives two moles of free chloride ions.

Deduce the formulae of the complex cations present in **T** and **U**. (2)

**(ii)** I. When 0.0200 mol of an orange compound **V** was completely decomposed by heating, *without the involvement of any other substance*, $0.448\ dm^3$ (273 K and 1 atmosphere) of nitrogen gas, 3.04 g of chromium(III) oxide and 1.44 g of $H_2O(g)$ were formed as the only products.

Calculate the numbers of moles of nitrogen molecules, chromium(III) oxide and $H_2O(g)$ formed from 1 mole of **V** and hence find the molecular formula of **V**.

[The molar volume of an ideal gas at 273 K and 1 atmosphere pressure is $22.4\ dm^3$.]

II. Solid **V** reacts with aqueous sodium hydroxide to evolve ammonia. **V** dissolves in water to give an orange solution which

1. turns green on the addition of excess aqueous iron(II) sulphate acidified with sulphuric acid,

2. turns yellow on the addition of aqueous sodium hydroxide.

Identify **V** and explain the changes, writing **ionic** equations for the reactions. (9)

WJEC (20)

# Volumetric analysis

8

## 8.1 CONCENTRATION

### Mass/Volume

One way of stating the concentration of a solution is to state the **mass** of solute present in 1 cubic decimetre of solution. The mass of solute is usually expressed in grams. A solution made by dissolving 5 grams of solute and making up to 1 cubic decimetre of solution has a concentration of $5 \text{ g dm}^{-3}$ (5 grams per cubic decimetre).

Other units of volume are the cubic centimetre, $cm^3$, the cubic metre, $m^3$, and the litre, l. The litre has the same volume as the cubic decimetre.

$$10^3 \text{ cm}^3 = 1 \text{ dm}^3 = 1 \text{ l} = 10^{-3} \text{ m}^3 \quad (10^3 = 1000; \; 10^{-3} = 0.001)$$

### Amount/Volume

A more common way of stating concentration in chemistry is to state the **molar concentration** of a solution. This is the **amount in moles** of a substance present per cubic decimetre (litre) of solution.

What is the **molar concentration** of a solution of 80 g of sodium hydroxide in 1 $dm^3$ of solution? The amount in moles of NaOH in 80 g of sodium hydroxide can be calculated from its molar mass.

Molar mass of NaOH $= 23 + 16 + 1 = 40 \text{ g mol}^{-1}$

$$\text{Amount of NaOH} = \frac{\text{Mass}}{\text{Molar mass}}$$

$$= \frac{80 \text{ g}}{40 \text{ g mol}^{-1}}$$

$$= 2 \text{ mol}$$

The concentration of the solution is given by:

$$\text{Concentration of solution (mol dm}^{-3}) = \frac{\text{Amount of solute (mol)}}{\text{Volume of solution (dm}^3)}$$

For this solution,

$$\text{Concentration} = \frac{2 \text{ mol}}{1 \text{ dm}^3} = 2 \text{ mol dm}^{-3}$$

If 3 moles of sodium hydroxide are dissolved in 500 $cm^3$ of solution,

$$\text{Concentration} = \frac{3 \text{ mol}}{0.5 \text{ dm}^3} = 6 \text{ mol dm}^{-3}$$

The symbol M is often used for mol dm$^{-3}$. This solution can be described as a 6 M sodium hydroxide solution.

The concentration in mol dm$^{-3}$ used to be referred to as the **molarity** of a solution. (In strict SI units, concentration is expressed in mol m$^{-3}$.)

Figure 8.1 gives more examples.

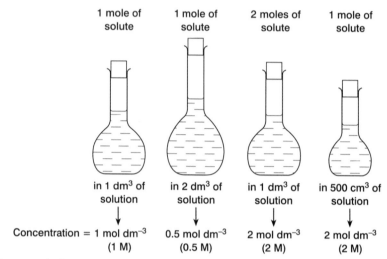

Fig. 8.1 *How to calculate concentration*

A useful rearrangement of the expression in the box on the previous page is:

$$\text{Amount of solute (mol)} = \text{Volume (dm}^3) \times \text{Concentration (mol dm}^{-3})$$

For example, the amount in moles of solute in 2.5 dm$^3$ of a 2.0 mol dm$^{-3}$ solution is given by:

Amount of solute = 2.5 dm$^3$ × 2.0 mol dm$^{-3}$ = 5.0 mol

**Reminder :** Rearranging equations § 1.1.
Does the triangle help in rearranging?
Cover up the quantity you want, e.g. volume.

Then you see $\dfrac{\text{Amount}}{\text{Concentration}}$

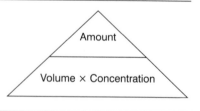

**EXAMPLE 1**

Calculate the concentration in mol dm$^{-3}$ of a solution containing 36.5 g of hydrogen chloride in 4.00 dm$^3$ of solution.

**METHOD**

Molar mass of HCl = (35.5 + 1.0) = 36.5 g mol$^{-1}$

Amount (mol) present in 36.5 g = 1.00 mol

Volume = 4.00 dm$^3$

$$\text{Concentration of solution} = \frac{\text{Amount of solute (mol)}}{\text{Volume of solution (dm}^3)}$$

$$= \frac{1.00 \text{ mol}}{4.00 \text{ dm}^3}$$

$$= 0.25 \text{ mol dm}^{-3}$$

**ANSWER**      The concentration is $0.25$ mol dm$^{-3}$. (It is a $0.25$ M solution.)

**EXAMPLE 2**      Calculate the amount of solute (mol) in $250$ cm$^3$ of a solution of sodium hydroxide which has a concentration of $2.00$ mol dm$^{-3}$.

**METHOD**      Concentration of solution $= 2.00$ mol dm$^{-3}$
Volume of solution $= 250$ cm$^3$ $= 0.250$ dm$^3$
Amount of solute $= Volume \times Concentration$
$\qquad\qquad\qquad$ (dm$^3$) $\qquad$ (mol dm$^{-3}$)

$$= 0.250 \text{ dm}^3 \times 2.00 \text{ mol dm}^{-3}$$

$$= 0.500 \text{ mol}$$

**ANSWER**      The solution contains $0.500$ mol of solute.

**EXERCISE 8.1**      **Problems on concentration**

Reminder : Rearranging equations § 1.1. Ratio calculations § 1.2.
$\qquad\qquad\qquad$ Amount = Mass/Molar mass § 4.1.

**1**      Calculate the concentration in mol dm$^{-3}$ of

    **a** $3.65$ g of hydrogen chloride in $2.00$ dm$^3$ of solution

    **b** $73.0$ g of hydrogen chloride in $2.00$ dm$^3$ of solution

    **c** $49.0$ g of sulphuric acid in $2.00$ dm$^3$ of solution

    **d** $49.0$ g of sulphuric acid in $250$ cm$^3$ of solution

    **e** $2.80$ g of potassium hydroxide in $500$ cm$^3$ of solution

    **f** $28.0$ g of potassium hydroxide in $4.00$ dm$^3$ of solution

    **g** $5.30$ g of anhydrous sodium carbonate $Na_2CO_3$ in $200$ cm$^3$ of solution

    **h** $53.0$ g of anhydrous sodium carbonate in $2.50$ dm$^3$ of solution

**2**      Calculate the amount in moles of solute in

    **a** $250$ cm$^3$ of sodium hydroxide solution containing $1.00$ mol dm$^{-3}$

    **b** $500$ cm$^3$ of sodium hydroxide solution containing $0.250$ mol dm$^{-3}$

   **c** 250 cm$^3$ of 0.0200 M calcium hydroxide solution

   **d** 2.00 dm$^3$ of 1.25 M sulphuric acid (1.25 mol dm$^{-3}$)

   **e** 125 cm$^3$ of aqueous nitric acid, having a concentration of 0.400 mol dm$^{-3}$

   **f** 200 cm$^3$ of ammonia solution, having a concentration of 0.125 mol dm$^{-3}$

   **g** 120 cm$^3$ of aqueous hydrochloric acid of concentration 3.00 mol dm$^{-3}$

   **h** 1500 cm$^3$ of potassium hydroxide solution of concentration 0.200 mol dm$^{-3}$

**3** What mass of the solute must be used in order to prepare the solutions listed below?

   **a** 500 cm$^3$ of 0.100 mol dm$^{-3}$ NaOH(aq) from NaOH(s)

   **b** 250 cm$^3$ of 0.200 mol dm$^{-3}$ Na$_2$CO$_3$(aq) from Na$_2$CO$_3$(s)

   **c** 750 cm$^3$ of 0.100 mol dm$^{-3}$ H$_2$C$_2$O$_4$(aq) from H$_2$C$_2$O$_4$ . 2H$_2$O(s)

   **d** 2.50 dm$^3$ of 0.200 mol dm$^{-3}$ NaHCO$_3$(aq) from NaHCO$_3$(s)

   **e** 500 cm$^3$ of 0.100 mol dm$^{-3}$ KI(aq) from KI(s)

**4** What volumes of the following concentrated solutions are required to give the stated volumes of the more dilute solutions?

   **a** 2.00 dm$^3$ of 0.500 mol dm$^{-3}$ H$_2$SO$_4$(aq) from 2.00 mol dm$^{-3}$ H$_2$SO$_4$(aq)

   **b** 1.00 dm$^3$ of 0.750 mol dm$^{-3}$ HCl(aq) from 10.0 mol dm$^{-3}$ HCl(aq)

   **c** 250 cm$^3$ of 0.250 mol dm$^{-3}$ NaOH(aq) from 5.00 mol dm$^{-3}$ NaOH(aq)

   **d** 500 cm$^3$ of 1.25 mol dm$^{-3}$ HNO$_3$(aq) from 5.00 mol dm$^{-3}$ HNO$_3$(aq)

   **e** 250 cm$^3$ of 2.00 mol dm$^{-3}$ KOH(aq) from 2.50 mol dm$^{-3}$ KOH(aq)

## 8.2 TITRATION

A solution of known concentration is called a **standard solution.** Such a solution can be used to find the concentrations of solutions of other reagents.

In **volumetric analysis,** the concentration of a solution is found by measuring the volume of solution that will react with a known volume of a standard solution. The procedure of adding one solution to another in a measured way until the reaction is complete is called **titration.** Volumetric analysis is often referred to as **titrimetric analysis.**

## 8.3 ACID–BASE TITRATIONS

A standard solution of acid can be used to find the concentration of a solution of alkali. A known volume of alkali is taken by pipette, a suitable indicator is added, and the alkali is titrated against the standard acid until the equivalence point is reached. The number of moles of acid used can be calculated and the equation used to give the number of moles of alkali neutralised.

**EXAMPLE 1**  Standardising sodium hydroxide solution

Of a solution of sodium hydroxide, 25.0 cm$^3$ requires 20.0 cm$^3$ of hydrochloric acid of concentration 0.100 mol dm$^{-3}$ for neutralisation. What is the concentration of the sodium hydroxide solution?

**METHOD**
**STEPS TO TAKE**

In tackling this calculation,

**a** Write the equation. Find the number of moles of acid needed to neutralise one mole of alkali.

$$NaOH(aq) + HCl(aq) \longrightarrow NaCl(aq) + H_2O(l)$$

1 mole of NaOH needs 1 mole of HCl for neutralisation.

**b** Use the expression

$$\text{Amount of solute (mol)} = \text{Volume (dm}^3) \times \text{Concentration (mol dm}^{-3})$$

to find the number of moles of the reagent of known concentration, in this case HCl.

Amount (mol) of HCl = Volume (dm$^3$) × Concn (mol dm$^{-3}$)
$$= 20.0 \times 10^{-3} \text{ dm}^3 \times 0.100 \text{ mol dm}^{-3} = 2.00 \times 10^{-3} \text{ mol}$$

From equation: No. of moles of NaOH = No. of moles of HCl
$$= 2.00 \times 10^{-3} \text{ mol}$$

But: Amount (mol) of NaOH = Volume (dm$^3$) × Concn (mol dm$^{-3}$)
$$= 25.0 \times 10^{-3} \text{ dm}^3 \times c$$

(where $c$ = concn)
These two amounts are equal: $2.00 \times 10^{-3}$ mol $= 25.0 \times 10^{-3}$ dm$^3 \times c$
$$c = (2.00 \times 10^{-3} \text{ mol})/(25.0 \times 10^{-3} \text{ dm}^3)$$
$$= 0.080 \text{ mol dm}^{-3}$$

**ANSWER**  The concentration of sodium hydroxide is 0.080 mol dm$^{-3}$.

**EXAMPLE 2**  **Standardising hydrochloric acid**

Sodium carbonate (anhydrous) is used as a primary standard in volumetric analysis. A solution of sodium carbonate of concentration 0.100 mol dm$^{-3}$ is used to standardise a solution of hydrochloric acid. 25.0 cm$^3$ of the standard solution of sodium carbonate require 35.0 cm$^3$ of the acid for neutralisation. Calculate the concentration of the acid.

**METHOD**
**STEPS TO TAKE**

**a** Write the equation:
$$Na_2CO_3(aq) + 2HCl(aq) \longrightarrow 2NaCl(aq) + CO_2(g) + H_2O(l)$$

1 mole of Na$_2$CO$_3$ neutralises 2 moles of HCl.

**b** Find the amount (mol) of the standard reagent used.

Amount (mol) of Na$_2$CO$_3$(aq) = Volume (dm$^3$) × Concn (mol dm$^{-3}$)
$$= 25.0 \times 10^{-3} \text{ dm}^3 \times 0.100 \text{ mol dm}^{-3}$$
$$= 2.50 \times 10^{-3} \text{ mol}$$

From equation: No. of moles of HCl = 2 × No. of moles of Na$_2$CO$_3$
$$= 5.00 \times 10^{-3} \text{ mol}$$

But: Amount (mol) of HCl(aq) = Volume $(dm^3)$ × Concn $(mol\,dm^{-3})$

$$= 35.0 \times 10^{-3}\,dm^3 \times c$$

(where $c$ = concn)

These two amounts are equal: $5.00 \times 10^{-3}\,mol = 35.0 \times 10^{-3}\,dm^3 \times c$

$$c = (5.00 \times 10^{-3}\,mol)/(35.0 \times 10^{-3}\,dm^3)$$
$$= 0.143\,mol\,dm^{-3}$$

**ANSWER**   The concentration of hydrochloric acid is $0.143\,mol\,dm^{-3}$.

**EXAMPLE 3**   **Calculating the percentage of sodium carbonate in washing soda crystals**

5.125 g of washing soda crystals are dissolved and made up to $250\,cm^3$ of solution. A $25.0\,cm^3$ portion of the solution requires $35.8\,cm^3$ of $0.0500\,mol\,dm^{-3}$ sulphuric acid for neutralisation. Calculate the percentage of sodium carbonate in the crystals.

**METHOD**
**STEPS TO TAKE**

**a**  Write the equation:

$$Na_2CO_3(aq) + H_2SO_4(aq) \longrightarrow Na_2SO_4(aq) + CO_2(g) + H_2O(l)$$

1 mole of $Na_2CO_3$ neutralises 1 mole of $H_2SO_4$.

**b**  Calculate the amount, in moles, of the standard reagent.

Amount (mol) of $H_2SO_4 = 35.8 \times 10^{-3}\,dm^3 \times 0.0500\,mol\,dm^{-3} = 1.79 \times 10^{-3}\,mol$

Amount (mol) of $Na_2CO_3 = 1.79 \times 10^{-3}\,mol$

But: Amount of $Na_2CO_3 = 25.0 \times 10^{-3}\,dm^3 \times c$

(where $c$ = concn)

Equate these two values: $1.79 \times 10^{-3}\,mol = 25.0 \times 10^{-3}\,dm^3 \times c$

$$c = (1.79 \times 10^{-3}\,mol)/(25.0 \times 10^{-3}\,dm^3)$$
$$= 0.0716\,mol\,dm^{-3}$$

Amount (mol) of $Na_2CO_3$ in whole solution = Volume × Concn

$$= 250 \times 10^{-3}\,dm^3 \times 0.0716\,mol\,dm^{-3}$$
$$= 0.0179\,mol$$

Mass of $Na_2CO_3$ = Amount (mol) × Molar mass = $0.0179\,mol \times 106\,g\,mol^{-1}$

$$= 1.90\,g$$

$$\% \text{ of } Na_2CO_3 = \frac{\text{Mass of sodium carbonate}}{\text{Mass of crystals}} \times 100\%$$

$$= \frac{1.90\,g}{5.125\,g} \times 100\% = 37.1\%$$

**ANSWER**   Washing soda crystals are 37.1% sodium carbonate.

**EXAMPLE 4\***   **Estimating ammonium salts**

A sample containing ammonium sulphate was warmed with $250\,cm^3$ of $0.800\,mol\,dm^{-3}$ sodium hydroxide solution. After the evolution of ammonia had ceased, the excess of sodium hydroxide solution was neutralised by $85.0\,cm^3$ of hydrochloric acid of concentration $0.500\,mol\,dm^{-3}$. What mass of ammonium sulphate did the sample contain?

**METHOD**
**STEPS TO TAKE**

**a** There are two reactions taking place:

   **(i)** the reaction between the ammonium salt and the alkali:

$$(NH_4)_2SO_4(s) + 2NaOH(aq) \longrightarrow 2NH_3(g) + Na_2SO_4(aq) + 2H_2O(l)$$

   **(ii)** the reaction between the excess alkali and the hydrochloric acid:

$$NaOH(aq) + HCl(aq) \longrightarrow NaCl(aq) + H_2O(l)$$

**b** Pick out the substance for which you have the information you need to calculate the number of moles. As you know its volume and concentration, you can calculate the number of moles of HCl. This will tell you the number of moles of NaOH left over after reaction **(i)**. Subtract this from the number of moles of NaOH added to the ammonium salt to obtain the number of moles of NaOH used in reaction **(i)**. This will give you the number of moles of $(NH_4)_2SO_4$ with which it reacted.

Amount (mol) of HCl = $85.0 \times 10^{-3}$ $dm^3 \times 0.500$ mol $dm^{-3}$ = 0.0425 mol

Amount (mol) of NaOH left over from reaction **(i)** = 0.0425 mol

Amount (mol) of NaOH added = $250 \times 10^{-3}$ $dm^3 \times 0.800$ mol $dm^{-3}$ = 0.200 mol

Amount (mol) of NaOH used in reaction **(i)** = 0.200 mol – 0.0425 mol

$$= 0.1575 \text{ mol}$$

No. of moles of $(NH_4)_2SO_4$ = $0.5 \times$ No. of moles of NaOH

$$= 0.0788 \text{ mol}$$

Molar mass of $(NH_4)_2SO_4$ = 132 g $mol^{-1}$

Mass of ammonium sulphate = 0.0788 mol $\times$ 132 g $mol^{-1}$ = 10.4 g

**ANSWER**

The sample contained 10.4 g of ammonium sulphate.

**EXERCISE 8.2**  **Problems on neutralisation**

Section 1

> **Reminder :** Percentage § 1.3. Check your answer § 1.12. 1 mg = $1 \times 10^{-3}$ g
>
>         Amount = Mass/Molar mass § 4.1. Amount = Volume $\times$ Concentration § 8.1.

The following are problems on neutralisation. Show, giving your working, whether each of these statements is true or false.

**1** 1 mol of HCl will neutralise

  **a** 5 $dm^3$ of KOH(aq) of concentration 0.2 mol $dm^{-3}$. True or False?

  **b** 2 $dm^3$ of NaOH(aq) of concentration 0.2 mol $dm^{-3}$

  **c** 2 $dm^3$ of KOH(aq) of concentration 0.5 mol $dm^{-3}$

  **d** 0.5 $dm^3$ of NaOH(aq) of concentration 1 mol $dm^{-3}$

  **e** 250 $cm^3$ of $Na_2CO_3$(aq) of concentration 2 mol $dm^{-3}$

  **f** 200 $cm^3$ of $Na_2CO_3$(aq) of concentration 4 mol $dm^{-3}$

Hint : First write the equation. What is the ratio of moles of acid: moles of base? Then calculate the amount of base. Is it the amount of base that will exactly neutralise 1 mol of HCl?

**2** 1 mol of $H_2SO_4$ will neutralise

a 500 $cm^3$ of NaOH(aq) of concentration 4 mol $dm^{-3}$. True or False?

b 1 $dm^3$ of KOH(aq) of concentration 1 mol $dm^{-3}$

c 400 $cm^3$ of NaOH(aq) of concentration 5 mol $dm^{-3}$

d 500 $cm^3$ of $Na_2CO_3$(aq) of concentration 1 mol $dm^{-3}$

e 2 $dm^3$ of $Na_2CO_3$(aq) of concentration 0.5 mol $dm^{-3}$

f 4 $dm^3$ of KOH(aq) of concentration 0.25 mol $dm^{-3}$

**3** 5 mol of NaOH will neutralise

a 2 $dm^3$ of HCl(aq) of concentration 2 mol $dm^{-3}$. True or False?

b 250 $cm^3$ of HCl(aq) of concentration 10 mol $dm^{-3}$

c 250 $cm^3$ of $H_2SO_4$(aq) of concentration 10 mol $dm^{-3}$

d 500 $cm^3$ of $H_2SO_4$(aq) of concentration 5 mol $dm^{-3}$

e 2500 $cm^3$ of $HNO_3$(aq) of concentration 2 mol $dm^{-3}$

f 2 $dm^3$ of $HNO_3$(aq) of concentration 2 mol $dm^{-3}$

**4** 0.5 mol of $Na_2CO_3$ will neutralise

a 1 $dm^3$ of HCl(aq) of concentration 1 mol $dm^{-3}$. True or False?

b 1 $dm^3$ of $H_2SO_4$(aq) of concentration 1 mol $dm^{-3}$

c 500 $cm^3$ of HCl(aq) of concentration 1 mol $dm^{-3}$

d 250 $cm^3$ of $HNO_3$(aq) of concentration 2 mol $dm^{-3}$

e 200 $cm^3$ of $H_2SO_4$(aq) of concentration 2.5 mol $dm^{-3}$

f 500 $cm^3$ of $HNO_3$(aq) of concentration 2 mol $dm^{-3}$

**5** Sodium hydroxide is sold commercially as solid *lye*. A 1.20 g sample of lye required 45.0 $cm^3$ of 0.500 mol $dm^{-3}$ hydrochloric acid to neutralise it. Calculate the percentage by mass of NaOH in lye.

**6** Vinegar is a solution of ethanoic acid. A 10.0 cm$^3$ portion of a certain brand of vinegar needed 55.0 cm$^3$ of 0.200 mol dm$^{-3}$ sodium hydroxide solution to neutralise the ethanoic acid in it.

$$\text{Ethanoic acid + Sodium hydroxide} \longrightarrow \text{Sodium ethanoate + Water}$$

$$CH_3CO_2H(aq) + NaOH(aq) \longrightarrow CH_3CO_2Na(aq) + H_2O(l)$$

Calculate the concentration of ethanoic acid in the vinegar in mol dm$^{-3}$.

**7** Salt is a necessary ingredient of our diet. In certain illnesses, the salt balance can be lost, and a doctor or nurse must give salt intravenously. They inject *normal saline*, which is a 0.85% solution of sodium chloride in water (0.85 g of solute per 100 g of water). What is the molar concentration of normal saline?

**8** A chip of marble weighing 2.50 g required 28.0 cm$^3$ of 1.50 mol dm$^{-3}$ hydrochloric acid to react with all the calcium carbonate it contained. What is the percentage of calcium carbonate in this sample of marble?

**STEPS TO TAKE**   **a** Write the balanced equation for the reaction.

**b** Find how many moles of HCl were used … then how many moles of CaCO$_3$ reacted … what mass of CaCO$_3$ … and finally the percentage of CaCO$_3$.

**9** A mixture of gases coming from a coke-producing plant contains ammonia. The mixture is bubbled through dilute sulphuric acid to remove the ammonia.

**a** Write a balanced equation for the reaction which occurs.

**b** What volume of ammonia (at r.t.p.) could be removed by 50 dm$^3$ of 1.50 mol dm$^{-3}$ sulphuric acid?

**c** What use could be made of the product?

**10** If a person's blood sugar level falls below 60 mg per 100 cm$^3$, insulin shock can occur. The density of blood is 1.2 g cm$^{-3}$.

**a** What is the percentage by mass of sugar in the blood at this level?

**b** What is the molar concentration of sugar, C$_6$H$_{12}$O$_6$, in the blood?

**11** A blood alcohol level of 150–200 mg alcohol per 100 cm$^3$ of blood produces intoxication. A blood alcohol level of 300–400 mg per 100 cm$^3$ produces unconsciousness. At a blood alcohol level above 500 mg per 100 cm$^3$, a person may die. What is the molar concentration of alcohol (ethanol, C$_2$H$_5$OH) at the lethal level?

**12** An experiment was done to find the percentage composition of an alloy of sodium and lead. The alloy reacts with water:

$$\text{Alloy + Water} \longrightarrow \text{Sodium hydroxide + Hydrogen + Lead}$$

$$2Na \cdot Pb(s) + 2H_2O(l) \longrightarrow 2NaOH(aq) + H_2(g) + 2Pb(s)$$

3.00 g of the alloy were added to about 100 cm$^3$ of water. When the reaction was complete, the sodium hydroxide formed was titrated against 1.00 mol dm$^{-3}$ hydrochloric acid. The volume of acid required to neutralise the sodium hydroxide was 12.0 cm$^3$. Calculate:

**a** the amount in moles of HCl used

**b** the amount in moles of NaOH neutralised

**c** the amount in moles of Na in 3.00 g of the alloy

**d** the mass in grams of Na in 3.00 g of alloy

**e** the percentage composition by mass.

Section 2

**Reminder :** Percentage § 1.3. Units § 1.11. Check your answers § 1.12.
Amount = Mass/Molar mass § 4.1. Amount = Volume × Concentration § 8.1.
Empirical formulae § 6.1.

**1** 0.500 g of impure ammonium chloride is warmed with an excess of sodium hydroxide solution. The ammonia liberated neutralised 22.20 cm$^3$ of 0.200 mol dm$^{-3}$ sulphuric acid. Calculate the percentage of ammonium chloride in the sample.

STEPS TO TAKE

**a** Write the equation for the reaction between ammonia and sulphuric acid.

**b** Calculate **(i)** the amount of sulphuric acid neutralised **(ii)** the amount of ammonia liberated.

**c** Calculate the mass of ammonium chloride that will give this amount of ammonia.

**d** Find the percentage of ammonium chloride in the sample.

**2** The problem is to find the percentage of calcium carbonate in a sample of limestone. A 1.00 g sample of limestone is allowed to react with 100 cm$^3$ of 0.200 mol dm$^{-3}$ hydrochloric acid. Some acid was left unused. The excess of acid was neutralised by 24.8 cm$^3$ of 0.100 mol dm$^{-3}$ sodium hydroxide solution.

STEPS TO TAKE

**a** Balance the equation

$$CaCO_3(s) + HCl(aq) \longrightarrow CaCl_2(aq) + CO_2(g) + H_2O(l)$$

**b** What amount (moles) of sodium hydroxide was used in the titration?

**c** What amount of unused hydrochloric acid did it neutralise?

**d** What amount of hydrochloric acid was **(i)** added originally **(ii)** used in the reaction with limestone?

**e** Calculate **(i)** the amount **(ii)** the mass of calcium carbonate with which it reacted.

**f** Now find the percentage of calcium carbonate in the sample of limestone.

**3** An impure sample of barium hydroxide Ba(OH)$_2$ of mass 1.6524 g was added to 100 cm$^3$ of 0.200 mol dm$^{-3}$ hydrochloric acid. All the barium hydroxide reacted. The excess of acid needed 10.9 cm$^3$ of 0.200 mol dm$^{-3}$ sodium hydroxide solution for neutralisation.

Calculate the percentage purity of the sample of barium hydroxide.

**STEPS TO TAKE**  a  Complete and balance the equation

$$Ba(OH)_2(s) + HCl(aq) \longrightarrow$$

b  You want to find the amount of acid used. To do this, find the amount of acid left over. You can do this by finding the amount of sodium hydroxide used to neutralise it.

c  What amount of acid was left unused after the reaction with barium hydroxide?

d  What amount of acid was added originally?

e  What amount of acid reacted with barium hydroxide?

f  Now you can find **(i)** the amount **(ii)** the mass **(iii)** the percentage of barium hydroxide in the sample.

g  Now calculate the percentage of barium hydroxide in the sample.

**4**  The problem is to find the percentage of ammonia in a household cleaner. A 23.80 g sample of the cleaner is dissolved in water and made up to 250 cm$^3$. A 25.0 cm$^3$ sample of this solution needs 35.0 cm$^3$ of 0.360 mol dm$^{-3}$ sulphuric acid for neutralisation.

**STEPS TO TAKE**  a  Write the equation for the reaction between ammonia and sulphuric acid to form $(NH_4)_2SO_4$.

b  Find the amount of sulphuric acid used in the titration.

c  Find **(i)** the amount **(ii)** the mass of ammonia in 25.0 cm$^3$ of solution.

d  Find the mass of ammonia in the sample of cleaner and from this the percentage.

**5**  A fertiliser contains ammonium sulphate and potassium sulphate. A sample of 0.225 g of fertiliser was warmed with sodium hydroxide solution.

$$(NH_4)_2SO_4(s) + 2NaOH(aq) \longrightarrow 2NH_3(g) + Na_2SO_4(aq)$$

The ammonia evolved was neutralised by 15.7 cm$^3$ of 0.100 mol dm$^{-3}$ hydrochloric acid. Calculate the percentage of ammonium sulphate in the sample.

**STEPS TO TAKE**  a  Find the amount of hydrochloric acid used in the titration.

b  Find the amount of ammonia neutralised.

c  This will give you **(i)** the amount and **(ii)** the mass of ammonium sulphate in the sample.

d  Now work out the percentage.

**6**  Calculate the number of carboxyl groups per molecule in the compound $C_6H_8O_6$, given that 0.440 g of it neutralised 37.5 cm$^3$ of 0.200 mol dm$^{-3}$ sodium hydroxide.

$$RCO_2H(aq) + NaOH(aq) \longrightarrow RCO_2Na(aq) + H_2O(l)$$

**STEPS TO TAKE**  a  Where do you start? You have data on sodium hydroxide so first find the amount of sodium hydroxide used.

b  How many moles of —CO$_2$H groups did it react with?

**c** How many moles of $C_6H_8O_6$ are present in 0.440 g?

**d** From **b** and **c** you now know how many $-CO_2H$ groups per molecule.

**7** Sodium carbonate crystals have the formula $Na_2CO_3 . nH_2O$. A 27.8230 g sample of crystals was dissolved in water and made up to 1.00 $dm^3$. 25.0 $cm^3$ of the solution were neutralised by 48.8 $cm^3$ of 0.1000 $mol\ dm^{-3}$ hydrochloric acid. Find $n$ in the formula for the crystals.

**STEPS TO TAKE**

**a** Complete and balance the equation

$$Na_2CO_3(aq) + HCl(aq) \longrightarrow$$

**b** Find the amount of hydrochloric acid used.

**c** What amount of sodium carbonate did it neutralise?

**d** Find **(i)** the amount **(ii)** the mass of sodium carbonate present in 1.00 $dm^3$ of solution.

**e** You now know the mass of $Na_2CO_3$ in the sample and by subtraction, the mass of $H_2O$. From the masses and the molar masses you can find the formula of the crystals.

**8** An excess of sodium hydroxide, 30.0 $cm^3$ of 1.00 $mol\ dm^{-3}$ solution, reacted with 2.38 g of an impure sample of ethyl ethanoate.

$$CH_3CO_2C_2H_5(l) + NaOH(aq) \longrightarrow CH_3CO_2Na(aq) + C_2H_5OH(aq)$$

The remaining sodium hydroxide required 25.0 $cm^3$ of 0.100 $mol\ dm^{-3}$ sulphuric acid for neutralisation. Calculate the percentage purity of ethyl ethanoate.

**9** A mass 0.010 g of a monoprotic acid **HA** needed 5.0 $cm^3$ of 0.010 $mol\ dm^{-3}$ sodium hydroxide for neutralisation. What is the relative formula mass of the acid?

## 8.4 OXIDATION–REDUCTION REACTIONS

Oxidation–reduction (or 'redox') reactions involve a transfer of electrons. The oxidising agent accepts electrons, and the reducing agent gives electrons. In working out the equation for a redox reaction, a good method is to work out the 'half-reaction equation' for the oxidising agent and the 'half-reaction equation' for the reducing agent, and then add them together.

### Examples of half-reaction equations

**a** Iron(III) salts are reduced to iron(II) salts. The equation is

$$Fe^{3+} \longrightarrow Fe^{2+}$$

For the equation to balance, the charge on the right-hand side (RHS) must equal the charge on the left-hand side (LHS). This can be done by inserting an electron on the LHS:

$$Fe^{3+} + e^- \longrightarrow Fe^{2+}$$

**b** When chlorine acts as an oxidising agent, it is reduced to chloride ions:

$$Cl_2 \longrightarrow 2Cl^-$$

To obtain a balanced half-reaction equation, $2e^-$ must be inserted on the LHS:

$$Cl_2 + 2e^- \longrightarrow 2Cl^-$$

**c** Sulphites can be oxidised to sulphates:

$$SO_3^{2-} \longrightarrow SO_4^{2-}$$

To balance the equation with respect to mass, an extra oxygen atom is needed on the LHS. If $H_2O$ is introduced on the LHS to supply this oxygen, the equation becomes

$$SO_3^{2-} + H_2O \longrightarrow SO_4^{2-} + 2H^+$$

To balance the equation with respect to charge, $2e^-$ are needed on the RHS:

$$SO_3^{2-} + H_2O \longrightarrow SO_4^{2-} + 2H^+ + 2e^-$$

**d** Potassium manganate(VII) is an oxidising agent. In acid solution, it is reduced to a manganese(II) salt:

$$MnO_4^- + H^+ \longrightarrow Mn^{2+}$$

To balance the equation with respect to mass, $8H^+$ are needed to combine with 4 oxygen atoms:

$$MnO_4^- + 8H^+ \longrightarrow Mn^{2+} + 4H_2O$$

To balance the equation with respect to charge, $5e^-$ are needed on the LHS:

$$MnO_4^- + 8H^+ + 5e^- \longrightarrow Mn^{2+} + 4H_2O$$

It is a good idea to make a final check. Charge on LHS $= -1 + 8 - 5 = +2$. Charge on RHS $= +2$. The equation is balanced.

**e** Potassium dichromate(VI) is an oxidising agent in acid solution, being reduced to a chromium(III) salt:

$$Cr_2O_7^{2-} + H^+ \longrightarrow Cr^{3+}$$

To balance the equation for mass, $14H^+$ are needed:

$$Cr_2O_7^{2-} + 14H^+ \longrightarrow 2Cr^{3+} + 7H_2O$$

To balance the equation for charge, $6e^-$ are needed on the LHS:

$$Cr_2O_7^{2-} + 14H^+ + 6e^- \longrightarrow 2Cr^{3+} + 7H_2O$$

A final check shows that the charge on the LHS $= -2 + 14 - 6 = +6$.

Charge on RHS $= 2(+3) = +6$. The equation is balanced.

You may like to practise with the half-reaction equations in Exercise 8.4.

### Using half-reaction equations to obtain the equation for a redox reaction

**EXAMPLE 1**

#### Sodium thiosulphate and iodine

In the reaction between iodine and thiosulphate ions, the two half-reaction equations are

$$I_2 + 2e^- \longrightarrow 2I^- \tag{1}$$

$$2S_2O_3^{2-} \longrightarrow S_4O_6^{2-} + 2e^- \tag{2}$$

Adding equations [1] and [2] gives

$$I_2 + 2e^- + 2S_2O_3^{2-} \longrightarrow 2I^- + S_4O_6^{2-} + 2e^-$$

Deleting the $2e^-$ term from both sides of the equation gives

$$I_2 + 2S_2O_3^{2-} \longrightarrow 2I^- + S_4O_6^{2-}$$

A check shows that the charges on the LHS and the RHS are both $-4$.

**EXAMPLE 2**

### Potassium manganate(VII) and iron(II) salts

When potassium manganate(VII) oxidises an iron(II) salt to an iron(III) salt, the equations for the half-reactions are

$$MnO_4^- + 8H^+ + 5e^- \longrightarrow Mn^{2+} + 4H_2O \qquad [3]$$

$$Fe^{2+} \longrightarrow Fe^{3+} + e^- \qquad [4]$$

One manganate(VII) ion needs 5 electrons, and one iron(II) ion gives only one. Equation [4] must therefore be multiplied by 5:

$$5Fe^{2+} \longrightarrow 5Fe^{3+} + 5e^- \qquad [5]$$

Equations [3] and [5] can now be added to give

$$MnO_4^- + 8H^+ + 5Fe^{2+} \longrightarrow Mn^{2+} + 4H_2O + 5Fe^{3+}$$

**EXAMPLE 3**

### Potassium manganate(VII) and sodium ethanedioate

When potassium manganate(VII) oxidises sodium ethanedioate, the equation for the manganate(VII) half-reaction is [3] as in Example 2, and the equation for the reduction of ethanedioate is

$$C_2O_4^{2-} \longrightarrow 2CO_2 + 2e^- \qquad [6]$$

One manganate(VII) ion needs $5e^-$, and one ethanedioate ion gives $2e^-$. Multiplying equation [3] by 2 gives

$$2MnO_4^- + 16H^+ + 10e^- \longrightarrow 2Mn^{2+} + 8H_2O$$

Multiplying equation [6] by 5 give

$$5C_2O_4^{2-} \longrightarrow 10CO_2 + 10e^-$$

Adding the new equations gives

$$2MnO_4^- + 16H^+ + 5C_2O_4^{2-} \longrightarrow 2Mn^{2+} + 10CO_2 + 8H_2O$$

**EXAMPLE 4**

### Potassium dichromate(VI) and iron(II) salts

Potassium dichromate(VI) oxidises iron(II) salts to iron(III) salts. The equations for the two half-reactions are

$$Cr_2O_7^{2-} + 14H^+ + 6e^- \longrightarrow 2Cr^{3+} + 7H_2O \qquad [7]$$

$$Fe^{2+} \longrightarrow Fe^{3+} + e^- \qquad [8]$$

One dichromate ion will oxidise six iron(II) ions:

$$Cr_2O_7^{2-} + 14H^+ + 6Fe^{2+} \longrightarrow 2Cr^{3+} + 6Fe^{3+} + 7H_2O$$

You may like to try the problem on balancing equations in Exercise 8.4 before going on to tackle the numerical problems.

## 8.5 OXIDATION NUMBERS

It is helpful to discuss oxidation–reduction reactions in terms of the change in the **oxidation number** of each reactant. In the reaction

$$Cu(s) + \tfrac{1}{2}O_2(g) \longrightarrow Cu^{2+}O^{2-}(s)$$

copper is oxidised and oxygen is reduced. It is said that the oxidation number of copper increases from zero to +2, and the oxidation number of oxygen decreases from zero to −2. The following rules are followed in assigning oxidation numbers:

1   The oxidation number of an uncombined element is zero.

2   In ionic compounds, the oxidation number of each element is the charge on its ion. In NaCl, the oxidation number of Na = +1, and that of Cl = −1.

3   The sum of the oxidation numbers of all the elements in a compound is zero. In $AlCl_3$, the oxidation numbers are: Al = +3; Cl = −1, so that the sum of the oxidation numbers is $+3 + 3(-1) = 0$.

4   The sum of the oxidation numbers of all the elements in an ion is equal to the charge on the ion. In $SO_4^{2-}$, the oxidation numbers are S = +6, O = −2. The sum of the oxidation numbers for all the atoms is $+6 + 4(-2) = -2$, the same as the charge on the $SO_4^{2-}$ ion.

5   In a covalent compound, one element must be given a positive oxidation number and the other a negative oxidation number, such that the sum of the oxidation numbers for all the atoms is zero. The following elements always have the same oxidation numbers in all their compounds. A knowledge of their oxidation numbers helps us to assign oxidation numbers to the other elements combined with them:

| | | | |
|---|---|---|---|
| Na, K | + 1 | H | + 1, except in metal hydrides |
| Mg, Ca | + 2 | F | − 1 |
| Al | + 3 | Cl | − 1, except in compounds with O and F |
| | | O | − 2, except in peroxides and compounds with F |

**EXAMPLE 1**   What is the oxidation number of germanium in $GeCl_4$?

**METHOD**   Chlorine is one of the elements with a constant oxidation number of −1.

(Oxidation number of Ge) + 4(−1) = 0.

**ANSWER**   Oxidation number of Ge = +4.

**EXAMPLE 2**   What is the oxidation number of manganese in $Mn_2O_7$?

**METHOD**   Oxygen is one of the elements with a constant oxidation number of −2.

2(Oxidation number of Mn) + 7(−2) = 0.

**ANSWER**   Oxidation number of Mn = +7.

**EXAMPLE 3**    What is the oxidation number of iron in $Fe(CN)_6^{3-}$?

**METHOD**      Since the cyanide ion is $CN^-$, it has an oxidation number of $-1$.

(Oxidation number of Fe) $+ 6(-1) = -3$.

**ANSWER**      Oxidation number of Fe $= +3$.


**EXERCISE 8.3**    **Problems on oxidation numbers**

---

**Reminder :** Negative numbers § 1.4.

---

**1**    What is the oxidation number of the named element in the following compounds?

| | | |
|---|---|---|
| **a** Ba in $BaCl_2$ | **j** C in $CCl_4$ | **s** Cr in $CrO_4^{2-}$ |
| **b** Fe in $Fe(CN)_6^{4-}$ | **k** I in $I_2$ | **t** N in $NO_2$ |
| **c** Cl in $Cl_2$ | **l** Cl in $Cl_2O_7$ | **u** H in LiH |
| **d** Li in $Li_2O$ | **m** C in CO | **v** Cr in $Cr_2O_7^{2-}$ |
| **e** Fe in $Fe(CN)_6^{3-}$ | **n** I in $I^-$ | **w** N in $N_2O_4$ |
| **f** Cl in $ClO^-$ | **o** Cl in $Cl_2O_3$ | **x** H in HBr |
| **g** P in $P_2O_3$ | **p** Cr in $CrO_3$ | **y** S in $SO_3^{2-}$ |
| **h** Br in $BrO_3^-$ | **q** I in $IO_3^-$ | **z** P in $PO_4^{3-}$ |
| **i** Cl in $ClO_3^-$ | **r** O in $H_2O_2$ | |

**2**    **a** Calculate the oxidation numbers of tin and lead on each side of the equation

$$PbO_2(s) + 4H^+(aq) + Sn^{2+}(aq) \longrightarrow Pb^{2+}(aq) + Sn^{4+}(aq) + 2H_2O(l)$$

and state which element has been oxidised and which has been reduced.

**b** In the redox reaction

$$2Mn^{2+}(aq) + 5BiO_3^-(aq) + 14H^+(aq) \longrightarrow 2MnO_4^-(aq) + 5Bi^{3+}(aq) + 7H_2O(l)$$

calculate the oxidation numbers of all the elements, and state which have been oxidised and which have been reduced.

**c** Calculate the oxidation numbers of arsenic and manganese in each of the species in the reaction:

$$5As_2O_3(s) + 4MnO_4^-(aq) + 12H^+(aq) \longrightarrow 5As_2O_5(s) + 4Mn^{2+}(aq) + 6H_2O(l)$$

State which element has been oxidised and which has been reduced.

**3** In each of the following equations, one element is underlined. Calculate its oxidation number in each species, and state whether an oxidation or a reduction has occurred.

**a** $2\underline{F}_2(g) + 2OH^-(aq) \longrightarrow \underline{F}_2O(g) + 2\underline{F}^-(aq) + H_2O(l)$

**b** $3\underline{Cl}_2(g) + 6OH^-(aq) \longrightarrow \underline{Cl}O_3^-(aq) + 5\underline{Cl}^-(aq) + 3H_2O(l)$

**c** $\underline{N}H_4^+\underline{N}O_3^-(s) \longrightarrow \underline{N}_2O(g) + 2H_2O(l)$

**d** $\underline{Cr}_2O_7^{2-}(aq) + 14H^+(aq) + 6e^- \longrightarrow 2\underline{Cr}^{3+}(aq) + 7H_2O(l)$

**e** $\underline{C}_2O_4^{2-}(aq) \longrightarrow 2\underline{C}O_2(g) + 2e^-$

**4 a** Only N and I alter in oxidation number in the reaction

$$N_2H_6O(aq) + IO_3^-(aq) + 2H^+(aq) + Cl^-(aq) \longrightarrow N_2(g) + ICl(aq) + 4H_2O(l)$$

Calculate the oxidation number of N in $N_2H_6O$.

**b** In the reaction below, only S and Br change in oxidation number.

$$Na_2H_{10}S_2O_8(aq) + 4Br_2(aq) \longrightarrow 2H_2SO_4(aq) + 2NaBr(aq) + 6HBr(aq)$$

Calculate the oxidation number of S in $Na_2H_{10}S_2O_8$.

## 8.6 POTASSIUM MANGANATE(VII) TITRATIONS

When potassium manganate(VII) acts as an oxidising agent in acid solution, it is reduced to a manganese(II) salt:

$$MnO_4^-(aq) + 8H^+(aq) + 5e^- \longrightarrow Mn^{2+}(aq) + 4H_2O(l)$$

Potassium manganate(VII) is not sufficiently pure to be used as a primary standard, and solutions of the oxidant are standardised by titration against a primary standard such as sodium ethanedioate. This reductant can be obtained in a high state of purity as crystals of formula $Na_2C_2O_4 . 2H_2O$, which are neither deliquescent nor efflorescent, and can be weighed out exactly to make a standard solution.

Once it has been standardised, a solution of potassium manganate(VII) can be used to titrate reducing agents such as iron(II) salts. No indicator is needed as the oxidant changes from purple to colourless at the end-point.

**EXAMPLE 1**

**Standardising potassium manganate(VII) against the primary standard, sodium ethanedioate**

A 25.0 cm$^3$ portion of sodium ethanedioate solution of concentration 0.200 mol dm$^{-3}$ is warmed and titrated against a solution of potassium manganate(VII). If 17.2 cm$^3$ of potassium manganate(VII) are required, what is its concentration?

**METHOD**

Let $M$ = concentration of the manganate(VII) solution.

Amount (mol) of ethanedioate = 25.0 × 10$^{-3}$ dm$^3$ × 0.200 mol dm$^{-3}$

Amount (mol) of manganate(VII) = 17.2 × 10$^{-3}$ dm$^3$ × $M$ mol dm$^{-3}$

The equations for the half-reactions are

$$MnO_4^-(aq) + 8H^+(aq) + 5e^- \longrightarrow Mn^{2+}(aq) + 4H_2O(l) \qquad [1]$$

$$C_2O_4^{2-}(aq) \longrightarrow 2CO_2(g) + 2e^- \qquad [2]$$

Multiplying [1] by 2 and [2] by 5, and adding the two equations gives

$$2MnO_4^-(aq) + 16H^+(aq) + 5C_2O_4^{2-}(aq) \longrightarrow 2Mn^{2+}(aq) + 8H_2O(l) + 10CO_2(g)$$

No. of moles of $MnO_4^- = \dfrac{2}{5} \times$ No. of moles of $C_2O_4^{2-}$.

$$\therefore \quad 17.2 \times 10^{-3} \text{ dm}^3 \times M = \frac{2}{5} \times 25.0 \times 10^{-3} \text{ dm}^3 \times 0.200 \text{ mol dm}^{-3}$$

$$M = \frac{2 \times 25.0 \times 10^{-3} \text{ dm}^3 \times 0.200 \text{ mol dm}^{-3}}{5 \times 17.2 \times 10^{-3} \text{ dm}^3} = 0.116 \text{ mol dm}^{-3}$$

**ANSWER**

The potassium manganate(VII) solution has a concentration of 0.116 mol dm$^{-3}$.

**EXAMPLE 2**

**Oxidising iron(II) compounds**

Ammonium iron(II) sulphate crystals have the following formula:
$(NH_4)_2SO_4 . FeSO_4 . nH_2O$. In an experiment to determine $n$, 8.492 g of the salt were dissolved and made up to 250 cm$^3$ of solution with distilled water and dilute sulphuric acid. A 25.0 cm$^3$ portion of the solution was further acidified and titrated against potassium manganate(VII) solution of concentration 0.0150 mol dm$^{-3}$. A volume of 22.5 cm$^3$ was required.

**METHOD**

The equations for the two half-reactions are

$$MnO_4^-(aq) + 8H^+(aq) + 5e^- \longrightarrow Mn^{2+}(aq) + 4H_2O(l) \qquad [1]$$
$$Fe^{2+}(aq) \longrightarrow Fe^{3+}(aq) + e^- \qquad [2]$$

Multiplying [2] by 5 and then adding it to [1] gives

$$MnO_4^-(aq) + 8H^+(aq) + 5Fe^{2+}(aq) \longrightarrow Mn^{2+}(aq) + 5Fe^{3+}(aq) + 4H_2O(l)$$

Amount (mol) of manganate(VII) $= 22.5 \times 10^{-3} \text{ dm}^3 \times 0.0150 \text{ mol dm}^{-3}$

$$= 0.338 \times 10^{-3} \text{ mol}$$

No. of moles of iron(II) $= 5 \times$ No. of moles of manganate(VII)

$$= 5 \times 0.338 \times 10^{-3} \text{ mol} = 1.69 \times 10^{-3} \text{ mol}$$

$$\text{Concn of iron(II)} = \frac{1.69 \times 10^{-3} \text{ mol}}{25.0 \times 10^{-3} \text{ dm}^3} = 0.0674 \text{ mol dm}^{-3}$$

$$\text{Concn of } (NH_4)_2SO_4.FeSO_4.nH_2O = \frac{\text{Mass in 1 dm}^3 \text{ of solution}}{\text{Molar mass}}$$

$$= \frac{4 \times 8.492 \text{ g dm}^{-3}}{\text{Molar mass}}$$

$$0.0674 \text{ mol dm}^{-3} = \frac{4 \times 8.492 \text{ g dm}^{-3}}{\text{Molar mass}}$$

$$\text{Molar mass} = 503.9 \text{ g mol}^{-1}$$

Molar mass of $(NH_4)_2SO_4.FeSO_4.nH_2O = 284 + 18n = 504 \text{ g mol}^{-1}$

$$\therefore \quad n = 12$$

**ANSWER**

The formula of the crystals is $(NH_4)_2SO_4 . FeSO_4 . 12H_2O$

**EXAMPLE 3**

## Oxidising hydrogen peroxide

A solution of hydrogen peroxide was diluted 20.0 times. A 25.0 cm$^3$ portion of the diluted solution was acidified and titrated against 0.0150 mol dm$^{-3}$ potassium manganate(VII) solution. 45.7 cm$^3$ of the oxidant were required. Calculate the concentration of the hydrogen peroxide solution **a** in mol dm$^{-3}$ and **b** the 'volume concentration'. (This means the number of volumes of oxygen obtained from one volume of the solution.)

**METHOD**

The equations for the half-reactions are

$$MnO_4^-(aq) + 8H^+(aq) + 5e^- \longrightarrow Mn^{2+}(aq) + 4H_2O(l) \qquad [1]$$

$$H_2O_2(aq) \longrightarrow O_2(g) + 2H^+(aq) + 2e^- \qquad [2]$$

Multiplying [1] by 2 and [2] by 5, and adding the two equations gives

$$2MnO_4^-(aq) + 6H^+(aq) + 5H_2O_2(aq) \longrightarrow 2Mn^{2+}(aq) + 8H_2O(l) + 5O_2(g)$$

Amount (mol) of $MnO_4^-(aq) = 45.7 \times 10^{-3}$ dm$^3 \times 0.0150$ mol dm$^{-3}$

$$= 0.685 \times 10^{-3} \text{ mol}$$

No. of moles of $H_2O_2 = \dfrac{5}{2} \times$ No. of moles of $MnO_4^-$

$$= \dfrac{5}{2} \times 0.685 \times 10^{-3} \text{ mol} = 1.71 \times 10^{-3} \text{ mol}$$

Concn of $H_2O_2 = (1.71 \times 10^{-3} \text{ mol})/(25.0 \times 10^{-3} \text{ dm}^3) = 0.0684$ mol dm$^{-3}$

Concn of original solution $= 20.0 \times 0.0684$ mol dm$^{-3} = 1.37$ mol dm$^{-3}$.

When hydrogen peroxide decomposes,

$$2H_2O_2(aq) \longrightarrow 2H_2O(l) + O_2(g)$$

2 moles of hydrogen peroxide form 1 mole of oxygen. Therefore a solution of hydrogen peroxide of concentration 2 mol dm$^{-3}$ is a 22.4 volume solution (the volume of 1 mole of oxygen at s.t.p.).

A solution of $H_2O_2$ of concentration 1.37 mol dm$^{-3}$ is a $22.4 \times 1.37/2 = 15.4$ volume solution.

**ANSWER**

The concentration of hydrogen peroxide is: **a** 1.37 mol dm$^{-3}$ and **b** 15.4 volume.

## 8.7 POTASSIUM DICHROMATE(VI) TITRATIONS

Potassium dichromate(VI) can be obtained in a high state of purity, and its aqueous solutions are stable. It is used as a primary standard. The colour change when chromium(VI) changes to chromium(III) in the reaction

$$Cr_2O_7^{2-}(aq) + 14H^+(aq) + 6e^- \longrightarrow 2Cr^{3+}(aq) + 7H_2O(l)$$

is from orange to blue. As it is not possible to see a sharp change in colour, an indicator is used. Barium N-phenylphenylamine-4-sulphonate gives a sharp colour change, from blue-green to violet, when a slight excess of potassium dichromate has been added. Phosphoric(V) acid must be present to form a complex with the $Fe^{3+}$ ions formed during the oxidation reaction; otherwise $Fe^{3+}$ ions affect the colour change of the indicator.

Since dichromate(VI) has a slightly lower redox potential than manganate(VII), it can be used in the presence of chloride ions, without oxidising them to chlorine.

**EXAMPLE**

Determination of the percentage of iron in iron wire

A piece of iron wire of mass 2.225 g was put into a conical flask containing dilute sulphuric acid. The flask was fitted with a bung carrying a Bunsen valve, to allow the hydrogen generated to escape but prevent air from entering. The mixture was warmed to speed up reaction. When all the iron had reacted, the solution was cooled to room temperature and made up to 250 $cm^3$ in a graduated flask. With all these precautions, iron is converted to $Fe^{2+}$ ions only, and no $Fe^{3+}$ ions are formed. 25.0 $cm^3$ of the solution were acidified and titrated against a 0.0185 $mol\ dm^{-3}$ solution of potassium dichromate(VI). The volume required was 31.0 $cm^3$. Calculate the percentage of iron in the iron wire.

**METHOD**

Amount (mol) of $Cr_2O_7^{2-}(aq)$ used $= 31.0 \times 10^{-3}\ dm^3 \times 0.0185\ mol\ dm^{-3}$

$$= 0.574 \times 10^{-3}\ mol$$

The equations for the two half-reactions are

$$Cr_2O_7^{2-}(aq) + 14H^+(aq) + 6e^- \longrightarrow 2Cr^{3+}(aq) + 7H_2O(l) \qquad [1]$$

$$Fe^{2+}(aq) \longrightarrow Fe^{3+}(aq) + e^- \qquad [2]$$

Multiplying [2] by 6 and adding gives

$$Cr_2O_7^{2-}(aq) + 14H^+(aq) + 6Fe^{2+}(aq) \longrightarrow 2Cr^{3+}(aq) + 6Fe^{3+}(aq) + 7H_2O(l)$$

Amount (mol) of $Fe^{2+}$ in 25.0 $cm^3 = 6 \times 0.574 \times 10^{-3}\ mol$

$$= 3.45 \times 10^{-3}\ mol$$

Amount (mol) of $Fe^{2+}$ in the whole solution $= 3.45 \times 10^{-2}\ mol$

Mass of Fe in the whole solution $= 3.45 \times 10^{-2}\ mol \times 55.8\ g\ mol^{-1} = 1.93\ g$

Percentage of Fe in wire $= \dfrac{1.93\ g}{2.225\ g} \times 100\% = 86.7\%$

**ANSWER**

The wire is 86.7% iron.

## 8.8 SODIUM THIOSULPHATE TITRATIONS

Sodium thiosulphate reduces iodine to iodide ions, and forms sodium tetrathionate, $Na_2S_4O_6$

$$2S_2O_3^{2-}(aq) + I_2(aq) \longrightarrow 2I^-(aq) + S_4O_6^{2-}(aq)$$

Sodium thiosulphate, $Na_2S_2O_3 . 5H_2O$, is not used as a primary standard as the water content of the crystals is variable. A solution of sodium thiosulphate can be standardised against a solution of iodine, or a solution of potassium iodate(V) or potassium dichromate(VI) or potassium manganate(VII).

**EXAMPLE 1**

Standardisation of a sodium thiosulphate solution, using iodine

Iodine has a limited solubility in water. It dissolves in a solution of potassium iodide because it forms the very soluble complex ion, $I_3^-$.

$$I_2(s) + I^-(aq) \rightleftharpoons I_3^-(aq)$$

An equilibrium is set up between iodine and tri-iodide ions, and if iodine molecules are removed from solution by a reaction, tri-iodide ions dissociate to form more iodine molecules. A solution of iodine in potassium iodide can therefore be titrated as though it were a solution of iodine in water.

When sufficient of a solution of thiosulphate is added to a solution of iodine, the colour of iodine fades to a pale yellow. Then 2 cm$^3$ of starch solution are added to give a blue colour with the iodine. Addition of thiosulphate is continued drop by drop, until the blue colour disappears.

2.835 g of iodine and 6 g of potassium iodide are dissolved in distilled water and made up to 250 cm$^3$. A 25.0 cm$^3$ portion titrated against sodium thiosulphate solution required 17.7 cm$^3$ of the solution. Calculate the concentration of the thiosulphate solution.

**METHOD**

Molar mass of iodine = 2 × 127 = 254 g mol$^{-1}$

Concn of iodine solution = 2.835 g × 4/254 g mol$^{-1}$ = 0.0446 mol dm$^{-3}$

Amount (mol) of I$_2$ in 25.0 cm$^3$ = 25.0 × 10$^{-3}$ dm$^3$ × 0.0446 mol dm$^{-3}$
$$= 1.115 \times 10^{-3} \text{ mol}$$

From the equation

$$2S_2O_3{}^{2-}(aq) + I_2(aq) \longrightarrow 2I^-(aq) + S_4O_6{}^{2-}(aq)$$

No. of moles of 'thio' = 2 × No. of moles of I$_2$

Amount (mol) of 'thio' in volume used = 2.23 × 10$^{-3}$ mol

Concn of 'thio' = $\dfrac{2.23 \times 10^{-3} \text{ mol}}{17.7 \times 10^{-3} \text{ dm}^3}$ = 0.126 mol dm$^{-3}$

**ANSWER**

The concentration of the thiosulphate solution is 0.126 mol dm$^{-3}$.

**EXAMPLE 2**

**Standardisation of thiosulphate against potassium iodate(V)**

Potassium iodate(V) is a primary standard. It reacts with iodide ions in the presence of acid to form iodine:

$$IO_3{}^-(aq) + 5I^-(aq) + 6H^+(aq) \longrightarrow 3I_2(aq) + 3H_2O(l)$$

A standard solution of iodine can be prepared by weighing out the necessary quantity of potassium iodate(V) and making up to a known volume of solution. When a portion of this solution is added to an excess of potassium iodide in acid solution, a calculated amount of iodine is liberated.

1.015 g of potassium iodate(V) are dissolved and made up to 250 cm$^3$. To a 25.0 cm$^3$ portion are added an excess of potassium iodide and dilute sulphuric acid. The solution is titrated with a solution of sodium thiosulphate, starch solution being added near the end-point. 29.8 cm$^3$ of thiosulphate solution are required. Calculate the concentration of the thiosulphate solution.

**METHOD**

Molar mass of KIO$_3$ = 39.1 + 127 + (3 × 16.0) = 214 g mol$^{-1}$

Concn of KIO$_3$ solution = 1.015 × 4 g dm$^{-3}$/214 g mol$^{-1}$ = 0.0189 mol dm$^{-3}$

Amount (mol) of KIO$_3$ in 25 cm$^3$ = 25.0 × 10$^{-3}$ dm$^3$ × 0.0189 mol dm$^{-3}$
$$= 0.473 \times 10^{-3} \text{ mol}$$

Since

$$IO_3^-(aq) + 5I^-(aq) + 6H^+(aq) \longrightarrow 3I_2(aq) + 3H_2O(l)$$

and

$$2S_2O_3^{2-}(aq) + I_2(aq) \longrightarrow 2I^-(aq) + S_4O_6^{2-}(aq)$$

No. of moles of 'thio' = 6 × No. of moles of $IO_3^-$

$$= 6 \times 0.473 \times 10^{-3} \text{ mol} = 2.84 \times 10^{-3} \text{ mol}$$

Concn of 'thio' = $(2.84 \times 10^{-3} \text{ mol})/(29.8 \times 10^{-3} \text{ dm}^3) = 0.0950 \text{ mol dm}^{-3}$

**ANSWER**   The sodium thiosulphate solution has a concentration 0.0950 mol dm$^{-3}$.

**EXAMPLE 3**   **Estimation of chlorine**

Chlorine displaces iodine from iodides. The iodine formed can be found by titration with a standard thiosulphate solution. Chlorate(I) solutions are often used as a source of chlorine as they liberate chlorine readily on reaction with acid:

$$ClO^-(aq) + 2H^+(aq) + Cl^-(aq) \longrightarrow Cl_2(aq) + H_2O(l)$$

The amount of chlorine available in a domestic bleach which contains sodium chlorate(I) can be found by allowing the bleach to react with an iodide solution to form iodine, and then titrating with thiosulphate solution:

$$ClO^-(aq) + 2H^+(aq) + 2I^-(aq) \longrightarrow I_2(aq) + Cl^-(aq) + H_2O(l)$$

A domestic bleach in solution is diluted by pipetting 10.0 cm$^3$ and making this volume up to 250 cm$^3$. A 25.0 cm$^3$ portion of the solution is added to an excess of potassium iodide and ethanoic acid and titrated against sodium thiosulphate solution of concentration 0.0950 mol dm$^{-3}$, using starch as an indicator. The volume required is 21.3 cm$^3$. Calculate the percentage of available chlorine in the bleach.

**METHOD**   Amount (mol) of 'thio' = $21.3 \times 10^{-3} \text{ dm}^3 \times 0.0950 \text{ mol dm}^{-3} = 2.03 \times 10^{-3} \text{ mol}$

Since

$$2S_2O_3^{2-}(aq) + I_2(aq) \longrightarrow S_4O_6^{2-}(aq) + 2I^-(aq)$$

Amount (mol) of $I_2$ = 1.015 mol

Since iodine is produced in the reaction

$$ClO^-(aq) + 2I^-(aq) + 2H^+(aq) \longrightarrow I_2(aq) + Cl^-(aq) + H_2O(l)$$

Amount (mol) of $ClO^-$ in 25 cm$^3$ of solution = $1.015 \times 10^{-3}$ mol

Since chlorate(I) liberates chlorine in the reaction

$$ClO^-(aq) + 2H^+(aq) + Cl^-(aq) \longrightarrow Cl_2(aq) + H_2O(l)$$

No. of moles of $Cl_2$ = No. of moles of chlorate(I)

$$= 1.015 \times 10^{-3} \text{ mol}$$

Mass of chlorine = $71.0 \text{ g mol}^{-1} \times 1.015 \times 10^{-3} \text{ mol} = 0.0720 \text{ g}$

This is the mass of chlorine available in 25 cm$^3$ of solution

Percentage of available $Cl_2 = \dfrac{\text{Mass of } Cl_2 \text{ from 250 cm}^3 \text{ solution}}{\text{Mass of bleach solution used}} \times 100\%$

$$= \frac{0.0720 \times 10 \text{ g}}{10.0 \text{ g}} \times 100\% = 7.2\%$$

**ANSWER**   The percentage of available chlorine in the bleach is 7.2%.

| | |
|---|---|
| **EXAMPLE 4** | ### Estimation of copper(II) salts |

Copper(II) ions oxidise iodide ions to iodine. The iodine produced can be titrated with standard thiosulphate solution, and, from the amount of iodine produced, the concentration of copper(II) ions in the solution can be calculated.

A sample of 4.256 g of copper(II) sulphate-5-water is dissolved and made up to $250 \text{ cm}^3$. A $25.0 \text{ cm}^3$ portion is added to an excess of potassium iodide. The iodine formed required $18.0 \text{ cm}^3$ of a $0.0950 \text{ mol dm}^{-3}$ solution of sodium thiosulphate for reduction. Calculate the percentage of copper in the crystals.

**METHOD**

Amount (mol) of 'thio' $= 18.0 \times 10^{-3} \text{ dm}^3 \times 0.0950 \text{ mol dm}^{-3} = 1.71 \times 10^{-3} \text{ mol}$

No. of moles of $I_2 = \frac{1}{2} \times$ No. of moles of 'thio' $= 0.855 \times 10^{-3} \text{ mol}$

Since

$$2Cu^{2+}(aq) + 4I^-(aq) \longrightarrow Cu_2I_2(s) + I_2(aq)$$

No. of moles of Cu $= 2 \times$ No. of moles of $I_2 = 1.71 \times 10^{-3} \text{ mol}$

Mass of Cu $= 63.5 \text{ g mol}^{-1} \times 1.71 \times 10^{-3} \text{ mol} = 0.109 \text{ g}$

Mass of Cu in whole solution $= 1.09 \text{ g}$

Percentage of Cu $= \dfrac{1.09 \text{ g}}{4.256 \text{ g}} \times 100\% = 25.6\%.$

**ANSWER**

The percentage of copper in the crystals is 25.6%.

| | |
|---|---|
| **Exercise 8.4** | ### Problems on redox reactions |

**Reminder :** Percentages § 1.3. Units § 1.11. Check your answer § 1.12.

Amount = Mass/Molar mass § 4.1. Amount = Volume × Concentration § 8.1.

**Table:** Half-reaction equations

$S_2O_8^{2-}(aq) + 2e^- \rightleftharpoons 2SO_4^{2-}(aq)$

$H_2O_2(aq) + 2H^+(aq) + 2e^- \rightleftharpoons 2H_2O(l)$

$MnO_4^-(aq) + 8H^+(aq) + 5e^- \rightleftharpoons Mn^{2+}(aq) + 4H_2O(l)$

$Ce^{4+} + e^- \rightleftharpoons Ce^{3+}(aq)$

$Cl_2(aq) + 2e^- \rightleftharpoons 2Cl^-(aq)$

$ClO_3^-(aq) + 6H^+(aq) + 6e^- \rightleftharpoons Cl^-(aq) + 3H_2O(l)$

$Cr_2O_7^{2-}(aq) + 14H^+(aq) + 6e^- \rightleftharpoons 2Cr^{3+}(aq) + 7H_2O(l)$

$MnO_2(s) + 4H^+(aq) + 2e^- \rightleftharpoons Mn^{2+}(aq) + 2H_2O(l)$

$Br_2(aq) + 2e^- \rightleftharpoons 2Br^-(aq)$

$Fe^{3+}(aq) + e^- \rightleftharpoons Fe^{2+}(aq)$

$I_2(aq) + 2e^- \rightleftharpoons 2I^-(aq)$

$Sn^{4+}(aq) + 2e^- \rightleftharpoons Sn^{2+}(aq)$

$S_4O_6^{2-}(aq) + 2e^- \rightleftharpoons 2S_2O_3^{2-}(aq)$

$SO_4^{2-}(aq) + 2H^+(aq) + 2e^- \rightleftharpoons SO_3^{2-}(aq) + H_2O(l)$

$2CO_2(aq) + 2e^- \rightleftharpoons C_2O_4^{2-}(aq)$

## Section 1

**1**   Complete and balance the following equations, using the half-reaction equations in the table:

   **a**  $MnO_4^-(aq) + H_2O_2(aq) + H^+(aq) \longrightarrow$

   **b**  $MnO_2(s) + H^+(aq) + Cl^-(aq) \longrightarrow$

   **c**  $MnO_4^-(aq) + C_2O_4^{2-}(aq) + H^+ \longrightarrow$

   **d**  $Cr_2O_7^{2-}(aq) + C_2O_4^{2-}(aq) + H^+(aq) \longrightarrow$

   **e**  $Cr_2O_7^{2-}(aq) + I^-(aq) + H^+(aq) \longrightarrow$

**2**   How many moles of the following reductants will be oxidised by $1.0 \times 10^{-3}$ mol of potassium manganate(VII) in acidic solution?

   **a**  $Fe^{2+}$   **b**  $Sn^{2+}$   **c**  $C_2O_4^{2-}$   **d**  $H_2O_2$   **e**  $I^-$

**3**   How many moles of the following will be oxidised by $1.0 \times 10^{-3}$ moles of potassium dichromate(VI) in acidic solution?

   **a**  $Fe^{2+}$   **b**  $SO_3^{2-}$   **c**  $Br^-$   **d**  $C_2O_4^{2-}$   **e**  $Sn^{2+}$

**4**   How many moles of the following will be reduced by $1 \times 10^{-3}$ mol of $Sn^{2+}$?

   **a**  $Fe^{3+}$   **b**  $Cl_2$   **c**  $Mn^{4+}$ to $Mn^{2+}$   **d**  $Ce^{4+}$ to $Ce^{3+}$   **e**  $BrO_3^-$ to $Br^-$

**5**   The problem is to find the percentage of iron that had rusted in a sample of rusty iron. A few grams of the rusty iron were completely dissolved in $1.00 \ dm^3$ sulphuric acid. Iron reacted to form $Fe^{2+}$ ions and rust reacted to form $Fe^{3+}$ ions. A $25.0 \ cm^3$ sample of the solution was titrated with $0.0200 \ mol \ dm^{-3}$ aqueous potassium manganate(VII). It took $27.2 \ cm^3$ for complete reaction. Zinc was added to the remaining acidic solution to reduce $Fe^{3+}(aq)$ to $Fe^{2+}(aq)$. A $25.0 \ cm^3$ sample of this solution now took $29.0 \ cm^3$ of the same potassium manganate(VII) solution for titration.

**STEPS TO TAKE**   **a**  Write the half-reaction equations for the reduction of $MnO_4^-$ to $Mn^{2+}$ and the oxidation of $Fe^{2+}$ to $Fe^{3+}$.

   **b**  Write the overall equation for the reaction.

   **c**  Calculate the amount of $Fe^{2+}(aq)$ in $25.0 \ cm^3$ of solution ... and $[Fe^{2+}]$.

   **d**  What volume of potassium manganate(VII) solution was required by the *extra* $Fe^{2+}$ formed by reduction with zinc? This will give you the $[Fe^{3+}]$ before reduction.

   **e**  Now you know $[Fe^{2+}]$ and $[Fe^{3+}]$ you can find the percentage of iron that had rusted.

**6**   $25.0 \ cm^3$ of potassium iodate(V) solution, $KIO_3(aq)$ were added to an excess of potassium iodide solution dissolved in sulphuric acid. The iodine liberated needed $30 \ cm^3$ of $0.50 \ mol \ dm^{-3}$ sodium thiosulphate, $Na_2S_2O_3$. Which is the concentration of potassium iodate solution?

   **A**  $0.1 \ mol \ dm^{-3}$   **B**  $0.2 \ mol \ dm^{-3}$   **C**  $0.4 \ mol \ dm^{-3}$   **D**  $0.5 \ mol \ dm^{-3}$

**7** $200 \text{ cm}^3$ of chlorine at r.t.p. were passed into $100 \text{ cm}^3$ of $0.200 \text{ mol dm}^{-3}$ KI(aq). What mass of iodine was produced?

**A** $1.06 \text{ g}$    **B** $2.12 \text{ g}$    **C** $4.24 \text{ g}$    **D** $5.08 \text{ g}$

**8** A mass $1.249 \text{ g}$ $CuSO_4 . 5H_2O$ was dissolved and added to an excess of KI(aq). The iodine liberated was titrated against aqueous sodium thiosulphate. What volume of $0.100 \text{ mol dm}^{-3}$ $Na_2S_2O_3$(aq) would be needed to react with the iodine liberated?

**A** $10 \text{ cm}^3$    **B** $25 \text{ cm}^3$    **C** $50 \text{ cm}^3$    **D** $100 \text{ cm}^3$

**9** Sulphur dioxide is an atmospheric pollutant. The amount of sulphur dioxide in a sample of air can be found by reaction with hydrogen peroxide to form sulphuric acid.

$$SO_2(g) + H_2O_2(aq) \longrightarrow H_2SO_4(aq)$$

The sulphuric acid formed can be titrated against a standard alkali.

When $200 \text{ cm}^3$ of polluted air were bubbled slowly through $25.0 \text{ cm}^3$ of hydrogen peroxide solution, the resulting solution was neutralised by $15.3 \text{ cm}^3$ of $0.100 \text{ mol dm}^{-3}$ aqueous sodium hydroxide. Calculate the concentration of sulphur dioxide in

**a** $\text{mol m}^{-3}$ of air    **b** $\text{g m}^{-3}$ air

**10** $25.0 \text{ cm}^3$ of aqueous sodium sulphite require $45.0 \text{ cm}^3$ of $0.0200 \text{ mol dm}^{-3}$ potassium manganate(VII) for oxidation. What is the concentration of the sodium sulphite solution?

**STEPS TO TAKE**   **a** First write the equation, referring to the half-reaction equations in the table above.

**b** Find the amount of $MnO_4^-$.

**c** How many moles of $SO_3^{2-}$ will it oxidise?

**d** Now find the concentration of $SO_3^{2-}$.

**11** $35.0 \text{ cm}^3$ of potassium manganate(VII) solution are needed to oxidise a $0.2145 \text{ g}$ sample of ethanedioic acid-2-water, $H_2C_2O_4 . 2H_2O$. What is the concentration of the potassium manganate(VII) solution?

**STEPS TO TAKE**   **a** First combine the half-reaction equations for $MnO_4^-$ and $C_2O_4^{2-}$.

**b** Find the amount of $C_2O_4^{2-}$.

**c** How many moles of $MnO_4^-$ will it react with?

**d** Now find the concentration of $KMnO_4$.

**12** $37.5 \text{ cm}^3$ of cerium(IV) sulphate solution are required to titrate a $0.2245 \text{ g}$ sample of sodium ethanedioate, $Na_2C_2O_4$. What is the concentration of cerium(IV) sulphate solution?

**Hint :** Equation ... amount of $C_2O_4^{2-}$ ... amount of $Ce^{4+}$ ... concentration of $Ce^{4+}$.

**13** A piece of iron wire weighs 0.2756 g. It is dissolved in acid, reduced to the $Fe^{2+}$ state, and titrated with 40.8 cm$^3$ of 0.0200 mol dm$^{-3}$ potassium dichromate solution. What is the percentage purity of the iron wire?

**Hint :** Amount of $K_2Cr_2O_7$ ... amount of $Fe^{2+}$ ... mass of Fe ... % of Fe in iron wire.

## Section 2

**1** A solution of potassium dichromate is standardised by titration with sodium ethanedioate solution. If 47.0 cm$^3$ of the dichromate solution were needed to oxidise 25.0 cm$^3$ of ethanedioate solution of concentration 0.0925 mol dm$^{-3}$, what is the concentration of the potassium dichromate solution?

**2** A 25.0 cm$^3$ portion of a solution containing $Fe^{2+}$ ions and $Fe^{3+}$ ions was acidified and titrated against potassium manganate(VII) solution. 15.0 cm$^3$ of a 0.0200 mol dm$^{-3}$ solution of potassium manganate(VII) were required. A second 25.0 cm$^3$ portion was reduced with zinc and titrated against the same manganate(VII) solution. 19.0 cm$^3$ of the oxidant solution were required. Calculate the concentrations of **a** $Fe^{2+}$ and **b** $Fe^{3+}$ in the solution.

**3** **a** What volume of acidified potassium manganate(VII) of concentration 0.0200 mol dm$^{-3}$ is decolourised by 100 cm$^3$ of hydrogen peroxide of concentration 0.0100 mol dm$^{-3}$? The equation is

$$MnO_4^-(aq) + H_2O_2(aq) + H^+(aq) \longrightarrow Mn^{2+}(aq) + O_2(g) + H_2O(l)$$

**b** What volume of oxygen is evolved at s.t.p.? (GMV = 22.4 dm$^3$ mol$^{-1}$ at s.t.p.)

**4** A piece of steel of mass 0.200 g reacts with dilute sulphuric acid. The resulting solution requires 34.0 cm$^3$ of 0.0200 mol dm$^{-3}$ potassium manganate(VII) in acidic solution in a titration.

Calculate the percentage of iron in the steel.

**5** Brass is an alloy of copper and zinc. It reacts with nitric acid to give a solution containing $Cu^{2+}$ and $Zn^{2+}$ ions. The concentration of $Cu^{2+}$ can be found by reaction with an excess of potassium iodide

$$2Cu^{2+}(aq) + 4I^-(aq) \longrightarrow 2CuI(s) + I_2(aq)$$

followed by titration of iodine against thiosulphate

$$I_2(aq) + 2S_2O_3^{2-}(aq) \longrightarrow 2I^-(aq) + S_4O_6^{2-}(aq)$$

The iodine liberated by a 0.2055 g piece of brass required 13.7 cm$^3$ of 0.150 mol dm$^{-3}$ thiosulphate solution.

Calculate the percentage of copper in brass.

**6** A 0.6125 g sample of potassium iodate(V), $KIO_3$, is dissolved in water and made up to 250 $cm^3$. A 25.0 $cm^3$ portion of the solution is added to an excess of potassium iodide in acid solution. The iodine formed requires 22.5 $cm^3$ of sodium thiosulphate solution for titration. What is the concentration of the thiosulphate solution?

$$IO_3^-(aq) + 5I^-(aq) + 6H^+(aq) \longrightarrow 3I_2(aq) + 3H_2O(l)$$

**7** 25.0 $cm^3$ of a solution of $X_2O_5$ of concentration 0.100 mol $dm^{-3}$ is reduced by sulphur dioxide to a lower oxidation state. To reoxidise **X** to its original oxidation number required 50.0 $cm^3$ of 0.0200 mol $dm^{-3}$ potassium manganate(VII) solution. To what oxidation number was **X** reduced by sulphur dioxide?

**Hint :** Amount of $X_2O_5$ ... amount of **X** ... amount of $MnO_4^-$ ...
ratio amount of **X**/amount of $MnO_4$.

**8** A piece of impure copper was allowed to react with dilute nitric acid. The copper(II) nitrate solution formed liberated iodine from an excess of potassium iodide solution. The iodine was estimated by titration with a solution of sodium thiosulphate. If a 0.877 g sample of copper was used, and the volume required was 23.7 $cm^3$ of 0.480 mol $dm^{-3}$ thiosulphate solution, what is the percentage of copper in the sample?

**Hint :** Equation ... amount of thio ... amount of iodine ... amount of copper ...
mass of copper ... percentage of copper.

**9** A household bleach contains sodium chlorate(I), NaOCl. The chlorate(I) ion will react with potassium iodide to give iodine, which can be estimated with a standard thiosulphate solution.

**a** Write the equations for the reaction of $ClO^-$ and $I^-$ to give $I_2$ and for the reaction of iodine and thiosulphate ions.

**b** A 25.0 $cm^3$ sample of household bleach is diluted to 250 $cm^3$. A 25.0 $cm^3$ portion of the solution is added to an excess of potassium iodide solution and titrated against 0.200 mol $dm^{-3}$ sodium thiosulphate solution. The volume required is 18.5 $cm^3$. What is the concentration of sodium chlorate(I) in the bleach?

$$ClO_3^-(aq) + 2I^-(aq) + H^+(aq) \longrightarrow I_2(aq) + Cl^-(aq) + H_2O(l)$$
$$I_2(aq) + 2S_2O_3^{2-}(aq) \longrightarrow 2I^-(aq) + S_4O_6^{2-}(aq)$$

## 8.9 COMPLEXOMETRIC TITRATIONS

The complexes formed by a number of metal ions with

bis[bis(carboxymethyl)amino]ethane,

$(HO_2CCH_2)_2NCH_2CH_2N(CH_2CO_2H)_2$

which is usually referred to as edta (short for its old name) are very stable, and can be used to find the concentration of metal ions by titration. The end-point in the titration is shown by an indicator which forms a coloured complex with the metal ion being titrated. If Eriochrome Black T is used as indicator, the metal-indicator colour of red is seen at the beginning of the titration. As the titrant is added, the metal ions are removed from the indicator and complex with edta. At the end-point, the colour of the free indicator, blue, is seen:

Metal-indicator (red) + edta $\longrightarrow$ Metal-edta + Indicator (blue)

**EXAMPLE**

### Measuring the hardness of tap water

Hardness in water is caused by the presence of calcium ions and magnesium ions. Both these ions complex strongly with edta.

$$Ca^{2+}(aq) + edta^{4-}(aq) \rightleftharpoons Caedta^{2-}(aq)$$

100 cm$^3$ of tap water are measured into a flask. An alkaline buffer and a suitable indicator are added and the solution is titrated against 0.100 mol dm$^{-3}$ edta solution. The volume required is 2.10 cm$^3$.

**METHOD**

Amount of edta = $2.10 \times 10^{-3}$ dm$^3 \times 0.100$ mol dm$^{-3} = 2.10 \times 10^{-3}$ mol

Amount of metal present as hardness = $2.10 \times 10^{-3}$ mol dm$^{-3}$

**ANSWER**

The concentration of calcium and magnesium = $2.10 \times 10^{-3}$ mol dm$^{-3}$

**EXERCISE 8.5**

### Problems on complexometric titrations

1 'Health salts' contain magnesium sulphate. A sample of 0.1 g of health salts was dissolved in water, the pH was adjusted to 10 by means of a buffer. The solution was titrated against 0.020 mol dm$^{-3}$ edta(aq), using a suitable indicator. The titre was 23.3 cm$^3$.

Calculate the percentage of magnesium in the health salts.

2 Calculate the concentration of a solution of zinc sulphate from the following data. 25.0 cm$^3$ of the solution, when added to an alkaline buffer and Eriochrome Black T indicator, required 22.3 cm$^3$ of a $1.05 \times 10^{-2}$ mol dm$^{-3}$ solution of edta for titration. The equation for the reaction can be represented as

$$Zn^{2+}(aq) + edta^{4-}(aq) \longrightarrow Znedta^{2-}(aq)$$

3 To a 50.0 cm$^3$ sample of tap water were added a buffer and a few drops of Eriochrome Black T. On titration against a 0.0100 mol dm$^{-3}$ solution of edta, the indicator turned blue after the addition of 9.80 cm$^3$ of the titrant. Calculate the hardness of water in parts per million of calcium, assuming that the hardness is entirely due to the presence of calcium salts. (1 p.p.m. = 1 g in 10$^6$ g water.)

**4** A 0.2500 g sample of a mixture of magnesium oxide and calcium oxide was dissolved in dilute nitric acid and made up to 1.00 $dm^3$ of solution with distilled water. A 50.0 $cm^3$ portion was buffered and, after addition of indicator, was titrated against 0.0100 mol $dm^{-3}$ edta solution. 25.8 $cm^3$ of the titrant were required. Find the percentage by mass of calcium oxide and magnesium oxide in the mixture.

**5** Find $n$ in the formula $Al_2(SO_4)_3 . nH_2O$ from the following analysis. 1.000 g of aluminium sulphate hydrate was weighed out and made up to 250 $cm^3$. A 25.0 $cm^3$ portion was allowed to complex with edta by being boiled with 50.0 $cm^3$ of edta solution of concentration $1.00 \times 10^{-2}$ mol $dm^{-3}$. The excess of edta was determined by adding Eriochrome Black T and titrating against a solution of $1.115 \times 10^{-2}$ mol $dm^{-3}$ zinc sulphate. 17.9 $cm^3$ of zinc sulphate solution were required to turn the indicator from blue to red. The reactions taking place are

$$Al^{3+}(aq) + edta^{4-}(aq) \longrightarrow Aledta^{-}(aq)$$
$$Zn^{2+}(aq) + edta^{4-}(aq) \longrightarrow Znedta^{2-}(aq)$$

## EXERCISE 8.6    Questions from examination papers

**1** **a** Titanium(IV) chloride reacted with water as shown in the following equation.

$$TiCl_4(l) + 2H_2O(l) \longrightarrow 4HCl(aq) + TiO_2(s)$$

The reaction produced 200 $cm^3$ of a 1.20 M solution of hydrochloric acid. Calculate the number of moles of HCl in the solution and use your answer to find the original mass of $TiCl_4$. (4)

**b** Calculate the volume of 1.10 M sodium hydroxide solution which would be required to neutralise a 100 $cm^3$ portion of the 1.20 M solution of hydrochloric acid. (3)

**c** An excess of magnesium metal was added to a 100 $cm^3$ portion of the 1.20 M solution of hydrochloric acid. Calculate the volume of hydrogen gas produced at 98 kPa and 20 °C.

$$Mg(s) + 2HCl(aq) \longrightarrow MgCl_2(aq) + H_2(g)$$ (4)

*NEAB 99*   (11)

**2** **a** Define the term *relative molecular mass*. (1)

**b** How would you calculate the mass of one mole of atoms from the mass of a single atom? (1)

**c** Sodium hydride reacts with water according to the following equation.

$$NaH(s) + H_2O(l) \longrightarrow NaOH(aq) + H_2(g)$$

A 1.00 g sample of sodium hydride was added to water and the resulting solution was diluted to a volume of exactly 250 $cm^3$.

**(i)** Calculate the concentration, in mol $dm^{-3}$, of the sodium hydroxide solution formed.

**(ii)** Calculate the volume of hydrogen gas evolved, measured at 293 K and 100 kPa.

**(iii)** Calculate the volume of 0.112 M hydrochloric acid which would react exactly with a 25.0 cm$^3$ sample of the sodium hydroxide solution. (8)

*NEAB 98* (10)

**3** **a** The ammonium ion content of fertilisers can be found by heating the fertiliser with sodium hydroxide and passing the ammonia produced into an excess of hydrochloric acid.

**(i)** Write an equation for the reaction between ammonium sulphate and sodium hydroxide. (1)

**(ii)** 3.00 g of a fertiliser mixture containing ammonium sulphate was made up to 250 cm$^3$ of solution. 25.0 cm$^3$ portions of this were added to an excess of sodium hydroxide solution, and the ammonia produced passed into 50.0 cm$^3$ of 0.100 mol dm$^{-3}$ hydrochloric acid solution. The residual acid was then titrated with 0.100 mol dm$^{-3}$ sodium hydroxide solution, 25.4 cm$^3$ being required. Find the percentage by mass of ammonium sulphate in the fertiliser mixture. (4)

**b** Some of the nitrogen content of a fertiliser could be present as nitrate ions, which are not detected by sodium hydroxide solution. Nitrate ions can be reduced to ammonia by the use of aluminium in sodium hydroxide solution.

**(i)** Complete the half equation below for the reduction of nitrate ions:

$$NO_3^-(aq) + 6H_2O(l) + \cdots e^- \longrightarrow NH_3(g) + \cdots$$ (2)

**(ii)** The reaction occurs in strongly alkaline solution; suggest a formula for the aluminium-containing species present at the end of the reaction. (1)

**c** In cold aqueous sodium hydroxide ammonium salts produce the following equilibrium:

$$NH_4^+(aq) + OH^-(aq) \rightleftharpoons NH_3(aq) + H_2O(l)$$

**(i)** Identify the two acid-base conjugate pairs in the equilibrium. (2)

**(ii)** Explain the effect of raising the temperature on this equilibrium. (2)

**d** Pure **liquid** ammonia ionises as follows:

$$NH_3(l) + NH_3(l) \rightleftharpoons NH_4^+(am) + NH_2^-(am)$$

where (am) represents solutions in liquid ammonia. The Brønsted-Lowry theory of acid-base behaviour applies to solutions in liquid ammonia.

**(i)** Suggest why ammonium salts behave as acids in liquid ammonia. (1)

**(ii)** The salt sodium amide Na$^+$NH$_2^-$ reacts with ammonium chloride in liquid ammonia in an acid-base reaction. Write an equation to represent the reaction. (2)

*L* (15)

**4** Wines often contain a small amount of sulphur dioxide that is added as a preservative. The amount of sulphur dioxide added needs to be carefully calculated: too little and the wine readily goes bad, too much and the wine tastes of sulphur dioxide.

The sulphur dioxide content of a wine can be found by using its reaction with aqueous iodine.

$$SO_2(aq) + I_2(aq) + 2H_2O(l) \longrightarrow SO_4^{2-}(aq) + 2I^-(aq) + 4H^+(aq)$$

**a** **(i)** State the oxidation number of sulphur in $SO_2$ and in $SO_4^{2-}$.

**(ii)** State, with a reason, whether sulphur is oxidised or reduced in the conversion of $SO_2$ into $SO_4^{2-}$. (3)

**b** The sulphur dioxide content of a wine can be found by titration. An analyst found that the sulphur dioxide in 50.0 cm$^3$ of a sample of white wine reacted with exactly 16.4 cm$^3$ of 0.0100 mol dm$^{-3}$ aqueous iodine.

**(i)** How many moles of iodine, $I_2$, did the analyst use in the titration?

**(ii)** How many moles of sulphur dioxide were in the 50.0 cm$^3$ of wine?

**(iii)** What was the concentration, in mol dm$^{-3}$, of sulphur dioxide in the wine?

**(iv)** What was the concentration, in g dm$^{-3}$, of sulphur dioxide in the wine? (5)

**c** The generally accepted maximum concentration of sulphur dioxide in wine is 0.25 g dm$^{-3}$. A concentration of less than 0.01 g dm$^{-3}$ is insufficient to preserve the wine.

Comment on the effectiveness of the sulphur dioxide in the wine analysed in **b**. (1)

*OCR* (9)

**5** **a** Use the concept of oxidation states to deduce whether either of the reactions given below involves redox processes. Explain your answers and, where appropriate, identify the element which is being oxidised.

**(i)** $2NH_3 + 3Cl_2 \longrightarrow N_2 + 6HCl$

**(ii)** $CuO + 2HCl \longrightarrow CuCl_2 + H_2O$ (5)

**b** **(i)** Potassium manganate(VII), $KMnO_4$, can be used in the quantitative estimation of ethanedioate ions, $C_2O_4^{2-}$, in an acidified aqueous solution. In this reaction, ethanedioate ions are converted into carbon dioxide. Deduce half equations for the redox processes involved and hence derive an equation for the overall reaction. (3)

**(ii)** A 1.93 g sample of a crystalline ethanedioate salt was dissolved in water and made up to 250 cm$^3$. 25.0 cm$^3$ of this solution, after acidification, was found to react with 30.4 cm$^3$ of 0.0200 M $KMnO_4$. Calculate the percentage by mass of ethanedioate ions in the original salt.

**NB** If you are unable to deduce the overall ratio $C_2O_4^{2-} : MnO_4^-$ for this reaction you may assume the ratio 5 : 3 (This is not the correct ratio). (6)

*NEAB 97* (14)

**6** Commercial bleaches contain sodium hypochlorite (sodium chlorate (I)), which acts as an oxidising agent. The concentration of sodium hypochlorite in solution can be determined by reaction with acidified potassium iodide solution.

$$NaOCl + 2KI + H_2SO_4 \longrightarrow I_2 + H_2O + NaCl + K_2SO_4$$

The liberated iodine is titrated with standard sodium thiosulphate solution.

**a** Which species is oxidised in this equation? (1)

**b** Explain the term **standard** sodium thiosulphate solution. (1)

**c** Name a suitable indicator for this titration, stating the expected colour change at the end point. (3)

**d** Write the equation for the reaction between iodine and sodium thiosulphate. (2)

**e** 15.0 cm$^3$ of a bleach sample was diluted to 250 cm$^3$ with de-ionised water. 25.0 cm$^3$ portions of the solution were treated with excess acidified potassium iodide solution and then titrated with 0.1 M sodium thiosulphate solution. The average titre was found to be 25.2 cm$^3$. Calculate the concentration of sodium hypochlorite in the original bleach sample. (4)

*NI* (11)

**7 a (i)** Draw a diagram to show the bonding in the nitrile group. (1)

**(ii)** Give the reagents and necessary conditions for the conversion of a halogenoalkane to a nitrile. (1)

**(iii)** Give the reagents and necessary conditions for the conversion of a nitrile to a carboxylic acid. (1)

**b** Ethanedioic acid (oxalic acid) is the simplest dicarboxylic acid having the formula $(COOH)_2 . 2H_2O$ in its hydrated form.

At 70 °C acidified aqueous ethanedioic acid is oxidised by manganate(VII) ions, the reaction being catalysed by manganese(II) ions.

The ion/electron half-equation is

$$\begin{array}{c} COOH \\ | \\ COOH \end{array} (aq) \longrightarrow 2CO_2(g) + 2H^+(aq) + 2e^-$$

Simultaneously manganate(VII) ions are reduced according to the ion/electron half equation

$$MnO_4^-(aq) + 8H^+(aq) + 5e^- \longrightarrow Mn^{2+}(aq) + 4H_2O(l)$$

**(i)** Write an ionic equation for the oxidation of ethanedioic acid by acidified aqueous manganate(VII) ions. (1)

**(ii)** The reaction is catalysed by manganese(II) ions. When potassium manganate(VII) solution is added to an aqueous ethanedioic acid and sulphuric acid mixture at 70 °C, the rate of reaction steadily increases in the period after mixing. Explain why this rate increase occurs. (1)

**(iii)** 34.60 cm$^3$ of aqueous potassium manganate(VII) of concentration 0.0200 mol dm$^{-3}$ were required to oxidise 25.00 cm$^3$ of an aqueous solution of ethanedioic acid acidified with aqueous sulphuric acid. Calculate the concentration of the ethanedioic acid-2-water in **g dm**$^{-3}$. (3)

**c (i)** The aqueous manganate(VI) ion, $MnO_4^{2-}(aq)$, is unstable in neutral solution and reacts with water to form aqueous manganate(VII) ions, hydroxide ions and manganese(IV) oxide, $MnO_2$. Give an ionic equation for this change. (1)

**(ii)** Manganese(IV) oxide reacts with concentrated hydrochloric acid in the same way that lead(IV) oxide reacts with concentrated hydrochloric acid.

State what type of reaction occurs when manganese(IV) oxide reacts with hydrochloric acid and name the gaseous product. (1)

*WJEC* (10)

**8** The sulphite of a Group I metal was oxidised by aqueous potassium dichromate(VI), $K_2Cr_2O_7$, under acidic conditions. 1.90 g of anhydrous metal sulphite was made up to 250 cm$^3$ of solution with acid and water. 20.0 cm$^3$ portions of this solution were titrated with 0.0200 mol dm$^{-3}$ potassium dichromate(VI) from the burette.

The following burette readings were obtained;

| | Trial | 1 | 2 | 3 |
|---|---|---|---|---|
| Final reading/cm$^3$ | 26.0 | 42.0 | 17.6 | 33.6 |
| Start reading/cm$^3$ | 8.6 | 26.0 | 1.0 | 17.6 |

**a** **(i)** What colour change would be observed during the titration reaction? (1)

**(ii)** During the reaction, the sulphite ion is oxidised, and the dichromate(VI) ion is reduced to the chromium(III) ion. Give the ionic half-equations for *both* reactions. Hence show that 1 mole of dichromate(VI) ion oxidises 3 moles of sulphite ion. (2)

**(iii)** What titre value would you use for the calculation? (1)

**(iv)** Use the data to calculate the relative molecular mass of the metal sulphite. Identify the metal. (3)

**b** Excess potassium hydroxide was added to aqueous potassium dichromate(VI). Predict what observation would be made, and give the ionic equation for the reaction. (2)

*OCR* (9)

**9** **a** A pale green solid, **X**, dissolves in water to form a pale green solution. The following tests are carried out on three portions of this solution of **X**.

**A** sodium hydroxide solution is added and a green gelatinous precipitate forms;

**B** potassium hexacyanoferrate(III) solution is added and a blue precipitate forms;

**C** barium chloride solution is added and a white precipitate forms.

**(i)** Deduce the name of substance **X**. (2)

**(ii)** Write a balanced ionic equation, including state symbols, for the reaction occurring in test **A**. (2)

**b** Potassium manganate(VII) is an effective oxidizing agent when mixed with dilute sulphuric acid. It is able to oxidize iron(II) ions to iron(III) ions. Solutions containing $Fe^{2+}(aq)$ can be titrated against potassium manganate(VII) solution. This titration forms the basis of an analytical technique for the estimation of iron in substances such as flour.

1.00 g of a flour was vigorously shaken with dilute sulphuric acid and the volume of the solution was made up to 100 cm$^3$ in a volumetric flask. 10 cm$^3$ portions of the solution were titrated with $5.00 \times 10^{-6}$ M potassium manganate(VII). An average titre of 11.0 cm$^3$ was recorded.

**(i)** Copy and complete the balancing of the equation for the titration reaction.

$$MnO_4^-(aq) + 5Fe^{2+}(aq) + H^+(aq) \longrightarrow Mn^{2+}(aq) + Fe^{3+}(aq) + H_2O(l)$$ (1)

**(ii)** Calculate the number of moles of manganate(VII) ion used in each titration. (1)

(iii) Calculate the number of moles of iron(II) ions **in 1.00 g of flour.** (1)

(iv) Flour must contain 1.65 mg of iron per 100 g of flour, by law. By carrying out an appropriate further calculation, decide if this flour contains sufficient iron to fulfil this legal requirement. (2)

c What are the advantages of iron being present in flour? (2)

*L(N)* (11)

10 a The active ingredients present in a brand of 'health salts' are listed as sodium hydrogencarbonate, $NaHCO_3$, citric acid, $C_5H_7O_5COOH$, and magnesium sulphate dihydrate, $MgSO_4 . 2H_2O$. Explain, with the aid of ionic equations, why vigorous effervescence occurs on the addition of these 'health salts' to water. (5)

b The concentration of magnesium ions in solution can be found by titration with a solution of the compound known as edta. A complex ion is formed between the magnesium ions and the edta, in which the edta acts as a *ligand* and hydrogen ions are released. The titration works best when buffered with a solution containing ammonia and ammonium chloride.

(i) Explain the term *ligand* and hence describe the type of bonding in complex ions. (2)

(ii) Explain, with the aid of relevant equations, how the mixture of ammonia and ammonium chloride acts as a buffer in this titration. (6)

(iii) Aqueous magnesium ions, $Mg^{2+}(aq)$, react with edta in the mole ratio 1 : 1. 7.80 g of health salts were dissolved in distilled water and, after effervescence had stopped, the solution was made up to 250 $cm^3$. A 25.0 $cm^3$ portion of this solution was titrated against edta solution. 17.4 $cm^3$ of 0.0500 mol $dm^{-3}$ edta were required.

Calculate the percentage of $MgSO_4.2H_2O$ in the health salts. (5)

*L* (18)

11 a Potassium manganate(VII) is a common laboratory oxidising agent. It can be prepared in the laboratory, from $MnO_2$, by the following sequence of reactions.

$$3MnO_2 + 6KOH + KClO_3 \longrightarrow 3K_2MnO_4 + 3H_2O + KCl$$
$$2K_2MnO_4 + Cl_2 \longrightarrow 2KMnO_4 + 2KCl$$

Potassium manganate(VII) readily oxidises concentrated hydrochloric acid to chlorine at room temperature.

(i) Copy and complete the equation for preparation of chlorine.

$$2KMnO_4 + HCl \longrightarrow KCl + MnCl_2 + Cl_2 + H_2O$$

(ii) By reference to the equation, suggest why this is not an economical method of preparing $^{36}Cl_2$ from very expensive $H^{36}Cl(aq)$.

(iii) **Name** each of the manganese compounds involved in the above reactions, and in each case identify the oxidation state of manganese: $MnO_2$, $K_2MnO_4$, $MnCl_2$. (5)

**b** Potassium manganate(VII) is used to titrate acidified solutions of reducing agents.

    **(i)** Suggest why hydrochloric acid is not a suitable acid to use for acidification of the reducing agent.

    **(ii)** Calculate the mass of $KMnO_4$ required to prepare 250 $cm^3$ of 0.0200 $mol\,dm^{-3}$ solution.

    **(iii)** Describe briefly an investigation or experiment involving a titration with $KMnO_4$. You should name the reducing agent and the product of oxidation and describe how to detect the end point. (6)

*OCR* (11)

**12** Calcium oxalate, $CaC_2O_4$ is the main component of 'beer-stone' which precipitates from beers in storage tanks.

  **a** What is the oxidation number of carbon in calcium oxalate? (1)

  **b** Calcium oxalate may be converted to oxalic acid and calcium sulphate by reaction with dilute sulphuric acid:

$$CaC_2O_4 + H_2SO_4 \longrightarrow H_2C_2O_4 + CaSO_4$$

The calcium sulphate precipitated can be hydrated to make plaster of Paris, $CaSO_4 \cdot xH_2O$. Calculate the value of $x$ in the formula if 1.45 g of plaster of Paris is heated to dryness and 1.36 g of $CaSO_4$ remains. (3)

  **c** Oxalate ions, $C_2O_4^{2-}$, can be oxidised to carbon dioxide using manganate(VII) (permanganate) ions, $MnO_4^-$.

$$C_2O_4^{2-} \longrightarrow 2CO_2 + 2e^-$$
$$MnO_4^- + 8H^+ + 5e^- \longrightarrow Mn^{2+} + 4H_2O$$

Using these two half-equations, write a balanced redox equation for the reaction between oxalate ions and manganate(VII) ions. (2)

  **d** In the oxalate ion the two carbons are joined by a single bond and each carbon has two oxygen atoms attached. Suggest and explain the shape of the oxalate ion. (2)

*NI* (8)

**13** Hydrogen peroxide has largely replaced chlorine as an oxidising agent in the paper industry.

The concentration of solutions of hydrogen peroxide may be found by oxidising excess potassium iodide solution and titrating the iodine produced with sodium thiosulphate solution. The redox half-equations are:

$$H_2O_2 + 2H^+ + 2e^- \longrightarrow 2H_2O$$
$$2I^- \longrightarrow I_2 + 2e^-$$

If 10 $cm^3$ of a solution of hydrogen peroxide is made up with distilled water to 1 $dm^3$ and a 25 $cm^3$ sample reacted with excess iodide, it is found that 30 $cm^3$ of 0.2 $mol\,dm^{-3}$ sodium thiosulphate solution is needed to react completely with the iodine produced.

  **a** Name the indicator used to make the end-point clearer and describe the colour change. (3)

**b** Calculate the molarity of the original hydrogen peroxide solution. (4)

**c** Hydrogen peroxide decomposes according to the equation:

$$2H_2O_2 \longrightarrow 2H_2O + O_2$$

Calculate the volume of oxygen, at 20 °C and 1 atm pressure, which would be produced from the complete decomposition of 10 cm$^3$ of the original hydrogen peroxide solution. (2)

NI (9)

**14** In a textbook, the instructions for an experiment to find the value of $x$ in the formula $Na_2CO_3 . xH_2O$ in a sample of hydrated sodium carbonate crystals were as follows:

> Weigh out accurately about 5 grams of soda crystals, being careful to select well-defined translucent crystals rather than those covered in powder. Dissolve these in a beaker of water, transfer to a 250 cm$^3$ flask, shake well and make up to the mark. Pipette 25.0 cm$^3$ of this solution into a conical flask, add a few drops of methyl orange and titrate with standard 0.2 M hydrochloric acid until the first permanent colour change is observed. Repeat to obtain two results within 0.1 cm$^3$ of each other.

A student who follows these instructions weighs out 4.90 g of the crystals and finds that an accurate titre of 17.7 cm$^3$ is required for neutralisation.

**a** What is meant by **standard** 0.2 M hydrochloric acid? (1)

**b** **(i)** What colour change will be observed with the methyl orange? (2)

**(ii)** Why did the instructions suggest the use of translucent crystals rather than those covered in powder? (1)

**c** **(i)** How many moles of hydrochloric acid were used? (1)

**(ii)** Write an equation for the reaction between the hydrochloric acid and the sodium carbonate solution. (1)

**(iii)** Calculate the concentration of the sodium carbonate solution. (1)

**(iv)** Calculate the value of $x$. (2)

NI (9)

**15** Potassium iodate(V), $KIO_3$, is used in the standardisation of aqueous sodium thiosulphate.

Iodine is liberated from excess acidified aqueous potassium iodide according to

$$IO_3^-(aq) + 6H^+(aq) + 5I(aq) \longrightarrow 3I_2(aq) + 3H_2O(l)$$

and the liberated iodine is titrated against sodium thiosulphate.

**a** Write down the balanced equation for the reaction between iodine and sodium thiosulphate. (1)

**b** A mass of 0.1500 g of potassium iodate(V), $KIO_3$, was dissolved in water and made up to 250.0 cm$^3$ of solution. 25.00 cm$^3$ of this solution was added to excess aqueous potassium iodide and aqueous acid and the liberated iodine just reacted with 22.60 cm$^3$ of aqueous sodium thiosulphate. Calculate the concentration of the potassium iodate(V) solution and hence the concentration of the aqueous sodium thiosulphate, in mol dm$^{-3}$. (4)

WJEC(part) (5)

**16** **a** Discuss the structure of, and bonding in, crystals of sodium chloride, diamond and iodine. Predict the relative melting temperatures of these three crystals. (10)

**b** Explain, in terms of the motion of particles and the forces acting between them what happens when water, starting as ice, is heated from −10 °C to 110 °C. (8)

**c** A supply of sodium metal was contaminated with sodium oxide and sodium chloride. A sample of this metal, weighing 1.00 g, was dissolved in water, producing $4.95 \times 10^{-4}$ m$^3$ of hydrogen gas at 298 K and 98.0 kPa. The resulting solution was made up to a volume of 250 cm$^3$ with water and a 25.0 cm$^3$ portion of this final solution was exactly neutralised by 37.2 cm$^3$ of 0.112 M hydrochloric acid.

**(i)** Calculate the number of moles of hydrogen gas released.

**(ii)** Calculate the number of moles of sodium hydroxide produced by the reaction between water and the sample.

**(iii)** Use your answers to parts **(i)** and **(ii)**, together with the equations given below, to calculate the number of moles of sodium and sodium oxide in the sample and hence the percentages by mass of sodium and sodium oxide.

$$2Na + 2H_2O \longrightarrow 2NaOH + H_2$$
$$Na_2O + H_2O \longrightarrow 2NaOH$$ (12)

*NEAB 98* (30)

**17** **a** Vanadium(V) oxide is used as an industrial catalyst.

**(i)** Write an equation for a process which uses this catalyst.

**(ii)** What feature of vanadium chemistry makes vanadium(V) oxide suitable for use in this process? (2)

**b** In order to analyse an impure sample of vanadium(V) oxide, a 0.200 g sample was treated with dilute sulphuric acid and an excess of sodium sulphite. A blue solution was formed containing vanadium(IV) as $VO^{2+}$. This solution required 20.0 cm$^3$ of 0.0200 M potassium manganate(VII) to convert all the vanadium back into the vanadium(V) state as $VO_2^+$. The equations for the reactions which occurred were

$$V_2O_5 + 2H^+ \longrightarrow 2VO_2^+ + H_2O$$
$$2VO_2^+ + SO_3^{2-} + 2H^+ \longrightarrow 2VO^{2+} + SO_4^{2-} + H_2O$$
$$5VO^{2+} + MnO_4^- + H_2O \longrightarrow Mn^{2+} + 5VO_2^+ + 2H^+$$

Use these data to calculate the percentage by mass of vanadium in the sample of vanadium(V) oxide. (6)

*NEAB 98* (8)

**18** **a** Define the term *relative molecular mass*. (1)

**b** How would you calculate the mass of one mole of atoms from the mass of a single atom? (1)

**c** Sodium hydride reacts with water according to the following equation.

$$NaH(s) + H_2O(l) \longrightarrow NaOH(aq) + H_2(g)$$

A 1.00 g sample of sodium hydride was added to water and the resulting solution was diluted to a volume of exactly 250 cm$^3$.

   **(i)** Calculate the concentration, in mol dm$^{-3}$, of the sodium hydroxide solution formed.

   **(ii)** Calculate the volume of hydrogen gas evolved, measured at 293 K and 100 kPa.

   **(iii)** Calculate the volume of 0.112 M hydrochloric acid which would react exactly with a 25.0 cm$^3$ sample of the sodium hydroxide solution.
   (8)

                                        *NEAB 98*   (10)

**19**   Silicon tetrachloride is a liquid of formula $SiCl_4$. It reacts with water to form silicon dioxide and hydrochloric acid according to the equation:

$$SiCl_4(l) + 2H_2O(l) \longrightarrow SiO_2(s) + 4HCl(aq)$$

**a**   **(i)** Is the formula of silicon tetrachloride consistent with the position of silicon in the Periodic Table? Justify your answer.
   (2)

   **(ii)** Would you expect silicon tetrachloride to contain ionic or covalent bonds? Which two items of information are there in the question which support your answer?
   (3)

**b**  Hydrochloric acid is described as a strong acid. What is meant by a **strong** acid? Write an equation to illustrate your answer.
   (2)

**c**  1.50 g of an impure sample of silicon tetrachloride was weighed out on an accurate balance. It was cautiously added to water and when the reaction was complete, the mixture was made up to 250 cm$^3$ with pure water.

A 25.0 cm$^3$ portion was titrated, using a suitable indicator, with 0.100 M sodium hydroxide solution. 30.0 cm$^3$ was needed for neutralization.

   **(i)** What piece of apparatus would you use for

      **A** making up the solution to exactly 250 cm$^3$?

      **B** removing a 25.0 cm$^3$ portion for the titration?
   (2)

   **(ii)** Suggest a suitable indicator for this titration and give the colour change.
   (2)

   **(iii)** The procedure described in **c** is incomplete. Suggest TWO important steps that the student should have taken to ensure an accurate result.
   (2)

   **(iv)** Use the information given in **c** and the equation

$$SiCl_4(l) + 2H_2O(l) \longrightarrow SiO_2(s) + 4HCl(aq)$$

to calculate the percentage purity of the sample of silicon tetrachloride.
   (4)

                                        *L(N)*   (17)

# The atom

9

## 9.1 RELATIVE ISOTOPIC MASS

**Reminder :** Relative atomic mass § 3.1. Relative molecular mass § 3.2.

The definition of **relative atomic mass** in § 3.1 is:

$$\text{Relative atomic mass} = \frac{\text{Mass of one atom of an element}}{\frac{1}{12} \text{ Mass of one atom of carbon-12}}$$

We shall now take this further. Many elements consist of a mixture of isotopes, for example there are other isotopes of carbon with different atomic masses: carbon-13 and carbon-14. To take account of these we need another definition. The term **relative isotopic mass** is defined by:

$$\text{Relative isotopic mass} = \frac{\text{Mass of one atom of the isotope}}{\frac{1}{12} \text{ Mass of one atom of carbon-12}}$$

By definition, carbon-12 has a relative isotopic mass of 12.0. The relative isotopic masses of the other carbon isotopes are 13.0 and 14.0.

### Relative atomic mass

What is the relationship between relative atomic mass and relative isotopic mass? Chlorine consists of two isotopes, chlorine-35 and chlorine-37. The average relative isotopic mass of chlorine is 35.5. It is not 36 because there are three chlorine-35 atoms for each chlorine-37 atom. Therefore:

- sum of relative isotopic masses of 4 typical atoms of chlorine = $(3 \times 35) + (1 \times 37)$
- weighted average relative isotopic mass = $[(3 \times 35) + (1 \times 37)]/4 = 35.5$

This average is called the **weighted average relative isotopic mass** because it takes into account the **relative abundance** of the isotopes (the proportions of the isotopes in the element) as well as their relative isotopic masses. It is also called the **relative atomic mass** of chlorine.

**Relative atomic mass,** symbol $A_r$, is defined by

Relative atomic mass = Weighted average of relative isotopic masses
of element                 of the element

Naturally occurring carbon has a relative atomic mass of 12.011 11. Hydrogen has a relative atomic mass of 1.007 97.

### Atomic mass unit

The mass spectrometer measures mass in **atomic mass units**, u. One atom of carbon-12, with relative isotopic mass 12.0 has a mass of 12.0 atomic mass units, 12.0 u. One atom of chlorine-37, with relative isotopic mass = 37.0 has a mass of 37.0 mass units, 37.0 u.

One atomic mass unit, $1\ u = 1.661 \times 10^{-27}\ kg$

## 9.2 THE MASS SPECTROMETER

In a mass spectrometer, an element or compound is vaporised and then ionised. The ions are accelerated and focussed into a beam which is then deflected by a magnetic field. The angle of deflection depends on the ratio of mass/charge of the ion, as well as the values of the accelerating voltage and the magnetic field. The magnetic field is kept constant while the accelerating voltage is varied continuously to focus each species in turn into the ion detector. The detector records each species as a peak on a trace. The value of mass/charge, $m/z$, for each species can be read from the position of its peak on the trace (see Figure 9.1). Each ion has a charge of +1 so the ratio $m/z = m/1$ = the mass of the ion in atomic mass units. The heights of the peaks are proportional to the relative abundance of the different ions.

### Using a mass spectrum to obtain relative atomic mass

**EXAMPLE 1**

Figure 9.1 shows the mass spectrum of boron. Calculate the relative atomic mass of boron.

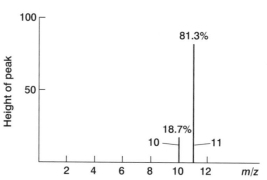

Fig. 9.1   *Mass spectrum of boron*

**METHOD**

There are peaks at $m/z = 10.0$ and at $m/z = 11.0$. Since each ion has a unit positive charge $m/z = m/1$ and the masses of the ions are 10.0 u and 11.0 u. The relative heights of the peaks show that the relative abundance of the two isotopes is 18.7% $^{10}$B: 81.3% $^{11}$B.

In 1000 typical atoms there are 187 atoms of mass 10.0 u = 1870 u
and 813 atoms of mass 11.0 u = 8943 u

The mass of 1000 atoms = 10 813 u, therefore weighted average isotopic mass = 10.8 u

**ANSWER**

The relative atomic mass of boron is 10.8

**EXAMPLE 2**

Figure 9.2 shows the mass spectrum of neon. The heights of the peaks are in the ratio 114 : 0.2 : 11.2. Calculate the relative atomic mass of neon.

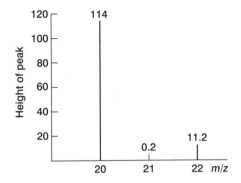

Fig. 9.2 *Mass spectrum of neon*

**METHOD**

The spectrum shows peaks at $m/z$ = 20.0, 21.0 and 22.0. These correspond to ions of mass 20.0 u, 21.0 u and 22.0 u. The relative heights of the peaks gives the relative abundance of the isotopes.

Multiplying the relative abundance by the isotopic mass of each isotope gives

Mass of neon-20 = 114 × 20.0 = 2280 u

Mass of neon-21 = 0.2 × 21.0 = 4.2 u

Mass of neon-22 = 11.2 × 22.0 = 246.4 u

Totals are relative abundance = 125.4 and mass = 2530.6 u

Weighted average isotopic mass = 2530.6/125.4 = 20.18 u

**ANSWER**

The relative atomic mass of neon is 20.2

**EXERCISE 9.1**

Problems on relative atomic mass

**Reminder :** Percentages § 1.3.

**1** The mass spectrum of rubidium consists of a peak at mass 85 u and a peak at mass 87 u. The relative abundance of the isotopes is 72 : 28. Calculate the weighted average isotopic mass of rubidium.

**2** If $^{69}$Ga and $^{71}$Ga occur in the proportions 60 : 40, calculate the weighted average isotopic mass of gallium.

**3** Figure 9.3 shows the mass spectrum of magnesium. The heights of the three peaks and the mass/charge ratios of the isotopes are shown in Figure 9.3. Calculate the relative atomic mass of magnesium.

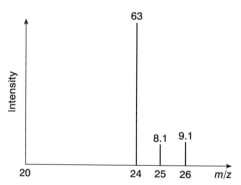

Fig. 9.3 *Mass spectrum of magnesium*

**4** The mass spectrum of chlorine shows peaks at masses 70, 72 and 74 u. The heights of the peaks are in the ratio of 9 to 6 to 1. What is the relative abundance of $^{35}Cl$ and $^{37}Cl$? What is the weighted average isotopic mass of chlorine?

**5** Calculate the relative atomic mass of lithium, which consists of 7.4% of $^6Li$ and 92.6% of $^7Li$.

**6** A sample of water containing $^1H$, $^2H$ and $^{16}O$ was analysed in a mass spectrometer. The trace showed peaks at mass numbers 1, 2, 3, 4, 17, 18, 19 and 20. Suggest which ions are responsible for these peaks.

**7** Calculate the weighted average isotopic mass of potassium, which consists of 93% $^{39}K$ and 7.0% $^{41}K$.

**8** Mass spectrometry of nickel shows the presence of three isotopes.

| $m/z$ | 58 | 60 | 62 |
|---|---|---|---|
| Relative abundance/% | 69 | 27 | 4 |

Calculate the weighted average relative isotopic mass of nickel.

## 9.3 MASS SPECTRA OF COMPOUNDS

### Using a mass spectrum to obtain relative molecular mass

When a compound is ionised in a mass spectrometer, a **molecular ion** is formed by the loss of one electron:

$$M \longrightarrow M^+ + e^-$$

In the mass spectra of volatile compounds, the peak with the highest value of $m/z$ is due to the molecular ion. Its mass gives the molecular mass (in u) of the compound. This is not the only peak because, as well as ionising molecules, the beam of electrons breaks chemical bonds in the molecule. Positively charged fragments are formed and each gives a separate peak. The largest peak is called the **base peak** and allocated a relative abundance of 100%. Other peak heights are expressed as a percentage of the base peak.

In Figure 9.4 the peak with $m/z = 32$ is due to the molecular ion, $CH_3OH^+$. The other peaks are $CH_3O^+$ at $m/z = 31$, $CHO^+$ at $m/z = 29$ and $CO^+$ at $m/z = 28$.

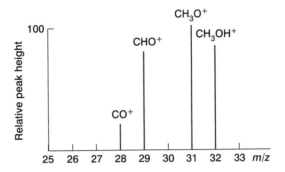

Fig. 9.4 *Mass spectrum of methanol*

If isotopes are present more than one molecular ion is formed; see below.

### Identifying compounds from mass spectra

The pattern of fragments obtained from a compound, together with its relative molecular mass, can be used to identify an unknown compound. The mass spectrum of an organic compound **X** is shown in Figure 9.5. A peak for the molecular ion is present at $m/z = 64$. An explanation has to be found for the second molecular peak at $m/z = 66$. Peaks due to fragments occur at $m/z = 29$, 35 and 37. The pair of fragments at $m/z = 35$ and $m/z = 37$ are probably $^{35}Cl^+$ and $^{37}Cl^+$ ions. This would explain the presence of two molecular ions, one containing $^{35}Cl$ and the other $^{37}Cl$. The fragment at $m/z = 29$ could be $C_2H_5^+$. By adding the fragments together we can deduce that the compound is chloroethane, $C_2H_5Cl$.

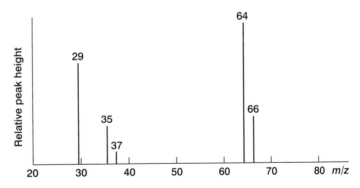

Fig. 9.5 *Mass spectrum of compound* **X**

### EXERCISE 9.2    Problems on mass spectra of compounds

For relative atomic masses see the table on p 243.

1   Bromine has two isotopes of mass 79 u and 81 u. The relative abundance is 1 : 1. The compound $C_3H_7Br$ therefore exists in two isotopic forms. State the $m/z$ values of the peaks which correspond to the molecular ion.

**2**   The mass spectrum of the aliphatic organic compound **A** is shown in Figure 9.6.

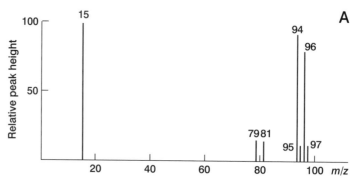

Fig. 9.6   *Mass spectrum of A*

**a**   There are four molecular peaks at $m/z$ = 94, 95, 96 and 97. This shows the presence of isotopes. Which element has isotopes of $m/z$ = 79 and 81?

**b**   The peak at $m/z$ = 15 is common in the mass spectra of organic compounds. Identify this fragment.

**c**   Deduce the formula of **A** and explain why there are **(i)** two molecular peaks at $m/z$ = 94 and 96 **(ii)** two smaller molecular peaks at $m/z$ = 95 and 97.

**3**   The mass spectrum of the aromatic compound **B** is shown in Figure 9.7.

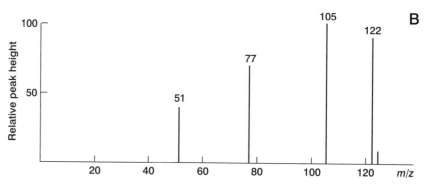

Fig. 9.7   *Mass spectrum of B*

**a**   Identify the molecular peak, and state the molecular mass of the compound.

**b**   The peak at $m/z$ = 77 occurs frequently in the mass spectra of aromatic compounds. Deduce the identity of the group responsible.

**c**   By subtracting the mass of the group in part **b** from the mass of the molecule in part **a**, suggest the identity of another group in the molecule.

**d**   Suggest the identity of the group responsible for the peak at $m/z$ = 105.

**4**    Figure 9.8 shows the mass spectrum of a compound of empirical formula $C_2H_6O$.

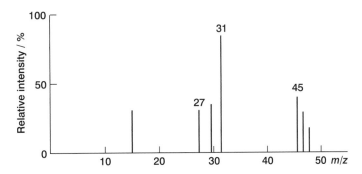

Fig. 9.8    *Mass spectrum of* $C_2H_6O$

**a**  Identify the two molecular peaks. Suggest why there are two of them.

**b**  Identify the groups responsible for the peaks at $m/z = 15$ and 29.

**c**  Identify the compound. Suggest which fragments give rise to the peaks at $m/z = 45$ and 31.

**5**    Chlorine consists of a mixture of isotopes, $^{35}Cl$ and $^{37}Cl$ which have a relative abundance of $3\,^{35}Cl : 1\,^{37}Cl$.

**a**  How many molecular ion peaks would you expect to see in the mass spectrum of 1,1,1-trichloroethane, $C_2H_3Cl_3$?

**b**  Write the isotopic composition of the ions.

**c**  Calculate the $m/z$ value of the heaviest molecular ion.

## 9.4 *NUCLEAR REACTIONS

In a nuclear reaction, a rearrangement of the protons and neutrons in the nuclei of the atoms takes place, and new elements are formed.

The **atomic number** or **proton number**, **Z**, of an element is the number of protons in the nucleus of an atom of the element. The **mass number** or **nucleon number**, **A**, is the number of protons and neutrons in the nucleus of an atom. Isotopes of an element differ in mass number but have the same atomic number. Isotopes are represented as $^A_Z$ Symbol, e.g. $^{12}_6C$. Protons are represented as $^1_1H$, electrons ($\beta$-particles) as $^0_{-1}e$, neutrons as $^1_0n$, and $\alpha$-particles as $^4_2He$.

In the equation for a nuclear reaction

● the sum of the mass numbers is the same on both sides,

● the sum of the atomic numbers is the same on both sides of the equation.

For practice in balancing nuclear equations, study the following examples.

**EXAMPLE 1**      Complete the equation

$$^{16}_{7}\text{N} \longrightarrow {}^{a}_{b}\text{O} + {}^{0}_{-1}\text{e}$$

**METHOD**      Consider mass numbers:      $16 = a + 0$      $\therefore \quad a = 16$
Consider atomic numbers:      $7 = b + (-1)$   $\therefore \quad b = 8$

**ANSWER**      $^{16}_{7}\text{N} \longrightarrow {}^{16}_{8}\text{O} + {}^{0}_{-1}\text{e}$

**EXAMPLE 2**      Find the values of $a$ and $b$ in the equation

$$^{27}_{13}\text{Al} + {}^{1}_{0}\text{n} \longrightarrow {}^{4}_{2}\text{He} + {}^{a}_{b}\text{X}$$

**METHOD**      Consider mass numbers:      $27 + 1 = 4 + a$      $\therefore \quad a = 24$
Consider atomic numbers:      $13 + 0 = 2 + b$      $\therefore \quad b = 11$

**ANSWER**      $a = 24$ and $b = 11$.

## EXERCISE 9.3      Problems on nuclear reactions

Complete the following equations, supplying values for the missing mass numbers (nucleon numbers) and atomic numbers (proton numbers).

**1**  **a**  $^{9}_{4}\text{Be} + \gamma \longrightarrow {}^{8}_{4}\text{Be} + {}^{a}_{b}\text{X}$

  **b**  $^{14}_{7}\text{N} + {}^{4}_{2}\text{He} \longrightarrow {}^{1}_{1}\text{H} + {}^{a}_{b}\text{Y}$

  **c**  $^{9}_{4}\text{Be} + {}^{1}_{1}\text{H} \longrightarrow {}^{6}_{3}\text{Li} + {}^{a}_{b}\text{Z}$

  **d**  $^{209}_{83}\text{Bi} + {}^{2}_{1}\text{D} \longrightarrow {}^{a}_{b}\text{X} + {}^{1}_{1}\text{H}$

  **e**  $^{16}_{8}\text{O} + {}^{1}_{0}\text{n} \longrightarrow {}^{13}_{6}\text{C} + {}^{a}_{b}\text{Y}$

  **f**  $^{10}_{5}\text{B} + {}^{a}_{b}\text{Y} \longrightarrow {}^{13}_{7}\text{N} + {}^{1}_{0}\text{n}$

  **g**  $^{14}_{7}\text{N} + {}^{1}_{0}\text{n} \longrightarrow {}^{a}_{b}\text{Q} + {}^{1}_{1}\text{H}$

  **h**  $^{19}_{9}\text{F} + {}^{1}_{0}\text{n} \longrightarrow {}^{a}_{b}\text{Z} + {}^{4}_{2}\text{He}$

  **i**  $^{207}_{82}\text{Pb} \longrightarrow {}^{a}_{b}\text{X} + {}^{0}_{-1}\text{e}$

  **j**  $^{27}_{13}\text{Al} + {}^{1}_{0}\text{n} \longrightarrow {}^{24}_{b}\text{Y} + {}^{a}_{2}\text{Z}$

  **k**  $^{35}_{17}\text{Cl} + {}^{p}_{q}\text{X} \longrightarrow {}^{r}_{16}\text{S} + {}^{1}_{1}\text{H}$

  **l**  $^{6}_{3}\text{Li} + {}^{1}_{0}\text{n} \longrightarrow {}^{4}_{2}\text{He} + {}^{c}_{d}\text{X}$

## EXERCISE 9.4      Questions from examination papers

**1**  **a**  A proton, a neutron and an electron all travelling at the same velocity enter a magnetic field. State which particle is deflected the most and explain your answer.      (2)

  **b**  Give two reasons why particles must be ionised before being analysed in a mass spectrometer.      (2)

  **c**  A sample of boron with a relative atomic mass of 10.8 gives a mass spectrum with two peaks, one at $m/z = 10$ and one at $m/z = 11$. Calculate the ratio of the heights of the two peaks.      (2)

**d** Compound **X** contains only boron and hydrogen. The percentage by mass of boron in **X** is 81.2%. In the mass spectrum of **X** the peak at the largest value of $m/z$ occurs at 54.

   **(i)** Use the percentage by mass data to calculate the empirical formula of **X**.

   **(ii)** Deduce the molecular formula of **X**.   (4)           *NEAB 98*   (10)

**2** **a** White phosphorus melts at 317 K. In a mass spectrometer, white phosphorus gives only one peak at $m/z = 124$ in the region of the molecular ion.

   **(i)** What can you deduce from the fact that there is only one peak in the region of the molecular ion? Give the formula of the species responsible for this peak.

   **(ii)** Explain in terms of its structure and bonding why white phosphorus has a low melting point.

   **(iii)** Explain why the melting point of silicon (1683 K) is higher than that of white phosphorus.           (10)

**b** When phosphorus is burned in a limited supply of air, it forms two oxides **A** and **B**. Oxide **A** contains 56.4% by mass of phosphorus.

   **(i)** Calculate the empirical formula of **A**.

   **(ii)** What additional information is required in order to calculate the molecular formula of **A**?

   **(iii)** Oxide **B** has the molecular formula $P_4O_{10}$. Write an equation for the formation of $P_4O_{10}$ from phosphorus and oxygen. Calculate the mass of $P_4O_{10}$ which would be formed by complete oxidation of 3.20 g of phosphorus.

   **(iv)** With the aid of an equation explain why $P_4O_{10}$ is classified as an acidic oxide.   (8)

**c** State the type of structure possessed by each of the following oxides. In each case, predict the resulting pH when the oxide is added to water and write an equation for any reaction which occurs.           (12)

$$Na_2O \quad MgO \quad SiO_2 \quad SO_2$$

*NEAB 97*   (30)

**3** **a** Give the symbol, including mass number and atomic number, for the isotope which has a mass number of 34 and which has 18 neutrons in each nucleus.           (2)

**b** Give the electronic configuration of the $F^-$ ion in terms of levels and sub-levels.           (1)

**c** Give a reason why it is unlikely that an $F^-$ ion would reach the detector in a mass spectrometer.           (1)

**d** Some data obtained from the mass spectrum of a sample of carbon are given below.

| Ion | $^{12}C^+$ | $^{13}C^+$ |
|---|---|---|
| Absolute mass of one ion/g | $1.993 \times 10^{-23}$ | $2.158 \times 10^{-23}$ |
| Relative abundance/% | 98.9 | 1.1 |

Use these data to calculate a value for the mass of one neutron, the relative atomic mass of $^{13}$C and the relative atomic mass of carbon in the sample. (6)

You may neglect the mass of an electron.

*NEAB 97* (10)

**4** **a** Describe, in terms of charge and mass, the properties of protons, neutrons and electrons. Explain fully how these particles are arranged in an atom of $^{14}$N. (6)

**b** Account for the existence of isotopes. (2)

**c** Isotopes can be separated in a mass spectrometer. Show how this is possible by describing the various parts of a mass spectrometer and by discussing the principles of operation of each part. (14)

**d** The mass spectrum of an element has peaks with relative intensity and $m/z$ values shown in the table below.

| $m/z$ | 80 | 82 | 83 | 84 | 86 |
|---|---|---|---|---|---|
| Relative intensity | 1 | 5 | 5 | 25 | 8 |

Identify this element and calculate its accurate relative atomic mass. (4)

**e** The mass spectrum of a compound has a molecular ion peak at $m/z = 168$. Elemental analysis shows it to contain 42.9% carbon, 2.4% hydrogen and 16.7% nitrogen by mass. The remainder is oxygen. Calculate the empirical and molecular formulae of this compound. (4)

*NEAB 99* (30)

# Gases and liquids

# 10

## 10.1 THE GAS LAWS

The behaviour of gases is described by the Gas laws: Boyle's law, Charles' law, the equation of state for an ideal gas, Avogadro's law, Dalton's law of partial pressures and the ideal gas equation.

**Boyle's law** states that the pressure of a fixed mass of gas at a constant temperature is inversely proportional to its volume:

$$PV = \text{Constant}$$

where $P$ = pressure, $V$ = volume.

**Charles' law** states that the volume of a fixed mass of gas at constant pressure is directly proportional to its temperature on the Kelvin scale:

$$\frac{V}{T} = \text{Constant}$$

where $T$ = temperature in kelvins.

Temperature on the Kelvin scale is obtained by adding 273 to the temperature on the Celsius scale.

$$\text{Temperature (K)} = \text{Temperature (°C)} + 273$$
$$273 \text{ K} = 0 \text{ °C}$$

### The equation of state for an ideal gas

Gases which obey Boyle's law and Charles' law are called **ideal** gases. By combining these two laws, the following equation can be obtained. It is called the **equation of state for an ideal gas:**

$$\frac{P_1 V_1}{T_1} = \frac{P_2 V_2}{T_2}$$

A gas has a volume $V_1$ at a temperature $T_1$ and pressure $P_1$. If the conditions are changed to a pressure $P_2$ and a temperature $T_2$, the new volume can be calculated from the equation. For example, if the pressure on a gas increases by a factor 5, the volume decreases to one fifth. It is usual to compare gas volumes at

- either standard temperature and pressure, 0 °C and 1 atm (s.t.p.)

- or room temperature and pressure, 20 °C and 1 atm (r.t.p.).

## 10.2 GAS MOLAR VOLUME

**Avogadro's law** states that equal volumes of gases, measured at the same temperature and pressure, contain equal numbers of molecules. It follows that the volume occupied by one mole of gas is the same for all gases. It is called the **gas molar volume**. It measures 22.4 dm³ at s.t.p. and 24.0 dm³ at r.t.p. If the volume occupied by a known mass of gas is known, the molar mass of the gas can be calculated.

**EXAMPLE**         11.0 g of a gas occupy 5.60 dm³ at s.t.p. What is the molar mass of the gas?

**METHOD**          Mass of 5.60 dm³ of gas = 11.0 g

Mass of 22.4 dm³ of gas = 11.0 g × 22.4 dm³/5.60 dm³ = 44.0 g

**ANSWER**          The molar mass of the gas is 44.0 g mol⁻¹.

**EXERCISE 10.1**   **Problems on gas molar volume**

---

Reminder : Ratio calculations § 1.2.

   GMV = 22.4 dm³ mol⁻¹ at s.t.p. = 24.0 dm³ mol⁻¹ at r.t.p.

---

**1**   Calculate the volume occupied by 0.250 mol of an ideal gas at **a** s.t.p. **b** r.t.p.

**2**   What amount (mol) of an ideal gas occupies 56.0 dm³ at s.t.p.?

**3**   A volume, 500 cm³ of krypton, measured at r.t.p. has a mass of 1.746 g. Calculate the molar mass of krypton.

**4**   An ideal gas occupies 1.50 dm³ at 101 kPa and 293 K. What amount of gas is present?

**5**   At s.t.p. 2.975 g of argon occupy 1.67 dm³. Calculate the molar mass of the gas.

## 10.3 MOLE FRACTION

One way of expressing the composition of a mixture of gases or a mixture of liquids is to state the **mole fraction** of each constituent. If a mixture consists of 4 moles of **A** and 3 moles of **B** the total number of moles is 7. Then,

$$\text{Mole fraction of } \mathbf{A} = \frac{\text{Number of moles of } \mathbf{A}}{\text{Total number of moles}} = \frac{3}{7}$$

$$\text{Mole fraction of } \mathbf{B} = \frac{\text{Number of moles of } \mathbf{B}}{\text{Total number of moles}} = \frac{4}{7}$$

and

$$\text{Mole fraction of A} + \text{Mole fraction of B} = 1$$

In general,

$$\text{Mole fraction of component} = \frac{\text{Number of moles of component}}{\text{Total number of moles}}$$

## 10.4 PARTIAL PRESSURE

Atmospheric pressure = 1 atm. All the gases in the air contribute to atmospheric pressure. Air is approximately $\frac{4}{5}$ nitrogen and $\frac{1}{5}$ oxygen by volume. Since the gas molar volume of all gases is the same, the mole fraction of nitrogen = $\frac{4}{5}$ and the mole fraction of oxygen = $\frac{1}{5}$. The nitrogen present contributes $\frac{4}{5}$ of atmospheric pressure and the oxygen contributes $\frac{1}{5}$ of atmospheric pressure. The contribution which each gas in the mixture makes to the total pressure is called its **partial pressure**.

Partial pressure of nitrogen = mole fraction of nitrogen × total pressure
$$= \tfrac{4}{5} \times 1 \text{ atm} = \tfrac{4}{5} \text{ atm}$$

Partial pressure of oxygen = mole fraction of oxygen × total pressure
$$= \tfrac{1}{5} \times 1 \text{ atm} = \tfrac{1}{5} \text{ atm}$$

In general, in a mixture of gases

> Partial pressure of gas = Mole fraction × Total pressure

### Dalton's law of partial pressure

> In a mixture of gases, the total pressure is the sum of the pressures that each of the gases would exert if it alone occupied the same volume as the mixture.

Imagine a container full of air at 1 atm. If all the oxygen were removed from it, the remaining nitrogen would fill the container. The pressure of nitrogen would be $\frac{4}{5}$ atm – the same as the partial pressure of nitrogen in the air. You will need to be able to calculate partial pressures when you study gaseous equilibria in § 14.2.

**EXAMPLE 1**

A container holds 2.00 mol of nitrogen and 5.00 mol of hydrogen at a pressure of $5.05 \times 10^5$ Pa. What is the partial pressure of each gas?

**METHOD**

Mole fraction of nitrogen = 2.00 mol/7.00 mol

Partial pressure of nitrogen = $5.05 \times 10^5$ Pa × 2.00/7.00 = $1.44 \times 10^5$ Pa

Mole fraction of hydrogen = 5.00 mol/7.00 mol
Partial pressure of hydrogen = $5.05 \times 10^5$ Pa × 5.00/7.00 = $3.61 \times 10^5$ Pa

Check: Do the partial pressures add up to the total pressure?

Sum = $(1.44 \times 10^5$ Pa$) + (3.61 \times 10^5$ Pa$) = 5.05 \times 10^5$ Pa

**ANSWER**

Partial pressures are nitrogen $1.44 \times 10^5$ Pa, hydrogen $3.61 \times 10^5$ Pa.

**EXAMPLE 2**

2.00 dm$^3$ of nitrogen at a pressure of 100 kPa and 5.00 dm$^3$ of hydrogen at a pressure of 500 kPa are injected into a 10.0 dm$^3$ container.

Calculate:

**a** the pressure of the mixture of gases

**b** the partial pressures of nitrogen and hydrogen.

**METHOD**
When the volume of nitrogen expands from $2.00 \text{ dm}^3$ to $10.0 \text{ dm}^3$ the pressure falls from $100 \text{ kPa}$ to $\frac{1}{5}$ of this pressure $= 20 \text{ kPa}$.

Hydrogen expands from $5.00 \text{ dm}^3$ to $10.0 \text{ dm}^3$ therefore its pressure decreases from $500 \text{ kPa}$ to $\frac{1}{2}$ this value $= 250 \text{ kPa}$

Total pressure $= 20 \text{ kPa} + 250 \text{ kPa} = 270 \text{ kPa}$

**ANSWER**
Total pressure $= 270 \text{ kPa}$, partial pressures are nitrogen $20 \text{ kPa}$, hydrogen $250 \text{ kPa}$.

**EXERCISE 10.2** **Problems on partial pressure**

**1** A mixture of gases at a pressure of $1.01 \times 10^5 \text{ N m}^{-2}$ contains $25.0\%$ by volume of oxygen. What is the partial pressure of oxygen in the mixture?

> **Hint :** Mole fraction of oxygen ... partial pressure of oxygen.

**2** $3.00 \text{ dm}^3$ nitrogen at a pressure of $101 \text{ kPa}$ and $7.00 \text{ dm}^3$ of hydrogen at a pressure of $101 \text{ kPa}$ are fed into a $10.0 \text{ dm}^3$ container. What is the partial pressure of each gas?

> **Hint :** Mole fraction of nitrogen ... total pressure ... partial pressure of nitrogen.

**3** A cylinder receives $2.50 \text{ dm}^3$ of methane, $7.50 \text{ dm}^3$ of ethane and $0.500 \text{ dm}^3$ of propane, all at the same temperature and pressure. The pressure inside the cylinder is $5.05 \times 10^5 \text{ Pa}$. What is the partial pressure of each gas?

> **Hint :** Mole fraction ... total pressure ... partial pressure.

**4** $3.0 \text{ dm}^3$ of carbon dioxide at a pressure of $200 \text{ kPa}$ and $1.0 \text{ dm}^3$ of nitrogen at a pressure of $300 \text{ kPa}$ are fed into a $1.5 \text{ dm}^3$ container. What is the total pressure in the container?

> **Hint :** $P \propto 1/V$

**5** A mixture of gases at a pressure $7.50 \times 10^4 \text{ N m}^{-2}$ has the volume composition $40\% \text{ N}_2$, $35\% \text{ O}_2$, $25\% \text{ CO}_2$.

    **a** What is the partial pressure of each gas?

    **b** What will the partial pressures of nitrogen and oxygen be if the carbon dioxide is removed by the introduction of some sodium hydroxide pellets?

**6** A mixture of gases at $1.50 \times 10^5 \text{ N m}^{-2}$ has the composition $40\% \text{ NH}_3$, $25\% \text{ H}_2$, $35\% \text{ N}_2$ by volume.

    **a** What is the partial pressure of each gas?

    **b** What will the partial pressures of the other gases become if the ammonia is removed by the addition of some solid phosphorus(V) oxide?

## 10.5 THE IDEAL GAS EQUATION

The equation of state for an ideal gas (§ 10.2) can be expressed:

$$\frac{P \times V}{T} = \text{Constant for a given mass of gas}$$

It follows from Avogadro's law that the constant is the same for one mole of all gases. It is called the **universal gas constant**, with symbol $R$. Then the equation becomes

$$PV = RT$$

This is called the **ideal gas equation**. For $n$ moles of gas the equation becomes

$$PV = nRT$$

The value of the universal gas constant is $R = 8.314 \text{ J K}^{-1} \text{ mol}^{-1}$.

**EXAMPLE**

Calculate the molar mass of a gas which has a density of $1.798 \text{ g dm}^{-3}$ at 298 K and $1.01 \times 10^5 \text{ N m}^{-2}$. The SI unit of mass is the kilogram.

**METHOD**

**Reminder :** Rearranging equations § 1.1.

Amount $n$ = mass/molar mass § 4.3.
Density = mass/volume

Since $PV = nRT$

$$PV = \frac{\text{mass} \times RT}{\text{molar mass}}$$

Rearranging,

$$\text{Molar mass} = \frac{\text{mass} \times RT}{PV} = \frac{\text{density} \times RT}{P}$$

Inserting the values $P = 1.01 \times 10^5 \text{ N m}^{-2}$

$$R = 8.314 \text{ J K}^{-1} \text{ mol}^{-1}$$
$$T = 298 \text{ K}$$

$$\text{density} = 1.798 \text{ g dm}^{-3} = 1.798 \text{ kg m}^{-3}$$

$$\text{Molar mass} = \frac{1.798 \text{ kg m}^{-3} \times 8.314 \text{ J K}^{-1} \text{mol}^{-1} \times 298 \text{ K}}{1.01 \times 10^5 \text{ N m}^{-2}} = 0.0441 \text{ kg mol}^{-1}$$

(In working out the unit, remember 1 Nm = 1 J)

**ANSWER**

Molar mass = $44.1 \text{ g mol}^{-1}$

## 10.6 LIQUIDS; FINDING THE MOLAR MASS

The molar mass of a liquid with a low boiling temperature can be found by **the gas syringe method**. A known mass of liquid is injected into a gas syringe. The liquid vaporises and the volume of the vapour is measured. Its temperature and pressure are noted. The measurements are inserted into the ideal gas equation to give the molar mass of the liquid.

1. The syringe is inside a furnace at a temperature 10 °C above the boiling temperature of the liquid.

2. The gas syringe contains $v_1$ cm³ of air.

Thermometer

4. The volume of air + vapour in the gas syringe is read, $v_2$ cm³
Volume of vapour = $(v_2 - v_1)$ cm³

3. Self-sealing rubber cap. Liquid is injected from a weighed syringe. The syringe is reweighed to find the mass of liquid injected, $m$ g.

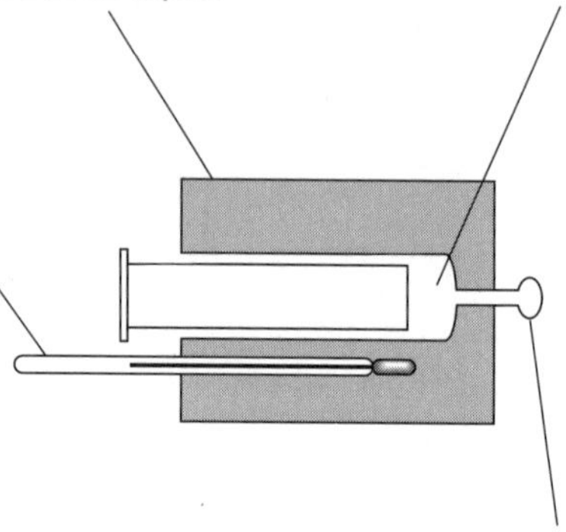

Fig. 10.1   *A gas syringe*

**EXAMPLE**

A mass 0.187 g of a volatile liquid is injected into a gas syringe at 57 °C and $1.01 \times 10^5 \, N \, m^{-2}$. The liquid formed 36.2 cm³ of vapour. Calculate the molar mass of the liquid.

**METHOD**

Using the values

$$P = 1.01 \times 10^5 \, N \, m^{-2}$$

$$V = 36.2 \, cm^3 = 36.2 \times 10^{-6} \, m^3$$

$$T = 273 + 57 = 330 \, K$$

$$R = 8.314 \, J \, K^{-1} \, mol^{-1}$$

In the equation $PV = nRT$, put $n = m/M$ (Amount = Mass/Molar mass)

Then $PV = mRT/M$

and, rearranging, $M = mRT/PV$

$$M = \frac{0.187 \, g \times 8.314 \, J \, K^{-1} \, mol^{-1} \times 330 \, K}{1.01 \times 10^5 \, N \, m^{-2} \times 36.2 \times 10^{-6} \, m^3}$$
$$= 140 \, g \, mol^{-1}$$

**ANSWER**

The molar mass is 140 g mol⁻¹.

The values of molar mass obtained by this method are not very accurate. If the empirical formula is known the result can be corrected. For example if a compound has empirical formula $CH_2O$ and an experimental value of 57 g mol⁻¹ for the molar mass, we can see that $C_2H_4O_2$ is the molecular formula and 60 g mol⁻¹ is the correct molar mass.

**EXERCISE 10.3**   **Problems on the ideal gas equation**

> **Reminder :** Rearranging equations § 1.1. Use $R = 8.31 \, \text{J K}^{-1} \, \text{mol}^{-1}$.
>
> Amount, $n = $ Mass/Molar mass § 4.3.

**1**   A gas has a density of 7.149 g dm$^{-3}$ at 50 °C and $3.00 \times 10^5$ N m$^{-2}$. What is its molar mass?

> **Hint :** In the equation $PV = nRT$, $P$ must be in N m$^{-2}$, $V$ in m$^3$ and $T$ in kelvins.

**2**   A volume 485 cm$^3$ of a gas measured at 40 °C and $2.00 \times 10^5$ N m$^{-2}$ has a mass of 1.715 g. Calculate the molar mass of the gas.

> **Hint :** In $PV = nRT$, put $n = $ Mass/ Molar mass. Use SI units.

**3**   0.104 g of a hydrocarbon gave 42.0 cm$^3$ of vapour measured at $1.01 \times 10^5$ N m$^{-2}$ and 150 °C. Calculate the relative molecular mass of the hydrocarbon.

> **Hint :** Use $PV = nRT$ and $n = m/M$

**4**   **a**   0.130 g of a liquid vaporised at 100 °C and 1 atm occupied a volume of 85.0 cm$^3$.

Calculate its relative molecular mass.

> **Hint :** Use the ideal gas equation again.

**b**   If the percentage composition of the liquid is C 52.2%, H 13.0%, O 34.8%, calculate the empirical formula of the compound. Can you suggest its identity?

**5**   A mass 0.3542 g of liquid was vaporised. The gas formed occupied 100 cm$^3$ at 20 °C and $1.01 \times 10^5$ Pa. What is the molar mass of the liquid?

**6**   A compound of phosphorus and fluorine contains 24.6% by mass of phosphorus. 1.000 g of this compound has a volume of 196 cm$^3$ at 300 K and $1.01 \times 10^5$ Pa. Deduce the molecular formula of the compound.

**7**   0.110 g of a liquid produced 42.0 cm$^3$ of vapour, measured at 147 °C and $1.01 \times 10^5$ N m$^{-2}$. What is the molar mass of the liquid?

**8**   0.228 g of liquid was injected into a gas syringe. The volume of vapour formed was 84.0 cm$^3$ at 17 °C and $1.01 \times 10^5$ N m$^{-2}$. Calculate the molar mass of the substance.

## 10.7 LIQUIDS; PARTITION COEFFICIENT

When two immiscible liquids are added, they form two separate layers. When a solid which dissolves in both of the liquids is added, it distributes itself between the two layers. When equilibrium is reached, it is found that the ratio of the concentrations in the two layers is always the same:

$$\frac{\text{Concentration in upper layer}}{\text{Concentration in lower layer}} = \frac{C_U}{C_L} = k$$

The constant $k$ is called the **partition coefficient** or **distribution coefficient**. It is constant for a particular temperature.

**EXAMPLE 1**

Figure 10.2 shows what happens when a substance **X** distributes itself between ethoxyethane and water.

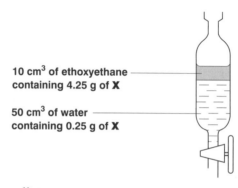

10 cm$^3$ of ethoxyethane
containing 4.25 g of **X**

50 cm$^3$ of water
containing 0.25 g of **X**

Fig. 10.2   *The partition coefficient*

What is the value of the partition coefficient, [**X** in ethoxyethane]/[**X** in water]?

**METHOD**

$C_U/C_L = k$

$C_U = 4.25\ \text{g}/10\ \text{cm}^3$

$C_L = 0.25\ \text{g}/50\ \text{cm}^3$

$k = (4.25\ \text{g}/10\ \text{cm}^3) \div (0.25\ \text{g}/50\ \text{cm}^3) = 85$

Both concentrations must be expressed in the same unit, in this case g/cm$^3$, so the concentration units will cancel.

**ANSWER**

Partition coefficient = 85.

**EXAMPLE 2**

The partition coefficient of **X** between ether and water is 25.0 at 20 °C. Calculate the mass of **X** extracted from a solution containing 10.0 g of **X** in 1.00 dm$^3$ of water by **a** 100 cm$^3$ of ether, **b** two successive portions of 50.0 cm$^3$ of ether.

**METHOD**

**a** Let the mass of **X** extracted by 100 cm$^3$ of ether be $m_1$.

Use $C_U/C_L = k$ ($C_U$ for ether layer, $C_L$ for water layer)

Concn of **X** in ether = $m_1/100\ \text{g cm}^{-3}$

Concn of $X$ in water $= (10.0 - m_1)/1000$ g cm$^{-3}$

$$\frac{m_1}{100}\bigg/\frac{(10.0 - m_1)}{1000} = 25.0$$

**ANSWER**

**a** giving $m_1 = 7.14$ g.

**b** Let $m_2 =$ mass of $X$ extracted by the first 50.0 cm$^3$ of ether, and $m_3 =$ mass of $X$ extracted by the second 50.0 cm$^3$ of ether.

Then

$$\frac{m_2}{50.0}\bigg/\frac{(10.0 - m_2)}{1000} = 25.0$$

giving $m_2 = 5.55$ g.

If 5.55 g of $X$ are extracted by ether, 4.45 g remain in the aqueous solution.

$$\therefore \qquad \frac{m_3}{50}\bigg/\frac{(4.45 - m_3)}{1000} = 25.0$$

giving $m_3 = 2.47$ g.

**ANSWER**

**b** Total mass of $X$ extracted by ether in two portions $= 5.55$ g $+ 2.47$ g $= 8.02$ g.

**Note :** that this is greater than the value of 7.14 g calculated for the mass of $X$ extracted by using all the ether at once.

## EXERCISE 10.4   Problems on partition

**1**   The partition coefficient of $Y$ between ethoxyethane and water is 4. What mass of $Y$ is extracted when 10 cm$^3$ of ethoxyethane is shaken with 100 cm$^3$ of water containing 5.00 g of $Y$?

**A** 1.25 g   **B** 1.43 g   **C** 3.57 g   **D** 4.00 g

**2**   $X$ is 12.0 times more soluble in trichloromethane than in water. What mass of $X$ will be extracted from 1.00 dm$^3$ of an aqueous solution containing 25.0 g by shaking with 100 cm$^3$ of trichloromethane?

**3**   The partition coefficient of $Y$ between ethoxyethane (ether) and water is 80. If 200 cm$^3$ of an aqueous solution containing 5.00 g of $Y$ is shaken with 50.0 cm$^3$ of ethoxyethane, what mass of $Y$ is extracted from the solution?

**4**   $Z$ is allowed to reach an equilibrium distribution between the liquids ethoxyethane and water. The ether layer is 50.0 cm$^3$ in volume and contains 4.00 g of $Z$. The aqueous layer is 250 cm$^3$ in volume and contains 1.00 g of $Z$. What is the partition coefficient of $Z$ between ethoxyethane and water?

**5**   500 cm$^3$ of an aqueous solution of concentration 0.120 mol dm$^{-3}$ is shaken with 50.0 cm$^3$ of ethoxyethane. The partition coefficient of the solute between ethoxyethane and water is 60.0. Calculate the amount (in mol) of solute which will be extracted by ethoxyethane.

**6** The distribution coefficient of **A** between ethoxyethane and water is 90. An aqueous solution of **A** with a volume of 500 cm$^3$ contains 5.00 g. What mass of **A** will be extracted by:

**a** 100 cm$^3$ of ethoxyethane, and

**b** two successive portions of 50.0 cm$^3$ of ethoxyethane?

**7** **a** The partition coefficient of **T** between ethoxyethane and water at room temperature is 19. A solution of 5.0 g of **T** in 100 cm$^3$ of water is extracted with 100 cm$^3$ of ethoxyethane at room temperature. What mass of **T** will be present in the ethoxyethane layer?

**b** What would be the total mass of **T** extracted if the ethoxyethane were used in two separate 50 cm$^3$ portions, instead of the single 100 cm$^3$ portion?

**8** An organic acid is allowed to reach an equilibrium distribution in a separating funnel containing 50.0 cm$^3$ of ethoxyethane and 500 cm$^3$ of water. On titration, 25.0 cm$^3$ of the ethoxyethane layer required 22.5 cm$^3$ of 1.00 mol dm$^{-3}$ sodium hydroxide solution, and 25.0 cm$^3$ of the aqueous layer required 9.0 cm$^3$ of 0.100 mol dm$^{-3}$ sodium hydroxide solution. Calculate the partition coefficient for the acid between ethoxyethane and water.

**9** A small amount of iodine is shaken in a separating funnel containing 50.0 cm$^3$ of tetrachloromethane and 500 cm$^3$ of water. On titration, 25.0 cm$^3$ of the aqueous layer require 6.7 cm$^3$ of a 0.0550 mol dm$^{-3}$ solution of sodium thiosulphate. 25.0 cm$^3$ of the organic solvent require 27.2 cm$^3$ of 1.15 mol dm$^{-3}$ sodium thiosulphate solution. Calculate the distribution coefficient for iodine between water and tetrachloromethane.

**10** A solid **A** is three times as soluble in solvent **X** as in solvent **Y**. **A** has the same relative molecular mass in both solvents. Calculate the mass of **A** that would be extracted from a solution of 4 g of **A** in 12 cm$^3$ of **Y** by extracting it with: **a** 12 cm$^3$ of **X**, and **b** three successive portions of 4 cm$^3$ of **X**.

**EXERCISE 10.5** **Questions from examination papers**

**1** **a** Define the term *relative molecular mass*. (2)

**b** The mass of one atom of $^{12}$C is $1.993 \times 10^{-23}$ g. Use this mass to calculate a value for the Avogadro constant ($L$) showing your working. (1)

**c** The following equation is not balanced.

$$MgI_2(s) + Fe^{3+}(aq) \longrightarrow Mg^{2+}(aq) + I_2(s) + Fe^{2+}(aq)$$

**(i)** In what way is the equation unbalanced?

**(ii)** Write the balanced equation. (2)

**d** A 153 kg sample of ammonia gas, $NH_3$, was compressed at 800 K into a cylinder of volume 3.00 m$^3$.

**(i)** Calculate the pressure in the cylinder assuming that the ammonia remained as a gas.

**(ii)** Calculate the pressure in the cylinder when the temperature is raised to 1000 K.

**(iii)** Calculate the molarity of the solution formed by dissolving this mass of ammonia in water to make 1.0 m$^3$ of solution. (7)

*NEAB 99* (12)

**2** Carbon disulphide, at 46 °C, is an excellent solvent for sulphur and other molecular substances. It is now rarely used in the laboratory as it is poisonous and highly flammable.

**a** The diagram below shows the preparation of a form of sulphur.

**(i)** What is the formula and structure of a sulphur molecule? (2)

**(ii)** Explain which form of sulphur is produced. (2)

**b** One of the dangers of using carbon disulphide is that it readily ignites.

$$CS_2 + 3O_2 \longrightarrow CO_2 + 2SO_2$$

**(i)** Calculate the total volume of gases produced (at 20 °C and 1 atmosphere pressure) if 25 cm$^3$ of liquid carbon disulphide were completely burnt. (4)

(The density of carbon disulphide at 20 °C is 1.29 g cm$^{-3}$)

**(ii)** How could you show that sulphur dioxide was one of the products of combustion? (2)

**c** The partition coefficient, $K_d$, for iodine between water and carbon disulphide at 20 °C is $2.4 \times 10^{-3}$. A 100 cm$^3$ sample of an aqueous solution of iodine ($I_2$) of concentration $0.50 \times 10^{-3}$ mol dm$^{-3}$ was shaken with 20.0 cm$^3$ of carbon disulphide.

**(i)** Write an expression for $K_d$. (2)

**(ii)** What mass of iodine would be extracted by the carbon disulphide? (3)

**(iii)** Using the same volume of carbon disulphide at the same temperature, how could more iodine have been extracted? (2)

*NI* (17)

# Electrochemistry

11

## 11.1 ELECTROLYSIS

Electrovalent compounds, when molten or in aqueous solution, conduct electricity. They are electrolysed in the process. When molten oxides $MO$ and halides $MX_2$ are electrolysed, the metal $M$ is deposited on the cathode (the negative electrode) and oxygen or a halogen is evolved at the anode (the positive electrode). When an aqueous solution of $MX_2$ is electrolysed, $M$ is deposited on the cathode if it is a metal low in the electrochemical series, but hydrogen is evolved if $M$ is a reactive metal. Oxygen or a halogen is evolved at the anode. Figure 11.1 shows the electrolysis of aqueous silver nitrate to give silver at the cathode and oxygen at the anode.

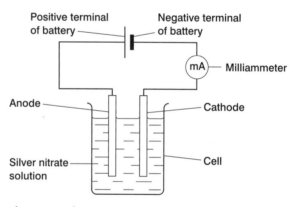

Fig. 11.1  *Electrolysis of aqueous silver nitrate*

The amount of each element liberated at the electrodes depends on the quantity of electric charge that passes. This is given by

$$\text{Charge (coulombs, C)} = \text{Current (amperes, A)} \times \text{Time (seconds, s)}$$

The equation shows that one mole of electrons is needed to discharge 1 mole of silver in the process

$$Ag^+(aq) + e^- \longrightarrow Ag(s)$$

By experiment it is found that the quantity of electric charge needed is 96 500 coulombs. This must be the charge on one mole of electrons. The ratio 96 500 coulombs mol$^{-1}$ (C mol$^{-1}$) is called the **Faraday constant**.

The charge on one mole of electrons is 96 500 C.

The Faraday constant is 96 500 C mol$^{-1}$.

There are two factors to consider in calculations on electrolysis:

**1**   How much charge flows through the electrolysis cell?

**2**   How many moles of electrons are needed to discharge 1 mole of the element? This may be 1, 2, 3 or 4 moles. Here are some examples:

1 mol of electrons $\longrightarrow$ 1 mol of element:  $Hg^+(aq) + e^- \longrightarrow Hg(l)$

2 mol of electrons $\longrightarrow$ 1 mol of element:  $Cu^{2+}(aq) + 2e^- \longrightarrow Cu(s)$

2 mol of electrons $\longrightarrow$ 1 mol of element:  $2H^+(aq) + 2e^- \longrightarrow H_2(g)$

2 mol of electrons $\longrightarrow$ 1 mol of element:  $2Cl^-(aq) \longrightarrow Cl_2(g) + 2e^-$

3 mol of electrons $\longrightarrow$ 1 mol of element:  $Al^{3+}(l) + 3e^- \longrightarrow Al(s)$

4 mol of electrons $\longrightarrow$ 1 mol of element:  $4OH^-(aq) \longrightarrow O_2(g) + 2H_2O(l) + 4e^-$

---

**Reminder :** Proportionality § 1.1. Ratio calculations § 1.2.

Amount = Mass/Molar mass § 4.3.

---

**EXAMPLE 1**   A direct current of 0.0100 A flows for 4.00 hours through three cells in series. They contain aqueous silver nitrate, copper(II) sulphate and gold(III) nitrate. Calculate the mass of metal deposited in each. $A_r(Ag) = 108$, $A_r(Cu) = 63.5$, $A_r(Au) = 197$.

**METHOD**   Coulombs = Amperes × Seconds = 0.0100 A × (4.00 × 60 × 60) s = 144 C

Charge passed = 144 C/96 500 C mol$^{-1}$ = $1.49 \times 10^{-3}$ mol of electrons

From the equation,

$$Ag^+(aq) + e^- \longrightarrow Ag(s)$$

Amount of Ag deposited = $1.49 \times 10^{-3}$ mol
Mass of Ag deposited = 108 g mol$^{-1}$ × $1.49 \times 10^{-3}$ mol = 0.161 g
From the equation,

$$Cu^{2+}(aq) + 2e^- \longrightarrow Cu(s)$$

Amount of Cu deposited = $\frac{1}{2} \times 1.49 \times 10^{-3}$ mol

Mass of Cu deposited = $\frac{1}{2} \times 63.5$ g mol$^{-1}$ × $1.49 \times 10^{-3}$ mol = 0.0474 g

From the equation,

$$Au^{3+}(aq) + 3e^- \longrightarrow Au(s)$$

Amount of Au deposited = $\frac{1}{3} \times 1.49 \times 10^{-3}$ mol

Mass of Au deposited = $\frac{1}{3} \times 197$ g mol$^{-1}$ × $1.49 \times 10^{-3}$ mol = 0.0980 g

**ANSWER**   Deposited are 0.161 g silver, 0.0474 g copper, 0.0980 g gold.

**EXAMPLE 2**   A metal of $A_r = 27$ is deposited by electrolysis. When 0.15 A flows for 3.5 hours, 0.176 g of metal is deposited. What is the charge on the cations of this metal?

**METHOD**     Coulombs = Amperes × Seconds = 0.15 A × (3.5 × 60 × 60) s = 1890 C

If 1890 C deposit 0.176 g of metal,

then 96 500 C deposit $\dfrac{96\ 500\ \text{C} \times 0.176\ \text{g}}{1890\ \text{C}} = 8.98$ g

1 mole of metal = 27 g

Since 8.98 g of metal are discharged by 1 mole of electrons,

27 g of metal are discharged by $\dfrac{27\ \text{g}}{8.98\ \text{g}} = 3$ moles of electrons

**ANSWER**     If 1 mole of metal needs 3 moles of electrons the charge on the cations is +3.

**EXERCISE 11.1   Problems on electrolysis**

> **Reminder :** Rearranging equations § 1.1. Ratio calculations § 1.2. Include the units § 1.11.
> Check your answer § 1.12.
>
> Coulombs = Amperes × Seconds, Faraday constant = 96 500 C mol$^{-1}$ § 11.1.

For relative atomic masses see p 243.

**1**   Calculate the mass of copper that would be deposited on a copper cathode from an aqueous solution of copper(II) sulphate if the same current, passed for the same time, liberated 0.900 g silver from aqueous silver nitrate, $AgNO_3$.

> **Hint :** Write the equations for the electrode processes.

**2**   A current of 0.100 A passed for 1.00 h through an aqueous solution of 1.00 mol dm$^{-3}$ silver nitrate. A silver cathode was used. Calculate the increase in mass of the cathode. State how the change in mass would be affected by:

**a**   passing a current of 0.200 A,

**b**   passing a current of 0.100 A for 2.00 h,

**c**   using a 2.00 mol dm$^{-3}$ solution of silver nitrate.

**3**   An electric current of 5.00 A was passed through molten anhydrous calcium chloride, $CaCl_2$, for 20.0 minutes between graphite electrodes. Calculate the mass of each product liberated.

**STEPS TO TAKE**     What quantity of charge passed? How many moles of electrons? How many moles of Ca and $Cl_2$? And their masses?

**4** A current of electricity deposited $6.00 \times 10^{-3}$ mol of silver from aqueous silver nitrate. What mass of gold would be deposited from aqueous gold(III) chloride, $AuCl_3$, by the same quantity of electricity?

**STEPS TO TAKE** Equations ... amount of gold ... mass of gold.

**5** A current is passed through three solutions in series, using platinum electrodes. In one, 0.216 g of silver is deposited from aqueous silver nitrate. In a second, hydrogen is evolved from a solution of sulphuric acid, and, in a third, lead is deposited from aqueous lead(II) nitrate. Calculate **a** the volume of hydrogen collected **b** the mass of lead deposited. (GMV = 24.0 $dm^3$ $mol^{-1}$ at r.t.p.)

**STEPS TO TAKE** Equations for electrode processes ... amount of Ag ... amount of $H_2$ ... volume ... amount of Pb ... mass.

**6** A current of 0.750 A passes through 250 $cm^3$ of aqueous copper(II) sulphate of concentration 1.12 mol $dm^{-3}$. How long (in minutes) will it take to deposit all the copper on the cathode?

**STEPS TO TAKE** Amount of copper ... no. of moles of electrons ... multiply by 96 500 C $mol^{-1}$ ... charge ... C = A × s

**7** How long (in hours) will it take a current of 0.100 A to deposit 1.00 g of silver from an unlimited source of silver ions?

**STEPS TO TAKE** Amount of Ag ... amount of electrons ... charge ... time.

**8** A current of 2.00 A passes through a solution of sulphuric acid for 5.00 h. Calculate the volumes of hydrogen and oxygen produced. (GMV = 24.0 $dm^3$ $mol^{-1}$ at r.t.p.)

**9** A steady current was passed through aqueous copper(II) sulphate until 6.05 g of copper were deposited on the cathode. How many coulombs of electricity passed?

**STEPS TO TAKE** Amount of copper ... amount of electrons ... multiply by Faraday constant ... charge in coulombs.

**10** In the electrolysis of aqueous potassium chloride 100 $cm^3$ of chlorine are produced (at r.t.p.). How many seconds has a current of 0.750 A been flowing to effect this?

**STEPS TO TAKE** Amount of $Cl_2$ ... amount of electrons ... multiply by Faraday constant ... charge in C ... C = A × s.

## 11.2 ACIDS AND BASES

The definition acids and bases according to the theory of T.M. Lowry and J.N. Brönsted is:

● An **acid** is a substance which can donate a proton to another substance.

● A **base** is a substance which can accept a proton from another substance.

No substance can act as an acid in solution unless a base is present to accept a proton. The reactions of acids are acid–base reactions. Similarly, all reactions of bases are acid–base reactions.

Acids react with water to give hydrogen ions. Acids such as hydrochloric acid are **monoprotic**: they give one hydrogen ion per formula unit.

$$HCl(aq) + H_2O(l) \longrightarrow H_3O^+(aq) + Cl^-(aq)$$

The base which accepts the hydrogen ion is a molecule of water. Complex ions of formula $H_3O^+$ are formed. Their proper name is **oxonium ions**, but in this text they will be called hydrogen ions. Sulphuric acid, $H_2SO_4$, is a **diprotic** acid: it gives two hydrogen ions per formula unit.

## 11.3 HYDROGEN ION CONCENTRATION

The hydrogen ion concentration of a solution can be indicated by means of a number on the pH scale.

The pH of a solution is the negative logarithm to the base 10 of the hydrogen ion concentration. pOH, $pK_a$ and $pK_b$ are defined below.

$$pH = -lg[H_3O^+/mol\ dm^{-3}] \qquad\qquad pK_a = -lg(K_a/mol\ dm^{-3})$$

$$pOH = -lg[OH^-/mol\ dm^{-3}] \qquad\qquad pK_b = -lg(K_b/mol\ dm^{-3})$$

### Strong acids and bases; calculation of pH and pOH

The pH of a solution of a strong acid or base is simply calculated. If the concentration of hydrochloric acid is 0.1 mol dm$^{-3}$,

$$[H_3O^+] = 0.1\ mol\ dm^{-3} = 10^{-1}\ mol\ dm^{-3}$$

$$lg[H_3O^+] = -1$$

$$pH = 1$$

If the concentration of a solution of sodium hydroxide is 0.01 mol dm$^{-3}$,

$$[OH^-] = 0.01\ mol\ dm^{-3} = 10^{-2}\ mol\ dm^{-3}$$

The product of the hydrogen ion concentration and the hydroxide ion concentration in a solution is called the **ionic product for water**, $K_w$. At 25 °C, the value of $K_w$ is $10^{-14}\ mol^2\ dm^{-6}$.

$$[H_3O^+][OH^-] = K_w = 10^{-14}\ mol^2\ dm^{-6}$$

Therefore, in this solution,

$$[H_3O^+] = \frac{10^{-14}}{10^{-2}} = 10^{-12}\ mol\ dm^{-2}$$

$$pH = -lg[H_3O^+] = 12$$

---

**Reminder :** Using a calculator to find logarithms § 1.9.

**EXAMPLE 1**        Calculate the pH of a solution of 0.020 mol dm$^{-3}$ hydrochloric acid.

**METHOD**        If the concentration of hydrochloric acid is 0.020 mol dm$^{-3}$,

$$[H_3O^+] = 2 \times 10^{-2} \text{ mol dm}^{-3}$$

**ANSWER**        $pH = -\lg[H_3O^+] = 1.70$

**EXAMPLE 2**        Calculate the pH of a 0.010 mol dm$^{-3}$ solution of calcium hydroxide.

**METHOD**        If the concentration of $Ca(OH)_2$ is $10^{-2}$ mol dm$^{-3}$,

$$[OH^-] = 2 \times 10^{-2} \text{ mol dm}^{-3}$$

$$pOH = -\lg(2 \times 10^{-2}) = 1.7$$

Since

$$[H_3O^+][OH^-] = K_w = 10^{-14} \text{ mol}^2 \text{ dm}^{-6}$$

$$\lg[H_3O^+] + \lg[OH^-] = \lg K_w = -14$$

$$\therefore \qquad pH + pOH = 14$$

Thus, if pOH = 1.7,

**ANSWER**        $pH = 14 - 1.7 = 12.3$.

### Weak acids and bases; calculation of pH and pOH

For calculating the pH of a solution of a weak acid or base, the concentration of the solution is not sufficient information. The extent to which the weak acid or base is ionised must be known. It may be that one molecule in a thousand is ionised. In such a case we say that the **degree of dissociation** is $1 \times 10^{-3}$. When a weak acid HA ionises,

$$HA + H_2O \rightleftharpoons H_3O^+ + A^-$$

The **dissociation constant** of the acid, $K_a$, is given by the expression

$$K_a = \frac{[H_3O^+] \, [A^-]}{[HA]}$$

The pH of a solution of the acid can be found from the concentration and the value of $K_a$. Conversely, if the pH of the solution is measured, the value of pH can be used to find the value of $K_a$.

**EXAMPLE 1**        Calculate the pH of a $1.00 \times 10^{-2}$ mol dm$^{-3}$ solution of ethanoic acid, which has a dissociation constant of $1.76 \times 10^{-5}$ mol dm$^{-3}$.

**METHOD**        When ethanoic acid ionises,

$$CH_3CO_2H + H_2O \rightleftharpoons H_3O^+ + CH_3CO_2^-$$

and

$$K_a = \frac{[H_3O^+][CH_3CO_2^-]}{[CH_3CO_2H]}$$

Since one $CH_3CO_2^-$ ion is formed for each $H_3O^+$ ion,

$$[H_3O^+] = [CH_3CO_2^-] \text{ and } [H_3O^+][CH_3CO_2^-] = [H_3O^+]^2$$

The degree of dissociation is so small that we can put $[CH_3CO_2H]$ equal to the total acid concentration, $1.00 \times 10^{-2}$ mol dm$^{-3}$. In fact it must be slightly less than this, but the error in putting $[CH_3CO_2H] = 1.00 \times 10^{-2}$ mol dm$^{-3}$ is smaller than the error in the practical measurements of pH and concentration. Therefore,

$$1.76 \times 10^{-5} \text{ mol dm}^{-3} = \frac{[H_3O^+]^2}{1 \times 10^{-2} \text{ mol dm}^{-3}}$$

$$[H_3O^+]^2 = 1.76 \times 10^{-7} \text{ mol}^2 \text{ dm}^{-6}$$

$$[H_3O^+] = 4.10 \times 10^{-4} \text{ mol dm}^{-3}$$

$$pH = -\lg[H_3O^+]$$

**ANSWER**       pH = 3.34

**EXAMPLE 2**       Calculate the dissociation constant of the weak acid phenol, given that a $1.00 \times 10^{-2}$ mol dm$^{-3}$ solution has a pH of 5.95.

**METHOD**       If pH = 5.95, $[H_3O^+] = $ antilg(−pH) $= 1.13 \times 10^{-6}$ mol dm$^{-3}$

If we write the dissociation of phenol

$$PhOH + H_2O \rightleftharpoons PhO^- + H_3O^+$$

$$K_a = \frac{[H_3O^+][PhO^-]}{[PhOH]} = \frac{[H_3O^+]^2}{[PhOH]}$$

$$K_a = \frac{(1.13 \times 10^{-6} \text{ mol dm}^{-3})^2}{1.00 \times 10^{-2} \text{ mol dm}^{-3}} = 1.28 \times 10^{-10} \text{ mol dm}^{-3}$$

**ANSWER**       The dissociation constant of phenol is $1.28 \times 10^{-10}$ mol dm$^{-3}$

**EXAMPLE 3**       Calculate the pH of a solution of ammonia (p$K_b$ = 4.75) of concentration $1.00 \times 10^{-2}$ mol dm$^{-3}$.

**Reminder :** Rearranging equations § 1.1.

**METHOD**       $K_b = $ antilg[−4.75] $= 1.78$ mol dm$^{-3}$

Since $K_b = \dfrac{[NH_4^+][OH^-]}{[NH_3]} = \dfrac{[OH^-]^2}{[NH_3]}$

$$[OH^-]^2 = K_b[NH_3]$$
$$= 1.78 \times 10^{-5} \text{ mol dm}^{-3} \times 1.00 \times 10^{-2} \text{ mol dm}^{-3}$$
$$[OH^-] = 4.22 \times 10^{-4} \text{ mol dm}^{-3}$$

**ANSWER**       pOH = 10.6

**EXERCISE 11.2**    **Problems on pH**

Section 1

**Reminder :** Finding logs on your calculator § 1.9.

Amount = Volume × Concentration § 8.1.

**1**    Calculate the pH of solutions with following $H_3O^+$ concentrations in mol $dm^{-3}$:

    **a**  $10^{-8}$        **d**  $6.8 \times 10^{-3}$      **g**  0.25        **j**  $9.9 \times 10^{-2}$

    **b**  $10^{-4}$        **e**  $3.2 \times 10^{-5}$      **h**  $5.4 \times 10^{-9}$

    **c**  $10^{-7}$        **f**  0.035         **i**  $7.1 \times 10^{-7}$

    Now calculate the pOH of each of the solutions.

**2**    Calculate the pH of solutions with the following $OH^-$ concentrations in mol $dm^{-3}$:

    **a**  $10^{-2}$        **d**  0.055         **g**  $7.6 \times 10^{-3}$      **j**  $3.7 \times 10^{-10}$

    **b**  $10^{-3}$        **e**  0.0010       **h**  $4.9 \times 10^{-5}$

    **c**  $10^{-8}$        **f**  0.083         **i**  $6.4 \times 10^{-8}$

**3**    Calculate the $H_3O^+$ concentrations in solutions with the following pH values:

    **a**  0.00        **d**  1.88         **g**  9.21        **j**  2.63

    **b**  4.30        **e**  4.15         **h**  13.7

    **c**  2.35        **f**  7.84         **i**  9.50

**4**    Calculate the pH of the solutions made by dissolving the following in distilled water and making up to 500 $cm^3$ of solution:

    **a**  3.00 g of hydrogen chloride

    **b**  4.50 g of chloric(VII) acid, $HClO_4$ (a strong acid)

    **c**  4.00 g of sodium hydroxide

    **d**  1.00 g of calcium hydroxide

    **e**  6.30 g of potassium hydroxide

**5**    Calculate the pH of:

    **a**  $1.00 \times 10^{-2}$ mol $dm^{-3}$ hydrochloric acid

    **b**  $1.00 \times 10^{-2}$ mol $dm^{-3}$ sodium hydroxide solution

    **c**  the solution obtained by diluting 5.00 $cm^3$ of 1.00 mol $dm^{-3}$ hydrochloric acid with distilled water to 1.00 $dm^3$.

**6**    **a**  Calculate the pH of $1.00 \times 10^{-3}$ mol $dm^{-3}$ hydrochloric acid.

    **b**  Calculate how many $cm^3$ of 1.00 mol $dm^{-3}$ hydrochloric acid are needed to decrease the pH of 1.00 $dm^3$ of hydrochloric acid by one unit. Ignore the small change in volume.

    **c**  Calculate how many $cm^3$ of 1.00 mol $dm^{-3}$ sodium hydroxide are needed to increase the pH of 1.00 $dm^3$ of hydrochloric acid by one unit. Ignore the small change in volume.

Section 2          **Reminder :** Rearranging equations § 1.1. Logarithms § 1.9. Check your answer § 1.12.

**1**   The pH of a 0.10 mol dm$^{-3}$ solution of HCl is 1.0, but the pH of a 0.10 mol dm$^{-3}$ solution of $CH_3CO_2H$ is 2.4. When 100 cm$^3$ of 0.10 mol dm$^{-3}$ $CH_3CO_2H$ reacts with an excess of zinc powder, the volume of hydrogen that is evolved is the same as when 100 cm$^3$ of 0.10 mol dm$^{-3}$ HCl is used. Suggest an explanation.

**2**                    $$H_3N^+CH_2CO_2^-(aq) \rightleftharpoons H_2NCH_2CO_2^-(aq) + H^+(aq)$$

The dissociation constant $K_a = 1.0 \times 10^{-12}$ mol dm$^{-3}$. What is the approximate pH of $1.0 \times 10^{-2}$ mol dm$^{-3}$ aminoethanoic acid?

**3**   A 0.10 mol dm$^{-3}$ solution of a weak monoprotic acid has pH = 4.8. The value of p$K_a$ is

   **A** 3.8   **B** 4.8   **C** 8.6   **D** 9.6

**4**   The ionic product of water $K_w$ varies with temperature.

| $K_w$/mol$^2$ dm$^{-6}$ | Temperature/°C |
|---|---|
| $6.8 \times 10^{-15}$ | 20 |
| $1.00 \times 10^{-14}$ | 25 |
| $5.47 \times 10^{-14}$ | 50 |

Write an expression for $K_w$. Calculate the pH of water  **a** at 25 °C  **b** at 50 °C.

**5**   Calculate the dissociation constants of the weak acids listed below:

   **a**   a solution of $1.00 \times 10^{-2}$ mol dm$^{-3}$ $CH_3CO_2H$ has a pH of 3.38

   **b**   a solution of 0.200 mol dm$^{-3}$ HCN has a pH of 5.05

   **c**   a solution of $1.00 \times 10^{-2}$ mol dm$^{-3}$ $CH_3CH_2CO_2H$ has a pH of 3.43

   **d**   a solution of $1.00 \times 10^{-2}$ mol dm$^{-3}$ HBrO has a pH of 5.35.

**6**   The pH of a 0.15 mol dm$^{-3}$ solution of a weak monoprotic acid HA is 2.69.

   **a**   Calculate  **(i)** [H$^+$] in the solution and  **(ii)** the value of the dissociation constant $K_a$ of the acid HA.

   **b**   A 25 cm$^3$ sample of the acid solution is titrated against 0.25 mol dm$^{-3}$ sodium hydroxide. What volume of sodium hydroxide is needed to reach the end point?

   **c**   Calculate the pH of the titration solution when half the HA has been neutralised and [HA] = [A$^-$].

   **d**   Calculate the pH of the titration solution when 25 cm$^3$ of 0.25 mol dm$^{-3}$ sodium hydroxide has been added.

**7** When sodium ethanoate dissolves in water an equilibrium is set up:

$$CH_3CO_2^-(aq) + H_2O(l) \rightleftharpoons CH_3CO_2H(aq) + OH^-(aq)$$

A 0.01 mol dm$^{-3}$ solution has a pH = 8.87.

**a** Calculate **(i)** [H$^+$] **(ii)** [OH$^-$] in the solution.

**b** Find [CH$_3$CO$_2$H] in the solution.

**c** Calculate the acid dissociation constant of CH$_3$CO$_2$H.

**8** The addition of 5.3 g of anhydrous sodium carbonate to 2.00 dm$^3$ of an aqueous solution of a strong acid exactly neutralises the acid.

$$2H^+(aq) + CO_3^{2-}(aq) \longrightarrow CO_2(g) + H_2O(l)$$

**a** Calculate the pH of the original solution.

**b** In a second experiment, 5.3 g of NaOH were added, instead of sodium carbonate, to 2.00 dm$^3$ of the original solution. Calculate the pH of the resulting solution.

## 11.4 BUFFER SOLUTIONS

A buffer solution is one which will resist changes in pH due to the addition of small amounts of acid and alkali. An effective buffer can be made by preparing a solution containing both a weak acid and also one of its salts with a strong base, e.g. ethanoic acid and sodium ethanoate. This will absorb hydrogen ions because they react with ethanoate ions to form molecules of ethanoic acid:

$$CH_3CO_2^- + H_3O^+ \rightleftharpoons CH_3CO_2H + H_2O$$

Hydroxide ions are absorbed by combining with ethanoic acid molecules to form ethanoate ions and water:

$$OH^- + CH_3CO_2H \rightleftharpoons CH_3CO_2^- + H_2O$$

A solution of a weak base and one of its salts formed with a strong acid, e.g. ammonia solution and ammonium chloride, will act as a buffer. If hydrogen ions are added, they combine with ammonia, and, if hydroxide ions are added, they combine with ammonium ions:

$$NH_3 + H_3O^+ \rightleftharpoons NH_4^+ + H_2O$$
$$OH^- + NH_4^+ \rightleftharpoons NH_3 + H_2O$$

The pH of a buffer solution consisting of a weak acid and its salt is calculated from the equation

$$K_a = \frac{[H_3O^+][A^-]}{[HA]}$$

$$[H_3O^+] = K_a \frac{[HA]}{[A^-]}$$

$$pH = pK_a + \lg \frac{[A^-]}{[HA]}$$

Since the salt is completely ionised and the acid only slightly ionised, we can assume that all the anions come from the salt, and put

$$[A^-] = [Salt]$$

$$[HA] = [Acid]$$

$$pH = pK_a + \lg \frac{[Salt]}{[Acid]}$$

For a buffer made from a base, **B**, and its salt with a strong acid, **BH⁺X⁻**, it can be shown in a similar way that

$$pH = pK_w - pK_b + \lg \frac{[Base]}{[Salt]}$$

**EXAMPLE**   Three solutions contain propanoic acid ($K_a = 1.34 \times 10^{-5}$ mol dm$^{-3}$) at a concentration of 0.10 mol dm$^{-3}$ and sodium propanoate at concentrations of **a** 0.10 mol dm$^{-3}$, **b** 0.20 mol dm$^{-3}$, **c** 0.50 mol dm$^{-3}$ respectively. Calculate the pH values of the three solutions.

**METHOD**   **a**                          $$pH = pK_a + \lg \frac{[Salt]}{[Acid]}$$

In solution **a**,            $$pH = 4.87 + \lg \frac{0.10}{0.10} = 4.87 + \lg 1.0$$

**ANSWER**   $$pH(\mathbf{a}) = 4.87$$

**b**  In solution **b**,      $$pH = 4.87 + \lg \frac{0.20}{0.10} = 4.87 + \lg 2.0$$

**ANSWER**   $$pH(\mathbf{b}) = 5.17$$

**c**  In solution **c**,      $$pH = 4.87 + \lg \frac{0.50}{0.10} = 4.87 + \lg 5.0$$

**ANSWER**   $$pH(\mathbf{c}) = 5.57$$

**EXERCISE 11.3**   **Problems on buffers**

**Reminder :** Rearranging equations § 1.1. Finding logs on your calculator § 1.9.

Amount = Volume × Concentration § 8.1.

**1**   What amount of sodium ethanoate must be added to 1.0 dm$^3$ of ethanoic acid of concentration 0.10 mol dm$^{-3}$ and $K_a = 2 \times 10^{-5}$ mol dm$^{-3}$ to produce a buffer solution of pH 4.7?

**A** 0.05 mol    **B** 0.1 mol    **C** 0.2 mol    **D** 0.5 mol

**2** Citric acid behaves as a weak monoprotic acid of $K_a = 7.24 \times 10^{-4}$ mol dm$^{-3}$. Calculate the pH of:

**a** a solution of 0.100 mol dm$^{-3}$ citric acid + 0.100 mol dm$^{-3}$ sodium citrate

**b** a solution of 0.100 mol dm$^{-3}$ citric acid + 0.050 mol dm$^{-3}$ sodium citrate.

**3** The pH of a 0.050 mol dm$^{-3}$ solution of propanoic acid is 3.10.

**a** Calculate the value of $K_a$ for the acid.

**b** Calculate the pH of a solution which is 0.050 mol dm$^{-3}$ with respect to propanoic acid and also to sodium propanoate.

**4** Calculate the pH of the solution formed when 40.0 cm$^3$ of 1.00 mol dm$^{-3}$ nitrous acid ($pK_a = 3.34$) is added to 20.0 cm$^3$ of 1.00 mol dm$^{-3}$ sodium hydroxide solution.

**5** What amount of sodium ethanoate must be added to 1.00 dm$^3$ of ethanoic acid of $pK_a = 4.73$, concentration 0.100 mol dm$^{-3}$, to produce a buffer of pH = 5.73?

## 11.5 ELECTRODE POTENTIALS

### Metals

If a strip of metal is placed in a solution of its ions, atoms of the metal may dissolve as positive ions, leaving a build-up of electrons on the metal:

$$M(s) \longrightarrow M^{2+}(aq) + 2e^-$$

The metal will become negatively charged. Alternatively, metal ions may take electrons from the strip of metal and be discharged as metal atoms:

$$M^{2+}(aq) + 2e^- \longrightarrow M(s)$$

In this case, the metal will become positively charged. The potential difference between the strip of metal and the solution depends on the nature of the metal and on the concentration of the ions. Zinc acquires a more negative potential than copper, since it has a greater tendency to dissolve as ions and a smaller tendency to be deposited as metal. In order to compare electrode potentials for different metals, **standard electrode potentials** are quoted at 25 °C with an ionic concentration of 1 mol dm$^{-3}$. The zero on the standard electrode potential scale is the potential of a strip of platinum in contact with hydrogen gas at 1 atm pressure and hydrogen ions at a concentration of 1 mol dm$^{-3}$.

### Redox systems

Metals are reducing agents. Other oxidation–reduction systems also have electrode potentials. The standard electrode potential of a redox system is the potential acquired by a piece of platinum immersed in a solution of the redox system in which the concentration of each dissolved component is 1 mol dm$^{-3}$. A powerful oxidising agent removes electrons and gives the platinum a high positive potential. When all the redox systems are arranged in order of their standard electrode potentials, the **electrochemical series** is obtained. Table 11.1 shows some of the redox systems in the series.

Table 11.1  *Values of standard electrode potential* $E^{\ominus}$ *at 298 K*

| Reaction | $E^{\ominus}/V$ |
|---|---|
| $K^+(aq) + e^- \longrightarrow K(s)$ | −2.92 |
| $Ca^{2+}(aq) + 2e^- \longrightarrow Ca(s)$ | −2.87 |
| $Na^+(aq) + e^- \longrightarrow Na(s)$ | −2.71 |
| $Mg^{2+}(aq) + 2e^- \longrightarrow Mg(s)$ | −2.36 |
| $Al^{3+}(aq) + 3e^- \longrightarrow Al(s)$ | −1.66 |
| $Zn^{2+}(aq) + 2e^- \longrightarrow Zn(s)$ | −0.76 |
| $Fe^{2+}(aq) + 2e^- \longrightarrow Fe(s)$ | −0.44 |
| $Cr^{3+}(aq) + e^- \longrightarrow Cr^{2+}(aq)$ | −0.41 |
| $Ni^{2+}(aq) + 2e^- \longrightarrow Ni(s)$ | −0.25 |
| $Sn^{2+}(aq) + 2e^- \longrightarrow Sn(s)$ | −0.14 |
| $Pb^{2+}(aq) + 2e^- \longrightarrow Pb(s)$ | −0.13 |
| $2H_3O^+(aq) + 2e^- \longrightarrow H_2(g) + 2H_2O(l)$ | 0.00 |
| $Sn^{4+}(aq) + 2e^- \longrightarrow Sn^{2+}(aq)$ | 0.15 |
| $Cu^{2+}(aq) + 2e^- \longrightarrow Cu(s)$ | 0.34 |
| $I_2(s) + 2e^- \longrightarrow 2I^-(aq)$ | 0.54 |
| $Fe^{3+}(aq) + e^- \longrightarrow Fe^{2+}(aq)$ | 0.77 |
| $Ag^+(aq) + e^- \longrightarrow Ag(s)$ | 0.80 |
| $Br_2(l) + 2e^- \longrightarrow 2Br^-(aq)$ | 1.09 |
| $MnO_2(s) + 4H^+(aq) + 2e^- \longrightarrow Mn^{2+}(aq) + 2H_2O(l)$ | 1.23 |
| $Cr_2O_7{}^{2-}(aq) + 14H^+(aq) + 6e^- \longrightarrow 2Cr^{3+}(aq) + 7H_2O(l)$ | 1.33 |
| $Cl_2(g) + 2e^- \longrightarrow 2Cl^-(aq)$ | 1.36 |
| $Ce^{4+}(aq) + e^- \longrightarrow Ce^{3+}(aq)$ (in $H_2SO_4(aq)$) | 1.44 |
| $PbO_2(s) + 4H^+(aq) + 2e^- \longrightarrow Pb^{2+}(aq) + 2H_2O(l)$ | 1.46 |
| $MnO_4{}^-(aq) + 8H^+(aq) + 5e^- \longrightarrow Mn^{2+}(aq) + 4H_2O(l)$ | 1.51 |
| $Ce^{4+}(aq) + e^- \longrightarrow Ce^{3+}(aq)$ (in $HNO_3(aq)$) | 1.61 |
| $H_2O_2(aq) + 2H^+(aq) + 2e^- \longrightarrow 2H_2O(l)$ | 1.78 |
| $F_2(g) + 2e^- \longrightarrow 2F^-(aq)$ | 2.85 |

## 11.6 ELECTROCHEMICAL CELLS

Figure 11.2 shows two metals inserted into solutions of their ions. The two solutions are joined by a salt bridge, and the two metal electrodes are connected by an external circuit. The cell has an electromotive force (e.m.f.) or voltage, which is equal to the difference between the standard electrode potentials of the two metals, and this e.m.f. makes a current flow through the external circuit.

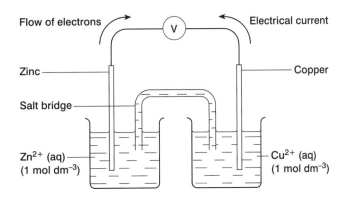

Fig. 11.2 *An electrochemical cell*

The cell shown in Figure 11.2 can be represented by

$$Zn(s) \mid Zn^{2+}(aq, 1 \text{ mol dm}^{-3}) \;\vdots\vdots\; Cu^{2+}(aq, 1 \text{ mol dm}^{-3}) \mid Cu(s)$$

By convention, the e.m.f. of the cell is taken as

$$E = E \text{ (RHS electrode)} - E \text{(LHS electrode)}$$

$$= E_{Cu}^{\ominus} - E_{Zn}^{\ominus} \text{ where } E^{\ominus} \text{ is the standard electrode potential}$$

$$= +0.34 \text{ V} - (-0.76 \text{ V}) = +1.10 \text{ V}$$

The flow of electrons is clockwise through the external circuit (from zinc to copper). Conventional electricity flows anticlockwise through the external circuit (from copper to zinc).

### Which electrode is which?

When two electrodes are combined, a cell reaction will take place if the e.m.f. for the cell is positive. The values of $E^{\ominus}$ for the two electrodes will tell you which will be the positive electrode and which the negative. For example,

$$Cu^{2+}(aq) + 2e^- \rightleftharpoons Cu(s); E^{\ominus} = +0.34 \text{ V}$$

$$Ag^+(aq) + e^- \rightleftharpoons Ag(s); E^{\ominus} = +0.80 \text{ V}$$

At one electrode reduction will occur, and the half-reaction will take place from left to right as written. At the other electrode, oxidation will take place and the half-reaction will take place from right to left. Which is which? For the cell $E^{\ominus}$ to be positive, the half-reaction at the silver electrode must go from left to right as a reduction, with $E^{\ominus} = +0.80$ V, that is, with the silver electrode being positive. Then the half-reaction at the copper electrode must go from right to left as an oxidation with $E^{\ominus} = -0.34$ V, that is, with the copper electrode being negative. For the cell, $E^{\ominus} = +0.80$ V $- 0.34$ V $= +0.46$ V. What would happen if the two half-reactions went in the opposite directions? Then $E^{\ominus} = -0.80$ V $+ 0.34$ V $= -0.46$ V. With a negative value of $E^{\ominus}$, this cell reaction does not happen.

## The anticlockwise rule

An alternative way to work out which reaction will happen is to use the anticlockwise rule. Write the two systems with the less positive or more negative value of $E^\ominus$ at the top. Then draw a circle anticlockwise.

$$Cu^{2+}(aq) + 2e^- \rightleftharpoons Cu(s); \quad E^\ominus = +0.34 \text{ V}$$
$$Ag^+(aq) + e^- \rightleftharpoons Ag(s); \quad E^\ominus = +0.80 \text{ V}$$

The circle tells you that the reaction that takes place is

$$Cu(s) + 2Ag^+(aq) \longrightarrow Cu^{2+}(aq) + 2Ag(s)$$

At the copper electrode electrons are released; the copper electrode is negative. At the silver electrode electrons are supplied to discharge silver ions; the silver electrode is positive.

## EXERCISE 11.4    Problems on standard electrode potentials

Refer to the standard electrode potentials in Table 11.1.

Section 1

**Reminder:** Negative numbers § 1.4. $E^\ominus$ must be positive for a cell reaction to happen. Or use the anticlockwise rule if you prefer.

**1**   Which of the following species are oxidised by manganese(IV) oxide?

$Br^-$     Ag     $I^-$     $Cl^-$

**2**   Which of the following species are reduced by $Sn^{2+}$?

$I_2$     $Ni^{2+}$     $Cu^{2+}$     $Fe^{3+}$

**3**   Calculate the standard e.m.f.s of the following cells at 298 K:

**a**  $Ni(s) \,|\, Ni^{2+}(aq) \,\vdots\, Sn^{2+}(aq), Sn^{4+}(aq) \,|\, Pt$

**b**  $Pt \,|\, I_2(s), I^-(aq) \,\vdots\, Ag^+(aq) \,|\, Ag(s)$

**c**  $Pt \,|\, Cl_2(g), Cl^-(aq) \,\vdots\, Br_2(l), Br^-(aq) \,|\, Pt$

**4**   The two half-reaction equations for the reduction of iodate(V) and chlorine are:

**a**  $IO_3^-(aq) + 6H^+(aq) + 5e^- \rightleftharpoons \frac{1}{2}I_2(aq) + 3H_2O(l); \quad E^\ominus = +1.19 \text{ V}$

**b**  $Cl_2(g) + 2e^- \rightleftharpoons 2Cl^-(aq); \quad E^\ominus = +1.36 \text{ V}$

Predict, with an explanation, whether a reaction will occur when chlorine is bubbled into a solution of potassium iodate(V).

**Hint:** A reaction will occur if $E^\ominus$ is positive, therefore half-reaction **b** must occur in the direction which gives $E^\ominus = 1.36$ V and half-reaction **a** in the reverse direction.

**5** A cell is represented by the diagram:

$$Zn(s) \,|\, ZnSO_4(aq) \,\vdots\, CuSO_4(aq) \,|\, Cu(s)$$

The standard electrode potentials are

**(i)** $Zn^{2+}(aq) + 2e^- \rightleftharpoons Zn(s); \quad E^\ominus = -0.76$ V

**(ii)** $Cu^{2+}(aq) + 2e^- \rightleftharpoons Cu(s); \quad E^\ominus = +0.34$ V

**a** State which electrode is the positive electrode.

**b** Write the equation for the overall cell reaction that occurs when the zinc and copper are connected by a conductor.

**Hint:** $E^\ominus$ must be positive, therefore half-reaction **(i)** must occur in the direction which gives $E^\ominus = +0.76$ V and half-reaction **(ii)** in the reverse direction.

**6** The standard electrode potentials for two systems are:

**a** $I_2(aq) + 2e^- \rightleftharpoons 2I^-(aq); \quad E^\ominus = +0.54$ V

**b** $Ce^{4+}(aq) + e^- \rightleftharpoons Ce^{3+}(aq); \quad E^\ominus = +1.70$ V

For the reaction between cerium(IV) ions and iodide ions, state the standard cell potential and write the ionic equation for the reaction.

**Hint:** To give a positive $E^\ominus$, **b** must have $E^\ominus = +1.70$ V (not $-1.70$ V).

**7** Calculate the standard e.m.f. of each of the cells:

**a** $Sn(s) \,|\, Sn^{2+}(aq) \,\vdots\, Ag^+(aq) \,|\, Ag(s)$

**b** $Ag(s) \,|\, Ag^+(aq) \,\vdots\, Cu^{2+}(aq) \,|\, Cu(s)$

**c** $Ce^{3+}(aq) \,|\, Ce^{4+}(aq) \,\vdots\, Fe^{3+}(aq) \,|\, Fe^{2+}(aq)$

**d** $Fe(s) \,|\, Fe^{2+}(aq) \,\vdots\, Cu^{2+}(aq) \,|\, Cu(s)$

**e** $Zn(s) \,|\, Zn^{2+}(aq) \,\vdots\, Pb^{2+}(aq) \,|\, Pb(s)$

Section 2

Refer to Table 11.1 for values of $E^\ominus$.

**1**   Iron filings are added to a solution containing the ions $Cu^{2+}$, $Fe^{2+}$, $Fe^{3+}$, $H_3O^+$ and $Zn^{2+}$, all at a concentration of 1 mol dm$^{-3}$. From the standard electrode potentials of the redox systems, deduce what reaction occurs, and write the equation.

**2**   A solution contains $Fe^{2+}$, $Fe^{3+}$, $Cr^{3+}$ and $Cr_2O_7^{2-}$ in their standard states and dilute sulphuric acid. Deduce what happens, and write the equation for the reaction.

**3**   Predict the reactions between:

    **a**  $Fe^{3+}(aq)$ and $I^-(aq)$         **d**  $Ag(s)$ and $Fe^{3+}(aq)$

    **b**  $Ag^+(aq)$ and $Cu(s)$         **e**  $Br_2(aq)$ and $Fe^{2+}(aq)$

    **c**  $Fe^{3+}(aq)$ and $Br^-(aq)$

**4**   From the standard electrode potentials, predict which of the halogens, $Cl_2$, $Br_2$, $I_2$, will oxidise **a** $Fe^{2+}$ to $Fe^{3+}$ **b** $Sn^{2+}$ to $Sn^{4+}$.

**5**   The nickel oxide–cadmium cell utilises the reactions:

$$NiO_2(s) + 2H_2O(l) + 2e^- \longrightarrow Ni(OH)_2(s) + 2OH^-(aq); \quad E^\ominus = +0.49 \text{ V}$$

$$Cd(OH)_2(s) + 2e^- \longrightarrow Cd(s) + 2OH^-(aq); \quad E^\ominus = -0.81 \text{ V}$$

    **a**  Write the equation for the overall cell reaction.

    **b**  Calculate the standard cell potential.

**6**   Some standard reduction potentials are listed:

| System | $E^\ominus/\text{V}$ |
|---|---|
| $Cu^{2+}(aq) + e^- \rightleftharpoons Cu^+(aq)$ | +0.15 |
| $Cu^{2+}(aq) + 2e^- \rightleftharpoons Cu(s)$ | +0.34 |
| $Cu^+(aq) + e^- \rightleftharpoons Cu(s)$ | +0.52 |
| $I_2(s) + 2e^- \rightleftharpoons 2I^-(aq)$ | +0.54 |
| $Cu^{2+}(aq) + I^-(aq) + e^- \rightleftharpoons CuI(s)$ | +0.92 |

    **a**  Explain why copper(I) compounds are unstable in aqueous solution.

    **b**  Predict what will happen when aqueous potassium iodide is added to aqueous copper(II) sulphate. Give an explanation, and write an equation for the reaction.

## EXERCISE 11.5   Questions from examination papers

**1**   **a**  The pH of a 0.15 M solution of a weak acid, HA, is 2.82 at 300 K.

        **(i)**  Write an expression for the acid dissociation constant, $K_a$, of HA, and determine the value of $K_a$ for this acid at 300 K, stating its units.

        **(ii)**  The dissociation of HA into its ions in aqueous solution is an endothermic process. How would its pH change if the temperature were increased? Explain your answer.

        (8)

**b** Solution **A** contains $n$ moles of a different weak acid, HX. The addition of some sodium hydroxide to **A** neutralises one third of the HX present to produce Solution **B**.

   **(i)** In terms of the amount, $n$, how many moles of HX are present in Solution **B**?

   **(ii)** Determine the ratio $\dfrac{[HX]}{[X^-]}$ in Solution **B**.

   **(iii)** Solution **B** has a hydrogen ion concentration of $4.2 \times 10^{-4}$ mol dm$^{-3}$. Use this information and your answer to part **b (ii)** to determine the value of the acid dissociation constant of HX. (5)

**c** Why is methyl orange **not** suitable as an indicator for the titration of HX with sodium hydroxide? (2)

**d** Solution **B** can act as a buffer. Explain what this means and write an equation that shows how Solution **B** acts as a buffer if a little hydrochloric acid is added. (3)

*NEAB 99* (18)

**2** Show details of **all** calculations in answering this question.

Solution **S** is 0.16 M hydrochloric acid, HCl, a strong acid.
Solution **W** is 0.16 M HX, a weak monoprotic acid ($K_a = 2.07 \times 10^{-5}$ mol dm$^{-3}$, pH = 2.74).
Solution **Z** is 0.12 M barium hydroxide, Ba(OH)$_2$, a strong base.

At a temperature of 25 °C, 18 cm$^3$ of solution **S** are titrated with solution **Z**. The titration is repeated using 18 cm$^3$ of solution **W**.

**a** Determine the equivalence volume (end-point) for these titrations and enter your result into the appropriate space in a copy of the incomplete table given below. Enter also the half-equivalence volume and the double-equivalence volume in the appropriate space.

**b** Calculate the missing pH values for each of the two titration solutions and enter these into a copy of the table.

   (You should assume that, at half-equivalence, [HX] = [X$^-$] for the weak acid HX, and that, for both acids at double-equivalence, only the alkali that is in excess contributes to the pH of the resulting solution.)

**c** Plot these results on a graph of pH against volume of Ba(OH)$_2$(aq) and use the points you have plotted to sketch the complete titration curves for solution **S** and solution **W** titrated with solution **Z**.

|  | Start | Half equivalence | Equivalence | Double equivalence |
|---|---|---|---|---|
| Volume/cm$^3$ Ba(OH)$_2$ solution added | 0.0 |  |  |  |
| pH for titration of **S** | 0.80 | 1.22 | 7 |  |
| pH for titration of **W** | 2.74 |  | 8.5 |  |

*NEAB 99* (15)

**3** **a** Write balanced equations for the reaction of gaseous chlorine with

    **(i)** aqueous potassium bromide, (1)

    **(ii)** cold dilute aqueous sodium hydroxide. (1)

**b** Name **two** different halogen compounds and give **one** use for **each** in **either** commerce or industry. (2)

**c** The table below contains standard electrode potential information which should be used, as appropriate, in answering the questions that follow.

| Half-equation | $E^{\ominus}/V$ |
|---|---|
| $I_2(aq) + 2e^- \rightleftharpoons 2I^-(aq)$ | +0.54 |
| $Fe^{3+}(aq) + e^- \rightleftharpoons Fe^{2+}(aq)$ | +0.77 |
| $VO_2^+(aq) + 2H^+(aq) + e^- \rightleftharpoons VO^{2+}(aq) + H_2O(l)$ | +1.00 |
| $Cl_2(g) + 2e^- \rightleftharpoons 2Cl^-(aq)$ | +1.36 |

    **(i)** A reaction occurs when gaseous chlorine is passed into iron(II) sulphate solution.

        I. State the change in oxidation state (number) of the iron. (1)

        II. Write an **ionic** equation for the reaction. (1)

        III. Calculate the value of the standard cell potential for the reaction. (1)

    **(ii)** A reaction occurs when acidified aqueous ammonium vanadate(V), $NH_4VO_3$, is added to aqueous potassium iodide. The addition of acid changes $VO_3^-(aq)$ ions to $VO_2^+(aq)$ ions.

        I. Write an **ionic** equation for the reaction between the vanadate(V) ion, $VO_3^-$, and acid to form the dioxovanadium(V) ion, $VO_2^+$. (1)

        II. Write an **ionic** equation for the subsequent reaction between the dioxovanadium(V) ion, $VO_2^+$, and iodide ions. (1)

        III. Calculate the standard cell potential for the reaction in **(ii)** II. (1)

                                          *WJEC* (10)

**4** One step in the extraction of gold from its ore uses oxygen in the presence of aqueous sodium cyanide:

**A**     $4Au(s) + 8CN^-(aq) + 2H_2O(l) + O_2(g) \longrightarrow 4Au(CN)_2^-(aq) + 4OH^-(aq)$

The resulting aqueous solution is then treated with zinc which re-precipitates the gold.

**a** The redox process in equation **A** may be divided up into two electrode systems. One of these is given as **B** below; copy and complete the other one as **C**: (2)

**B**                $Au(CN)_2^-(aq), 2CN^-(aq) \mid Au(s)$         $E^{\ominus} = +0.12$ V

**C**                [.............],............ $\mid$ Pt              $E^{\ominus} = +0.40$ V

**b** Draw a diagram of a half-cell based on **B**. No salt bridge or hydrogen electrode need be shown but the diagram should be labelled and concentrations given where appropriate. (4)

**c** **(i)** Combine **B** and **C** into a cell diagram with the electrode **B** as the left-hand electrode. (2)

   **(ii)** Calculate $E^\ominus$ for the cell you have represented in **c (i)**. Give a sign and units in your answer. (2)

   **(iii)** What does your answer suggest about the position of equilibrium of reaction **A** at 298 K? Justify your answer. (2)

**d** What shape would you expect for the complex ion $Au(CN)_2^-$? (1)

**e** One of the problems in extracting gold in this way is that the metal is trapped in a matrix of minerals such as arsenopyrite, FeAsS, which cannot easily be broken down. In some cases only 10% of the available gold is actually extracted.

This problem has recently been solved by using bacteria to catalyse reactions such as:

$$2FeAsS(s) + 7O_2(g) + 4H^+(aq) + 2H_2O(l) \longrightarrow 2Fe^{3+}(aq) + 2\overset{\bullet}{H}_3AsO_4(aq) + 2HSO_4^-(aq)$$

   **(i)** Assuming that arsenopyrite contains sulphide ions, $S^{2-}$, what oxidation numbers does this suggest for iron and arsenic? (2)

   **(ii)** Why would it be environmentally unacceptable simply to heat the arsenopyrite strongly in air in order to break it down? (2)

   *L(N)* (17)

**5** **a** At room temperature water is partially dissociated. The value of the ionic product of water, $K_w$, varies with temperature as shown below:

| Temperature/°C | $K_w$/mol$^2$ dm$^{-6}$ |
|---|---|
| 20 | $6.80 \times 10^{-15}$ |
| 25 | $1.00 \times 10^{-14}$ |
| 50 | $5.47 \times 10^{-14}$ |

   **(i)** Write an equation for the dissociation of water to produce hydroxonium ions. (1)

   **(ii)** Using the data above, explain whether this dissociation is endothermic or exothermic. (2)

   **(iii)** Write an expression for the ionic product of water, $K_w$. (1)

   **(iv)** The pH of water is 7.00 at 25 °C, but changes with temperature. Calculate the pH of water at 50 °C. (3)

**b** Water may be described as amphoteric since it acts as an acid and a base according to the Bronsted-Lowry theory.

   **(i)** Explain the term **Bronsted-Lowry acid**. (1)

   **(ii)** Ammonia reacts with water to form an alkaline solution.

$$NH_3 + H_2O \rightleftharpoons NH_4^+ + OH^-$$

Using the above equation, identify the **two** conjugate acid–base pairs. (2)

*NI* (10)

**6** Lemonade contains carbonated water, citric acid, sodium citrate and other ingredients. It has been suggested that the following equilibrium exists in carbonated water.

$$H_2CO_3(aq) + H_2O(l) \rightleftharpoons HCO_3^-(aq) + H_3O^+(aq)$$

**a** Use this equation to explain why carbonic acid, $H_2CO_3$, is classified as an acid. (1)

**b** **(i)** Write an expression for the acid dissociation constant of carbonic acid.

**(ii)** Calculate the pH of 0.010 mol dm$^{-3}$ carbonic acid, given that $K_a$ for carbonic acid is $2.0 \times 10^{-4}$ mol dm$^{-3}$. (4)

**c** The $pK_a$ of citric acid is 3.1. Suggest, with a reason, which is the stronger acid: citric acid or carbonic acid. (1)

**d** What is the name commonly given to a solution of a mixture of a weak acid and the salt of the weak acid? (1)

**e** Describe the behaviour of a solution containing citric acid and sodium citrate if small quantities of **(i)** acid, **(ii)** alkali are added separately. Use the formula **HA** to represent citric acid. (4)

*OCR* (11)

**7** **a** Values of electrode potentials, $E$, can vary with the experimental conditions used. State **three** specific conditions that must be fixed when determining **standard** electrode potentials. (3)

**b** Define the term *oxidation* in terms of electron transfer. (1)

**c** State the charge on the electrode at which oxidation occurs in an electrochemical cell. (1)

**d** A standard cell, conventionally written with a secondary standard reference electrode ($E^\ominus = +0.27$ V) as the left-hand electrode, has an emf of +1.03 V. Calculate the standard potential of the right-hand electrode. (2)

**e** The redox couples $Co^{2+}(aq)/Co(s)$ and $Ni^{2+}(aq)/Ni(s)$ have standard electrode potentials of −0.28 V and −0.23 V, respectively. Write an equation for the spontaneous reaction which will occur if two half-cells containing these two redox couples are connected. Use the standard electrode potentials to justify your choice of the direction of reaction. (3)

*NEAB 96* (10)

**8** **a** Two standard electrode potentials are given below.

$$E^\ominus/V$$

**A**  $Br_2(aq) + 2e^- \rightleftharpoons 2Br^-(aq)$   +1.07

**B**  $Co^{2+}(aq) + 2e^- \rightleftharpoons Co(s)$   −0.28

**(i)** Draw a labelled diagram of the standard cell formed from the half-cells corresponding to **A** and **B**.

**(ii)** Calculate the standard cell potential, $E^\ominus$, for this cell.

**(iii)** Write the equation for the overall cell reaction.

**(iv)** Identify the electrode at which reduction occurs. Explain your answer. (9)

**b** An environmental chemist investigated the chloride content of river water. He decided to change the chloride ions into chlorine.

Three standard electrode potentials are given below.

$$E^{\ominus}/V$$

$$MnO_4^-(aq) + 8H^+(aq) + 5e^- \rightleftharpoons Mn^{2+}(aq) + 4H_2O(l) \qquad +1.52$$

$$Cl_2(g) + 2e^- \rightleftharpoons 2Cl^-(aq) \qquad +1.36$$

$$Cr_2O_7^{2-}(aq) + 14H^+(aq) + 6e^- \rightleftharpoons 2Cr^{3+}(aq) + 7H_2O(l) \qquad +1.33$$

Suggest, with reasons, whether acidified manganate(VII) and/or dichromate(VI) would be suitable for this reaction. (2)

*OCR* (11)

**9** Use the data below to answer the questions that follow.

| Reaction at 298 K | $E^{\ominus}/V$ |
|---|---|
| $Ag^+(aq) + e^- \longrightarrow Ag(s)$ | +0.80 |
| $AgF(s) + e^- \longrightarrow Ag(s) + F^-(aq)$ | +0.78 |
| $AgCl(s) + e^- \longrightarrow Ag(s) + Cl^-(aq)$ | +0.22 |
| $AgBr(s) + e^- \longrightarrow Ag(s) + Br^-(aq)$ | +0.07 |
| $H^+(aq) + e^- \longrightarrow \frac{1}{2}H_2(g)$ | 0.00 |
| $D^+(aq) + e^- \longrightarrow \frac{1}{2}D_2(g)$ | −0.004 |
| $AgI(s) + e^- \longrightarrow Ag(s) + I^-(aq)$ | −0.15 |

The symbol D denotes deuterium, which is heavy hydrogen, $_1^2H$.

**a** By considering electron transfer, state what is meant by the term *oxidising agent*. (1)

**b** State which of the two ions, $H^+(aq)$ or $D^+(aq)$, is the more powerful oxidising agent. Write an equation for the spontaneous reaction which occurs when a mixture of aqueous $H^+$ ions and $D^+$ ions are in contact with a mixture of hydrogen and deuterium gas. Deduce the e.m.f. of the cell in which this reaction would occur spontaneously. (3)

**c** Write an equation for the spontaneous reaction which occurs when aqueous $F^-$ ions and $Cl^-$ ions are in contact with a mixture of solid AgF and solid AgCl. Deduce the e.m.f. of the cell in which this reaction would occur spontaneously. (4)

**d** Silver does not usually react with dilute solutions of strong acids to liberate hydrogen.
   **(i)** State why this is so.

   **(ii)** Suggest a hydrogen halide which might react with silver to liberate hydrogen in aqueous solution. Write an equation for the reaction and deduce the e.m.f. of the cell in which this reaction would occur spontaneously. (4)

*NEAB 99* (12)

**10** **a** Some data concerning three carboxylic acids are shown in the following table.

| Name of acid | Formula | Acid dissociation constant, $K_a$ /mol dm$^{-3}$ |
|---|---|---|
| Chloroethanoic | $CH_2ClCO_2H$ | $1.3 \times 10^{-3}$ |
| Methanoic | $HCO_2H$ | $1.6 \times 10^{-4}$ |
| Propanoic | $CH_3CH_2CO_2H$ | $1.3 \times 10^{-5}$ |

**(i)** Give an expression for the acid dissociation constant of chloroethanoic acid, including all appropriate symbols. (2)

**(ii)** Use the values of $K_a$ to decide which is the strongest of the three acids. Explain how you reached your decision. (2)

**b** **(i)** Calculate the pH of a 0.050 M solution of propanoic acid. Make the usual approximations in your calculation. (2)

**(ii)** Calculate the pH of a 0.050 M solution of sodium hydroxide.

(The ionization constant for water, $K_w = 1.0 \times 10^{-14}$ mol$^2$ dm$^{-6}$) (2)

**(iii)** Draw a graph, taking into account your values from **(i)** and **(ii)** to show how the pH changes when 0.050 M sodium hydroxide is added from a burette to 25.0 cm$^3$ of 0.050 M propanoic acid.

Assume a total of 40.0 cm$^3$ of alkali is added. (3)

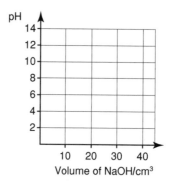

**(iv)** The names and pH ranges of three acid/alkali indicators are shown in the following table:

| Name of indicator | pH range |
|---|---|
| Bromophenol blue | 2.8–4.6 |
| Bromothymol blue | 6.0–7.6 |
| Phenolphthalein | 8.2–10.0 |

Which indicator would you use for the titration described in **(iii)**? Justify your decision. (2)

**c** **(i)** Methanoic acid reacts with sodium carbonate, liberating carbon dioxide gas. Write a balanced equation for this reaction. (1)

**(ii)** Draw the displayed formula of the organic product of this reaction. (1)

**d** Carboxylic acids and some of their derivatives react with alcohols, forming esters.

   **(i)** Draw the displayed formula of the ester produced when chloroethanoic acid reacts with methanol. (2)

   **(ii)** Why is a small quantity of concentrated sulphuric acid added to the mixture of chloroethanoic acid and methanol when carrying out this reaction? (1)

   **(iii)** Esters often contribute to the characteristic flavours and fragrances of fruit. They can be made synthetically for use in the food industry. Suggest ONE advantage and ONE disadvantage of using synthetically made esters in place of natural fruit flavourings. (2)

   *L(N)* (20)

**11** **a** State and explain the effect on the position of the equilibrium

$$H_2(g) + I_2(g) \rightleftharpoons 2HI(g) \qquad \Delta H = -9.6 \text{ kJ}$$

   **(i)** of an increase in the pressure (2)

   **(ii)** an increase in the temperature of the system. (2)

**b** Give the oxidation number (state) of the halogen in the ion $IO_4^-$. (1)

**c** Consider the following standard electrode potentials:

| Half-reaction | $E^\ominus/V$ |
|---|---|
| $S_4O_6^{2-}(aq) + 2e^- \rightleftharpoons 2S_2O_3^{2-}(aq)$ | +0.09 |
| $I_2(aq) + 2e^- \rightleftharpoons 2I^-(aq)$ | +0.54 |

   **(i)** An aqueous solution containing iodine is added, dropwise, to a small volume of a solution containing thiosulphate ions, $S_2O_3^{2-}$, until there is no further change. Describe what would be observed, and use the electrode potentials to explain your observations. (3)

   **(ii)** State the oxidation number (state) of sulphur in the product, $S_4O_6^{2-}$, of the reaction in **c (i)**. (1)

   **(iii)** Chlorine reacts with an aqueous solution containing $S_2O_3^{2-}$ ions as follows:

   $$4Cl_2(aq) + 5H_2O(l) + S_2O_3^{2-}(aq) \longrightarrow 8Cl^-(aq) + 2SO_4^{2-}(aq) + 10H^+(aq).$$

   State the oxidation number (state) of sulphur in the sulphur-containing product of this reaction. (1)

   **(iv)** What information do the sulphur oxidation number (state) values in **c (ii)** and **c (iii)** provide about the two halogens involved? Explain your reasoning. (2)

   *WJEC* (12)

**12** **a** **(i)** Write an equation to show the reaction that occurs when hydrogen fluoride dissolves in water. (1)

   **(ii)** Explain why hydrogen fluoride acts as a weak acid in dilute solution whilst all other hydrogen halides form strong acids. (2)

**b** The following reduction potentials are required in this question.

$$\begin{array}{ll} & E^{\ominus}/V \\ O_2(g) + 4H^+(aq) + 4e^- \rightleftharpoons 2H_2O(l) & +1.23 \\ Cl_2(aq) + 2e^- \rightleftharpoons 2Cl^-(aq) & +1.36 \\ F_2(g) + 2e^- \rightleftharpoons 2F^-(aq) & +2.87 \end{array}$$

**(i)** State and explain what occurs when fluorine is bubbled into water. (3)

**(ii)** Write an equation for the expected reaction of chlorine with water using the above data. (1)

**(iii)** In fact the reaction between chlorine and water is usually represented by the equation

$$Cl_2(g) + H_2O(l) \longrightarrow H^+(aq) + Cl^-(aq) + HClO(aq)$$

Suggest why this reaction occurs rather than that in **(ii)**. (2)

**c** **(i)** Write an equation to show the disproportionation of the chlorate(I) ion in solution and give the conditions necessary for this to occur. (2)

**(ii)** Use the equation to explain the term *disproportionation*. (2)

**d** When solid $KBrO_3$ is heated oxygen gas is evolved, and a white solid remains. If the white solid is dissolved in water and aqueous silver nitrate is added, a cream precipitate is formed which is soluble in concentrated ammonia solution.

**(i)** Suggest the name of the white solid and write an ionic equation for the reaction that occurs on the addition of the aqueous silver nitrate. (2)

**(ii)** Hence write an equation to represent the thermal decomposition of $KBrO_3$. (1)

**e** $KIO_3$ reacts with hydrogen peroxide according to the equation

$$2IO_3^-(aq) + 2H^+(aq) + 5H_2O_2(aq) \longrightarrow I_2(aq) + 6H_2O(l) + 5O_2(g)$$

**(i)** Deduce the ionic half equations for the reduction reaction and oxidation reaction. (3)

**(ii)** What would be seen if the reagents were mixed in a conical flask? (2)

*L* (21)

**13** The table below gives values for standard electrode potentials in acid solution. Use the data to answer the questions which follow.

$$\begin{array}{ll} & E^{\ominus}/V \\ MnO_4^-(aq) + 8H^+(aq) + 5e^- \rightleftharpoons Mn^{2+}(aq) + 4H_2O(l) & +1.52 \\ Cl_2(g) + 2e^- \rightleftharpoons 2Cl^-(aq) & +1.36 \\ Cr_2O_7^{2-}(aq) + 14H^+(aq) + 6e^- \rightleftharpoons 2Cr^{3+}(aq) + 7H_2O(l) & +1.33 \\ IO_3^-(aq) + 6H^+(aq) + 6e^- \rightleftharpoons I^-(aq) + 3H_2O(l) & +1.09 \\ I_2(s) + 2e^- \rightleftharpoons 2I^-(aq) & +0.54 \\ Cr^{3+}(aq) + e^- \rightleftharpoons Cr^{2+}(aq) & -0.41 \\ Zn^{2+}(aq) + 2e^- \rightleftharpoons Zn(s) & -0.76 \\ Cr^{2+}(aq) + 2e^- \rightleftharpoons Cr(s) & -0.91 \end{array}$$

**a** **(i)** Write the chemical equation for which $E^{\ominus}$ is arbitrarily taken as zero.

**(ii)** Hydrochloric acid and aqueous sodium chloride both contain chloride ions, but $KMnO_4$ liberates chlorine only from hydrochloric acid. Suggest a reason for this.

**(iii)** Write a balanced equation for the reaction of iodate(V) and iodide ions in acid solution.

**(iv)** Chromium(II) is a very unstable oxidation state of chromium. Use the data to predict a possible method of preparation of $Cr^{2+}(aq)$ and explain your reasoning. (5)

**b** **(i)** Use the data to identify a halide ion which will be oxidised by acidified $Cr_2O_7{}^{2-}(aq)$, giving a reason for your answer.

**(ii)** Calculate the mass of halogen liberated when excess of this halide reacts completely with 20.00 cm$^3$ of 0.0500 mol dm$^{-3}$ $K_2Cr_2O_7$ solution. (3)

OCR (8)

**14** In 1869 Mendeleev put chromium and lead in the same group of the Periodic Table; chromium in group 4A, and lead in group 4B. Group 4 now usually refers to the elements carbon to lead, and chromium is classified as a transition element. Nevertheless chromium and lead do have some properties in common. Both have amphoteric oxides; both have more than one oxidation state, and for both their highest oxidation state is strongly oxidising.

The following electrode potential data are required in this question:

$$
\begin{array}{lr}
 & E^{\ominus}/V \\
Cr^{3+} + 3e^- \rightleftharpoons Cr & -0.74 \\
O_2 + 2H^+ + 2e^- \rightleftharpoons H_2O_2 & +0.68 \\
Cr_2O_7{}^{2-} + 14H^+ + 6e^- \rightleftharpoons 2Cr^{3+} + 7H_2O & +1.33
\end{array}
$$

**a** Chromium(III) may be converted to chromium(VI) by heating a strongly alkaline solution of chromium(III) ions with hydrogen peroxide.

$$Cr_2(SO_4)_3(aq) \xrightarrow[\text{step 1}]{\text{KOH in excess}} \text{green solution} \xrightarrow[\text{step 2}]{H_2O_2,\ \text{heat}} K_2CrO_4(aq)$$

The yellow solution of potassium chromate(VI) is boiled until the excess of hydrogen peroxide has been destroyed, the solution then being made acidic with ethanoic acid to give a solution of potassium dichromate(VI).

$$K_2CrO_4(aq) \xrightarrow[\text{step 3}]{H^+} K_2Cr_2O_7(aq)$$

Show by means of equations the processes which are occurring in steps 1, 2 and 3 of the reaction scheme, and suggest in terms of the data given above why the excess of hydrogen peroxide must be destroyed before the solution is acidified in step 3. (8)

**b** Ammonium dichromate(VI), $(NH_4)_2Cr_2O_7$, decomposes on heating to chromium(III) oxide, nitrogen and water in a reaction which gives out enough heat to be self-sustaining. Write an equation representing the reaction, and by a consideration of the oxidation states of the atoms concerned show that the reaction is an oxidation of the cation by the anion. (4)

**c** Define the term *disproportionation*, and use the data given to show whether chromium(III) ions will disproportionate in acidic solution. (4)

**d** Explain why lead(IV) is an oxidising agent and give an example of a reaction in which a lead(IV) compound acts as an oxidising agent. (4)

**e** Illustrate by means of equations the amphoteric nature of chromium(III) oxide and lead(II) oxide. (5)

*L* (25)

**15 a** The structure of a substance influences its properties.

    **(i)** Silicon(IV) oxide melts at 1610 °C and is a good electrical insulator. Discuss how these properties are related to its structure.

    **(ii)** Describe the composition of brass. Suggest why it is used to make ornamental buttons.

    **(iii)** Suggest why aluminium is often used instead of copper in electrical cables. (7)

**b** In the UK, over 290 000 tonnes of aluminium are produced each year.

    **(i)** Describe the electrolytic extraction of aluminium from purified aluminium oxide. In your answer, include equations for relevant electrode processes.

    **(ii)** If a current of $1 \times 10^5$ A is passed for 8.0 hours, calculate the mass of aluminium produced.

    **(iii)** Describe the composition of duralumin. Suggest why it is used to make mountain bicycle frames.

    **(iv)** State the main purpose of recycling aluminium. (14)

*OCR* (21)

**16** A 20 cm$^3$ sample of a 0.1 M solution of a weak acid HA (p$K_a$ = 5.68) was titrated with 0.1 M sodium hydroxide. The temperature was kept constant at 25 °C.

**a** Calculate the pH of the acid at the beginning of the titration. (4)

**b** Calculate the pH after 40 cm$^3$ of sodium hydroxide had been added. (3)

**c**   **(i)** On a copy of the axes shown below, sketch the resulting titration curve.

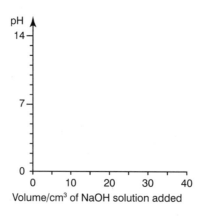

    **(ii)** Use your sketch to estimate the value of the pH at equivalence. (4)

**d** A 20 cm$^3$ sample of 0.1 M ethanedioic (*oxalic*) acid was titrated with 0.2 M sodium hydroxide.

    **(i)** What major difference is there between the titration curve for this acid–base pair and your sketch in part **c**.

    **(ii)** At what volumes, on this titration curve, would you expect the pH to alter rapidly? (2)

NEAB 97 (13)

**17** In 1854, Sinstede produced the first lead–acid battery to store electricity. One hundred and fifty years later, such batteries are playing a major part in powering the way to the global communications highway.

  **a**  **(i)** Explain the operation of the lead–acid battery, giving equations for the reactions taking place at the terminals as it discharges. (4)

    **(ii)** Explain why there is a progressive loss of capacity in such cells. (1)

  **b** The zinc–carbon dry cell is also based on older technology.

    **(i)** Copy and label the diagram below. (4)

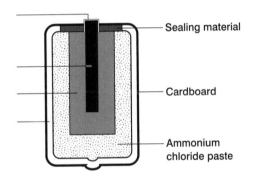

    **(ii)** Give equations for the reactions at the two electrodes indicating at which electrode each reaction occurs. (3)

  **c** In principle a fuel cell operates like a battery. However, unlike a battery, a fuel cell does not run down or require recharging. It will produce energy in the form of electricity as long as fuel is supplied.

    **(i)** Name a fuel used. (1)

    **(ii)** What is the function of oxygen in a fuel cell? (1)

    **(iii)** Alkanes cannot be used in fuel cells, yet they burn to produce energy. Write an equation for the combustion of butane. (2)

**d** Large amounts of energy are needed to power the diaphragm cell, used in the production of sodium hydroxide and chlorine.

    **(i)** On a copy of the diagram below, label the anode.     (1)

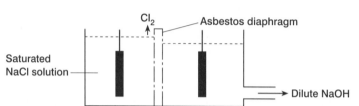

    **(ii)** Give an equation for the formation of chlorine.     (2)

    **(iii)** If a current of 5 A is passed through the cell for 2 hours, calculate the mass of chlorine which will be produced. (1 Faraday = 96 500 coulombs)     (3)

    **(iv)** Name the other gaseous product formed.     (1)

**e** Chlorine is used to make vinyl chloride (chloroethene), which is the monomer in the production of PVC.

    **(i)** What is meant by the term **monomer**?     (2)

    **(ii)** Explain whether vinyl chloride can exist as *cis-trans* isomers.     (2)

    **(iii)** Draw a two unit section of PVC.     (1)

**\*f** Corrosion of steel occurs by cell formation, but it can be prevented by the presence of zinc. Describe experiments to demonstrate the cells set up when steel rusts, including reference to the role of air and how zinc can protect the steel.

(Up to two marks may be awarded for quality of language in this part.)     (6)

                                                                   *NI*  (34)

# Thermochemistry

# 12

## 12.1 INTERNAL ENERGY AND ENTHALPY

Matter contains energy. You are familiar in physics with kinetic energy – the energy which enables an object to move – and potential energy – the energy which an object possesses due to its position. Matter possesses energy of both kinds. Matter possesses kinetic energy because the atoms and molecules of which it is composed are in motion. It possesses potential energy due to the positions which atoms and molecules occupy relative to one another; that is to the chemical bonds. During the course of a chemical reaction, one type of matter changes into another, and energy is either given out or taken in from the surroundings (see Figure 12.1). A reaction in which energy is given out is termed an **exothermic reaction**; a reaction in which energy is taken in is termed an **endothermic reaction**.

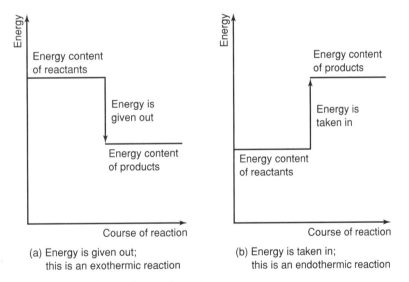

(a) Energy is given out;
   this is an exothermic reaction

(b) Energy is taken in;
   this is an endothermic reaction

Fig. 12.1 *Energy changes during chemical reactions*

The heat absorbed during a reaction is equal to the internal energy of the products minus the internal energy of the reactants plus any work done by the system on the surroundings. Since most laboratory work is done at constant pressure, any gases formed are allowed to escape into the atmosphere and work is done in expansion:

$$\begin{pmatrix} \text{Heat absorbed at} \\ \text{constant pressure} \end{pmatrix} = \begin{pmatrix} \text{Change in} \\ \text{internal energy} \end{pmatrix} + \begin{pmatrix} \text{Work done on} \\ \text{surroundings} \end{pmatrix}$$

The heat absorbed at constant pressure is given the name **change in enthalpy** and the symbol $\Delta H$.

## 12.2 STANDARD ENTHALPY CHANGES

The enthalpy of a substance is stated for a substance in its **standard state**. The standard state of a substance is 1 mole of the substance in a specified state (solid, liquid or gas) at 1 atmosphere pressure.

The value of an enthalpy change is stated for standard conditions; gases at 1 atmosphere, solutions at unit concentration and substances in their normal states at a stated temperature. $\Delta H_T^{\ominus}$ means the standard enthalpy change at a temperature $T$. $\Delta H_{298}^{\ominus}$ is often written as $\Delta H^{\ominus}$.

Definitions of some standard enthalpy changes follow.

**Standard enthalpy of formation,** $\Delta H_F^{\ominus}$ is the heat absorbed per mole when a substance is formed from its elements in their standard states at constant pressure. If the reaction is exothermic, the heat absorbed is negative, and $\Delta H_F^{\ominus}$ has a negative value.

It follows from this definition that the standard enthalpy of formation of an element in its standard state is zero.

**Standard enthalpy of combustion** $\Delta H_C^{\ominus}$ is the heat absorbed per mole when a substance decomposes into gaseous atoms at constant pressure.

**Standard enthalpy of neutralisation** $\Delta H_N^{\ominus}$ is the heat absorbed per mole of water formed when an acid and a base react at constant pressure to form water. It has a negative value.

**Standard enthalpy of reaction** $\Delta H_R^{\ominus}$ is the heat absorbed in a reaction at constant pressure between the numbers of moles of reactants shown in the equation for the reaction. In the reaction

$$3Fe(s) + 4H_2O(l) \longrightarrow Fe_3O_4(s) + 4H_2(g)$$

the standard enthalpy change refers to the reaction between 3 moles of iron and 4 moles of water.

**Standard enthalpy of solution** is the heat absorbed per mole when a substance is dissolved in a stated quantity of solvent at constant pressure. This may be 100 g or 1000 g of solvent or it may be an 'infinite' amount of solvent, that is, a volume so large that on further dilution there is no further heat change.

## 12.3 STANDARD ENTHALPY CHANGE FOR A CHEMICAL REACTION

It is worthwhile taking a little time to think about the fact that matter contains energy (§ 12.1). It makes calculations on thermochemistry very simple once you have really understood this idea. A substance **A** contains a certain quantity of energy per mole, $a$ kJ mol$^{-1}$. Substances **B**, **C** and **D** contain different quantities of energy per mole. When **A** + **B** are converted into **C** + **D**, the calculation of the change in energy is a simple matter. The first step is to write the energy content of each substance underneath it in the equation:

$$\begin{array}{ccccccc} A & + & B & \longrightarrow & C & + & D \\ a\ kJ\ mol^{-1} & & b\ kJ\ mol^{-1} & & c\ kJ\ mol^{-1} & & d\ kJ\ mol^{-1} \end{array}$$

Total energy of reactants = $(a + b)$ kJ mol$^{-1}$

Total energy of products $= (c + d)$ kJ mol$^{-1}$

Change in energy = Energy of products – Energy of reactants

$$= (c + d) - (a + b) \text{ kJ mol}^{-1}$$

Now consider three statements:

- If the reaction takes place at constant pressure under standard conditions, the change in energy is the change in standard enthalpy.

- The standard enthalpy of a substance is by definition (§ 12.2) its standard enthalpy of formation.

- The change in standard enthalpy is the standard enthalpy of reaction, $\Delta H_R^{\ominus}$.

The equation for change in energy then becomes:

$$\text{Standard enthalpy of reaction} = \begin{pmatrix} \text{Sum of standard enthalpies} \\ \text{of formation of products} \end{pmatrix} - \begin{pmatrix} \text{Sum of standard enthalpies} \\ \text{of formation of reactants} \end{pmatrix}$$

$$\Delta H_R^{\ominus} = \Delta H_F^{\ominus}(\text{products}) - \Delta H_F^{\ominus}(\text{reactants})$$

**EXAMPLE**

Hydrogen chloride adds to ethene to form chloroethane. Calculate the standard enthalpy of this reaction.

**METHOD**

$$CH_2{=}CH_2(g) + HCl(g) \longrightarrow C_2H_5Cl(g)$$

$$(+52.3) \qquad (-92.3) \qquad (-105)$$

The standard enthalpy of formation in kJ mol$^{-1}$ is shown under each substance.
The standard enthalpy of reaction $\Delta H_R^{\ominus}$ is given by

**ANSWER**

$$\Delta H_R^{\ominus} = (-105) - (52.3 + (-92.3)) = -65 \text{ kJ mol}^{-1}$$

**Reminder :** Negative numbers § 1.4.

The negative sign means that the products contain less enthalpy than the reactants.

The difference is the heat given out in the reaction: the reaction is exothermic.

## Hess's Law

The standard enthalpy of reaction depends only on the difference between the total standard enthalpy of the products and the total standard enthalpy of the reactants. It does not depend on the route by which the reaction occurs. This idea is contained in **Hess's law**, which states:

If a reaction can take place by more than one route, the overall change in enthalpy is the same, whichever route is followed.

## EXERCISE 12.1    Problems on standard enthalpy of reaction

**Reminder :** Negative numbers § 1.4.

**1**  Calculate the standard enthalpy change in the reaction

$$PbO(s) + CO(g) \longrightarrow Pb(s) + CO_2(g)$$

The standard enthalpies of formation of lead(II) oxide, carbon monoxide and carbon dioxide are $-219$, $-111$, and $-394$ kJ mol$^{-1}$, respectively.

**Hint :** Put values of $\Delta H_F^{\ominus}$ under all the substances in the equation.

**2**  Calculate the standard enthalpy change for the reaction

$$Fe_2O_3(s) + 2Al(s) \longrightarrow Al_2O_3(s) + 2Fe(s)$$

The standard enthalpies of formation of iron(III) oxide and aluminium oxide are $-822$ and $-1669$ kJ mol$^{-1}$. State whether the reaction is exothermic or endothermic.

**3**  Given the standard enthalpy change of formation of MgO = $-602$ kJ mol$^{-1}$ and of $Al_2O_3 = -1700$ kJ mol$^{-1}$, calculate the standard enthalpy change for the reaction

$$Al_2O_3 + 3Mg \longrightarrow 2Al + 3MgO$$

Does your answer tell you whether magnesium will reduce aluminium oxide?

**4**  The following are standard enthalpies of formation, $\Delta H_F^{\ominus}$, in kJ mol$^{-1}$ at 298 K:

$CH_4(g)$, $-76$; $CO_2(g)$, $-394$; $H_2O(l)$, $-286$; $H_2O(g)$, $-242$;
$NH_3(g)$, $-46.2$; $HNO_3(l)$, $-176$

**a**  Calculate the standard enthalpy change at 298 K for the reaction

$$CH_4(g) + 2O_2(g) \longrightarrow CO_2(g) + 2H_2O(l)$$

**b**  Calculate the standard enthalpy change for the reaction

$$\tfrac{1}{2}N_2(g) + \tfrac{3}{2}H_2O(g) \longrightarrow NH_3(g) + \tfrac{3}{4}O_2(g)$$

**c**  Calculate the standard enthalpy change for the reaction

$$\tfrac{1}{2}N_2(g) + \tfrac{1}{2}H_2O(g) + \tfrac{5}{4}O_2(g) \longrightarrow HNO_3(l)$$

## 12.4 EXPERIMENTAL METHODS

### Heat and temperature

Methods of finding standard enthalpy changes of chemical reactions depend on measuring a change in temperature and calculating the enthalpy change from it. The container used for measuring the change in temperature is called a **calorimeter**.

When an object receives heat its temperature rises. The rise in temperature depends on the **heat capacity** of the object:

$$\text{Heat capacity} = \frac{\text{Heat supplied}}{\text{Rise in temperature}}$$

The unit of heat capacity is $J\,K^{-1}$ (joules per kelvin).

The greater the mass of an object the more heat is needed to raise its temperature.

Heat capacity and mass, $m$, are related to a quantity called **specific heat capacity**, $c$, by the expression:

$$\text{Specific heat capacity} = \frac{\text{Heat capacity}}{\text{Mass}}$$
$$\text{Heat capacity} = m \times c$$

The unit of specific heat capacity is $J\,K^{-1}\,kg^{-1}$ (joules per kelvin per kilogram).

Putting these two expressions together gives

$$\text{Heat supplied} = \text{Mass} \times \text{Specific heat capacity} \times \text{Rise in temperature}$$
$$\text{Heat} = m \times c \times \Delta T$$

### Standard enthalpy of neutralisation

The standard enthalpy of neutralisation can be found by measuring the rise in temperature that occurs when solutions of an acid and a base are mixed. Sometimes a very simple type of calorimeter, consisting of an expanded polystyrene cup and lid, may be used. This material is a good insulator so the temperature rise can be measured before the loss of heat to the surroundings is serious.

**EXAMPLE 1**

250 $cm^3$ of sodium hydroxide solution of concentration 0.400 $mol\,dm^{-3}$ were added to 250 $cm^3$ of hydrochloric acid of concentration 0.400 $mol\,dm^{-3}$ in an expanded polystyrene cup. The temperature of the two solutions was 17.05 °C. The temperature rose to 19.97 °C. Assuming the specific heat capacity of all the solutions is 4200 $J\,kg^{-1}\,K^{-1}$, calculate the standard enthalpy of neutralisation.

**METHOD**

Mass of solutions = 500 g

Heat capacity of solutions = $m \times c$ = 0.500 kg × 4200 $J\,kg^{-1}\,K^{-1}$ = 2100 $J\,K^{-1}$

Rise in temperature = 2.74 °C = 2.74 K

Heat evolved = $m \times c \times \Delta T$ = 2100 $J\,K^{-1}$ × 2.74 K = 5750 J

Amount of water formed $= 250 \times 10^{-3}$ dm$^3 \times 0.400$ mol dm$^{-3} = 0.100$ mol

Heat evolved per mole $= \dfrac{5750 \text{ J}}{0.100 \text{ mol}} = 57\,500$ J mol$^{-1}$

**ANSWER**    Standard enthalpy of neutralisation $= -57.5$ kJ mol$^{-1}$

**EXAMPLE 2**    This time a copper calorimeter is used. Some of the heat liberated goes to raise the temperature of the copper calorimeter.

250 cm$^3$ of sodium hydroxide solution of concentration 0.400 mol dm$^{-3}$ were added to 250 cm$^3$ of hydrochloric acid of concentration 0.400 mol dm$^{-3}$ in a copper calorimeter of mass 500 g and specific heat capacity 400 J kg$^{-1}$ K$^{-1}$. The temperature of the two solutions was 17.05 °C. The temperature rose to 19.55 °C. Assuming the specific heat capacity of all the solutions is 4200 J kg$^{-1}$ K$^{-1}$, calculate the standard enthalpy of neutralisation.

**METHOD**    Mass of solution = 500 g

Heat capacity of solution = Mass of solution × Specific heat capacity of solution
$$= 0.500 \text{ kg} \times 4200 \text{ J kg}^{-1} \text{K}^{-1} = 2100 \text{ J K}^{-1}$$

Heat capacity of copper calorimeter = Mass of calorimeter × Specific heat capacity of copper
$$= 0.500 \text{ kg} \times 400 \text{ J kg}^{-1} \text{K}^{-1} = 200 \text{ J K}^{-1}$$

Rise in temperature = 2.50 °C

Heat evolved = (Heat capacity of solution + Heat capacity of copper) $\times \Delta T$
$$= (2100 + 200) \text{ J K}^{-1} \times 2.50 \text{ K} = 5750 \text{ J}$$

Amount of water formed $= 250 \times 10^{-3}$ dm$^3 \times 0.400$ mol dm$^{-3} = 0.100$ mol

Heat evolved per mole $= \dfrac{5750 \text{ J}}{0.100 \text{ mol}} = 57\,500$ J

**ANSWER**    Standard enthalpy of neutralisation $= -57.5$ kJ mol$^{-1}$

### Enthalpy of combustion

The enthalpy of combustion of a fuel can be found by burning a known mass of it and measuring the rise in temperature of a known mass of water. In order to know the mass of fuel burned, the fuel can be put into a spirit lamp and the lamp weighed before and after the experiment. These are not standard conditions and the result obtained is not the **standard** enthalpy of combustion.

**EXAMPLE 3**    Find the enthalpy of combustion of ethanol given that:

Mass of ethanol burned = 1.650 g

Mass of copper calorimeter = 295 g

Mass of water in copper calorimeter = 500 g

Initial temperature of water = 17.4 °C

Rise in temperature of water = 19.9 °C

Specific heat capacities: water 4200 J kg$^{-1}$ K$^{-1}$; copper 2100 J kg K$^{-1}$.

**METHOD** Heat absorbed by calorimeter = (Heat capacity of water + Heat capacity of calorimeter) $\times \Delta T$

$$= [(0.500 \text{ kg} \times 4200 \text{ J kg}^{-1}\text{ K}^{-1}) + (0.295 \text{ kg} \times 2100 \text{ J kg}^{-1}\text{ K}^{-1})] \times 19.9 \text{ K}$$

$$= 44.05 \text{ kJ}$$

$$\text{Amount of ethanol burned} = \frac{1.650 \text{ g}}{46 \text{ g mol}^{-1}} = 0.0359 \text{ mol}$$

$$\text{Heat given out per mol of ethanol} = \frac{44.05 \text{ kJ}}{0.0359 \text{ mol}} = 1230 \text{ kJ mol}^{-1}$$

**ANSWER** Enthalpy of combustion of ethanol = $-1230$ kJ mol$^{-1}$.

> **Note :** This result is lower than the standard enthalpy of combustion of ethanol = $-1371$ kJ mol$^{-1}$. Errors arise due to heat lost to the surroundings and incomplete combustion of ethanol. As mentioned above, conditions are not standard so the quantity obtained is not the standard enthalpy of combustion.

## EXERCISE 12.2 Problems on neutralisation and combustion

> **Reminder :** Significant figures § 1.7. Units § 1.11. Check your answer § 1.12.
>
>   Amount/mol = Volume × Concentration  Heat = $m \times c \times \Delta T$
>
>   Specific heat of water = 4.2 J K$^{-1}$ g$^{-1}$.

**1** 50.0 cm$^3$ of sodium hydroxide solution of concentration 0.400 mol dm$^{-3}$ required 20.0 cm$^3$ of sulphuric acid of concentration 0.500 mol dm$^{-3}$ for neutralisation. A temperature rise of 3.9 °C was observed if both solutions and the container were initially at the same temperature. Calculate the standard enthalpy of neutralisation of sodium hydroxide with sulphuric acid. (The specific heat capacity of all the solutions is 4.2 J K$^{-1}$ g$^{-1}$.) Assume no heat passes to the container.

**2** 100 cm$^3$ of potassium hydroxide solution of concentration 1.00 mol dm$^{-3}$ and 100 cm$^3$ of hydrochloric acid of concentration 1.00 mol dm$^{-3}$ were mixed in a container. All three were at the same temperature. The rise in temperature was 6.7 K. Calculate the standard enthalpy of neutralisation. (Specific heat capacity of water = 4.2 J K$^{-1}$ g$^{-1}$.) Assume no heat passes to the container.

**3** 100 cm$^3$ of 1.00 mol dm$^{-3}$ sodium hydroxide solution and 100 cm$^3$ of 1.00 mol dm$^{-3}$ ethanoic acid were mixed in a calorimeter. All three were at the same temperature. The rise in temperature was 5.9 K. Calculate the standard enthalpy of neutralisation. Assume no heat passes to the container.

**4** A calorimeter has a mass of 200 g and a specific heat capacity of 0.42 J g$^{-1}$. Into it are put 50 cm$^3$ of 1.25 mol dm$^{-3}$ hydrochloric acid and 50 cm$^3$ of 1.25 mol dm$^{-3}$ potassium hydroxide solution at the same temperature. The temperature of the calorimeter and contents rises by 7.0 °C. Calculate the standard enthalpy of neutralisation.

**Hint :** In this problem, Heat $= m \times c \times \Delta T$

Heat $= [(m$ of solution $\times c$ of solution) $+ (m$ of calorimeter $\times c$ of calorimeter$)] \times \Delta T$

**5** Figure 12.2 shows the results of a thermometric titration to find a value of the standard enthalpy of neutralisation. $50.0 \text{ cm}^3$ of a solution of sodium hydroxide of concentration $0.500 \text{ mol dm}^{-3}$ were titrated against a $0.568 \text{ mol dm}^{-3}$ solution of sulphuric acid. Calculate the standard enthalpy of neutralisation. Assume that the specific heat capacity of the solutions is $4.2 \text{ J K}^{-1} \text{ g}^{-1}$, and assume that no heat passes to the container.

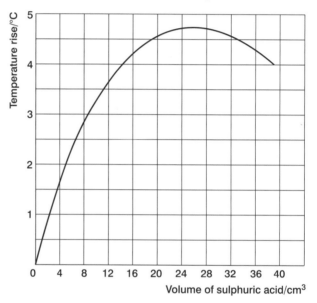

**6** A typical domestic boiler burns $10 \text{ m}^3$ of methane $CH_4$ per day.

   **a** Write the equation for the combustion.

   **b** Calculate the amount (moles) of methane burned per day in the domestic boiler (using GMV $= 25 \text{ dm}^3 \text{ mol}^{-1}$ under these conditions).

   **c** If $\Delta H_C^{\ominus}(CH_4) = -890 \text{ kJ mol}^{-1}$ calculate the heat evolved per day.

**7** When $1.60 \text{ g}$ of ethanol, $C_2H_6O$, are burned enough heat is produced to heat $250 \text{ g}$ of water from $22.0 \,°C$ to $67.0 \,°C$. Calculate the enthalpy of combustion of ethanol. (Specific heat capacity of water $= 4200 \text{ J K}^{-1} \text{ kg}^{-1}$)

**8** Hydrazine is used as a rocket fuel.

$$N_2H_4(l) + O_2(g) \longrightarrow N_2(g) + 2H_2O(g)$$

Values of $\Delta H^{\ominus}_{\text{Formation}}/\text{kJ mol}^{-1}$ are $N_2H_4(l)$ 50.6; $H_2O(g)$ −241.8.

Calculate $\Delta H^{\ominus}$ for the reaction.

**9** You are given the values of $\Delta H^{\ominus}_{\text{Formation}}/\text{kJ mol}^{-1}$: $CO_2(g)$ −394; $H_2O(l)$ −286; $C_4H_{10}(g)$ −125.

Which one is the standard enthalpy change of combustion of butane, $C_4H_{10}$?

**A** $-555$ kJ mol$^{-1}$  **B** $-805$ kJ mol$^{-1}$  **C** $-2881$ kJ mol$^{-1}$  **D** $-3131$ kJ mol$^{-1}$

**Hint :** Start with the balanced equation.

## 12.5 CALCULATING STANDARD ENTHALPY OF FORMATION

Sometimes, the standard enthalpy of formation of a compound can be measured directly by allowing known amounts of elements to combine and measuring the amount of heat evolved. Other reactions are difficult to study, and the standard enthalpy of reaction must be found indirectly.

To find the standard enthalpy of formation of ethyne from practical measurements is impossible as attempts to make ethyne from carbon and hydrogen,

$$2C(s) + H_2(g) \longrightarrow C_2H_2(g)$$

will result in the formation of a mixture of hydrocarbons. The standard enthalpy of combustion of ethyne can, however, be measured experimentally, and from it can be calculated the standard enthalpy of formation. The standard enthalpies of combustion of carbon and hydrogen are also required.

**EXAMPLE 1**

Find the standard enthalpy of formation of ethyne, given the standard enthalpies of combustion (in kJ mol$^{-1}$): $C_2H_2(g) = -1300$; $C(s) = -394$; $H_2(g) = -286$.

**Reminder :** Negative numbers § 1.4.

**METHOD 1**

The method of calculation is based on the three equations for the combustion of ethyne, carbon and hydrogen:

$$C(s) + O_2(g) \longrightarrow CO_2(g); \qquad \Delta H_1^\ominus = -394 \text{ kJ mol}^{-1} \qquad [1]$$
$$H_2(g) + \tfrac{1}{2}O_2(g) \longrightarrow H_2O(l) \qquad \Delta H_2^\ominus = -286 \text{ kJ mol}^{-1} \qquad [2]$$
$$C_2H_2(g) + 2\tfrac{1}{2}O_2(g) \longrightarrow 2CO_2(g) + H_2O(l) \qquad \Delta H_3^\ominus = -1300 \text{ kJ mol}^{-1} \qquad [3]$$

Looking at equation [1] we can see that the standard enthalpy of combustion of carbon is the same as the standard enthalpy of formation of carbon dioxide.

Likewise, equation [2] shows that the standard enthalpy of combustion of hydrogen is the same as the standard enthalpy of formation of water.

The standard enthalpy content of a substance is equal to the standard enthalpy of formation of the substance from its elements in their standard states.

Putting the standard enthalpy content of each substance into equation [3] gives

$$C_2H_2(g) + 2\tfrac{1}{2}O_2(g) \longrightarrow 2CO_2(g) + H_2O(l); \Delta H_3^\ominus = -1300 \text{ kJ mol}^{-1}$$

$$\Delta H_F^\ominus(C_2H_2) \quad 0 \qquad\qquad 2(-394) \quad (-286)$$

Since

$$\left(\begin{array}{c}\text{Standard}\\\text{enthalpy change}\end{array}\right) = \left(\begin{array}{c}\text{Standard}\\\text{enthalpy content}\\\text{of products}\end{array}\right) - \left(\begin{array}{c}\text{Standard}\\\text{enthalpy content}\\\text{of reactants}\end{array}\right)$$

$$\Delta H_3^{\ominus} = -1300 = 2(-394) + (-286) - \Delta H_F^{\ominus}(C_2H_2)$$

$$\Delta H_F^{\ominus}(C_2H_2) = 226 \text{ kJ mol}^{-1}$$

**ANSWER**

The standard enthalpy of formation of ethyne is 226 kJ mol$^{-1}$. Since $\Delta H_F^{\ominus}$ is positive, ethyne is referred to as an endothermic compound.

**METHOD 2**

Another method of tackling the problem is to construct an enthalpy diagram:

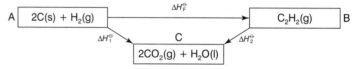

$\Delta H^{\ominus}$ for route A to C = $\Delta H_1^{\ominus}$

$\Delta H^{\ominus}$ for route A to B to C = $\Delta H_F^{\ominus} + \Delta H_2^{\ominus}$

According to Hess's Law, $\Delta H_1^{\ominus} = \Delta H_F^{\ominus} + \Delta H_2^{\ominus}$

Putting

$$\Delta H_1^{\ominus} = 2(\Delta H^{\ominus} \text{ for combustion of C}) + (\Delta H^{\ominus} \text{ for combustion of H}_2) \text{ gives}$$

$$\Delta H_1^{\ominus} = 2(-394) + (-286) = -1074 \text{ kJ mol}^{-1}$$

$$\Delta H_F^{\ominus} = \Delta H_1^{\ominus} - \Delta H_2^{\ominus} = -1074 - (-1300) \text{ kJ mol}^{-1}$$

**ANSWER**

$$\Delta H_F^{\ominus} = + 226 \text{ kJ mol}^{-1} \text{ (as before)}$$

**EXAMPLE 2**

Calculate the standard enthalpy of formation of propan-1-ol, given the standard enthalpies of combustion, in kJ mol$^{-1}$: $C_3H_7OH(l)$ –2010; C(s) –394; $H_2$(g) –286.

**METHOD 1**

Again, as the equation for combustion is the basis for the calculation, it must be carefully balanced:

$$C_3H_7OH(l) + 4\tfrac{1}{2}O_2(g) \longrightarrow 3CO_2(g) + 4H_2O(1); \qquad \Delta H^{\ominus} = -2010 \text{ kJ mol}^{-1}$$

Putting the standard enthalpies of formation of $CO_2$(g) and $H_2O$(l) into the equation, as in Example 1, gives

$$C_3H_7OH(l) + 4\tfrac{1}{2}O_2(g) \longrightarrow 3CO_2(g) + 4H_2O(l); \qquad \Delta H^{\ominus} = -2010 \text{ kJ mol}^{-1}$$

$$\Delta H_F^{\ominus}(C_3H_7OH) \quad 0 \qquad\qquad 3(-394) \quad 4(-286)$$

Since

$$\left(\begin{array}{c}\text{Standard}\\\text{enthalpy change}\\\text{for reaction}\end{array}\right) = \left(\begin{array}{c}\text{Standard}\\\text{enthalpy content}\\\text{of products}\end{array}\right) - \left(\begin{array}{c}\text{Standard}\\\text{enthalpy content}\\\text{of reactants}\end{array}\right)$$

$$-2010 = 3(-394) + 4(-286) - \Delta H_F^{\ominus}(C_3H_7OH(l))$$

$$\Delta H_F^{\ominus}(C_3H_7OH(l)) = -316 \text{ kJ mol}^{-1}$$

**ANSWER**

The standard enthalpy of formation of liquid propan-1-ol is –316 kJ mol$^{-1}$.

**METHOD 2**    The enthalpy diagram for the formation of propanol is

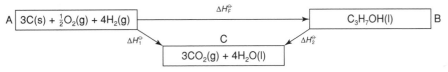

$$\Delta H_1^\ominus = 3(\Delta H^\ominus \text{ for combustion of C}) + 4(\Delta H^\ominus \text{ for combustion of H}_2)$$
$$= 3(-394) + 4(-286) = -2326 \text{ kJ mol}^{-1}$$

According to Hess's law,

$$\Delta H_1^\ominus = \Delta H_F^\ominus + \Delta H_2^\ominus$$
$$\Delta H_F^\ominus = \Delta H_1^\ominus - \Delta H_2^\ominus$$
$$\Delta H_F^\ominus = -2326 - (-2010) \text{ kJ mol}^{-1}$$

**ANSWER**    $$\Delta H_F^\ominus = -316 \text{ kJ mol}^{-1} \text{ (as before)}$$

You will have noticed in both Examples 1 and 2 that

$$\begin{pmatrix} \text{Standard enthalpy} \\ \text{of reaction} \end{pmatrix} = \begin{pmatrix} \text{Sum of standard} \\ \text{enthalpies of} \\ \text{combustion of} \\ \text{reactants} \end{pmatrix} - \begin{pmatrix} \text{Sum of standard} \\ \text{enthalpies of} \\ \text{combustion of} \\ \text{products} \end{pmatrix}$$

## 12.6 STANDARD ENTHALPY OF REACTION FROM STANDARD ENTHALPIES OF FORMATION

The standard enthalpies of formation of the reactants and products can be used to give the standard enthalpy of a reaction.

**EXAMPLE 1**    Calculate the standard enthalpy of the reaction

$$CH_2 = CH_2(g) + H_2(g) \longrightarrow CH_3 CH_3(g)$$

given that the standard enthalpies of formation are: ethene +52, ethane −85 kJ mol$^{-1}$.

**METHOD**    Put the standard enthalpy content of each species into the equation (units kJ mol$^{-1}$):

$$CH_2 = CH_2(g) + H_2(g) \longrightarrow CH_3 CH_3(g)$$
$$+52 \qquad\qquad 0 \qquad\qquad -85$$

$$\begin{pmatrix} \text{Standard enthalpy} \\ \text{of reaction} \end{pmatrix} = \begin{pmatrix} \text{Standard enthalpy} \\ \text{of product} \end{pmatrix} - \begin{pmatrix} \text{Standard enthalpy} \\ \text{of reactants} \end{pmatrix}$$
$$= -85 - (52 + 0) = -137 \text{ kJ mol}^{-1}$$

**ANSWER**    Standard enthalpy of reaction $= -137$ kJ mol$^{-1}$

The method of calculation is simply:

$$\begin{pmatrix} \text{Standard enthalpy} \\ \text{of reaction} \end{pmatrix} = \begin{pmatrix} \text{Sum of standard} \\ \text{enthalpies of} \\ \text{formation of} \\ \text{products} \end{pmatrix} - \begin{pmatrix} \text{Sum of standard} \\ \text{enthalpies of} \\ \text{formation of} \\ \text{reactants} \end{pmatrix}$$

**EXAMPLE 2**

Calculate the standard enthalpy change in the reaction

$$SO_2(g) + 2H_2S(g) \longrightarrow 3S(s) + 2H_2O(l)$$

The standard enthalpy of combustion of sulphur is $-297$ kJ mol$^{-1}$, and the standard enthalpies of formation of hydrogen sulphide and water are $-20.2$ kJ mol$^{-1}$ and $-286$ kJ mol$^{-1}$.

**METHOD**

This problem is tackled by putting the standard enthalpies of formation of each species into the equation (units kJ mol$^{-1}$):

$$SO_2(g) + 2H_2S(g) \longrightarrow 3S(s) + 2H_2O(l)$$
$$(-297) + 2(-20.2) \qquad 3 \times 0 \quad 2(-286)$$

$$\begin{pmatrix} \text{Standard enthalpy} \\ \text{of reaction} \end{pmatrix} = \begin{pmatrix} \text{Standard enthalpy} \\ \text{of products} \end{pmatrix} - \begin{pmatrix} \text{Standard enthalpy} \\ \text{of reactants} \end{pmatrix}$$

$$= -572 + 297 + 40.4 \text{ kJ mol}^{-1}$$

**ANSWER**

Standard enthalpy change $= -235$ kJ (mol of the equation)$^{-1}$

**EXERCISE 12.3**  **Problems on standard enthalpy of formation**

---

**Reminder** : Negative numbers § 1.4.

---

**1** Given the standard enthalpy changes for the reactions:

$$H_2(g) \longrightarrow 2H(g); \quad \Delta H^\ominus = 436 \text{ kJ mol}^{-1}$$
$$Br_2(g) \longrightarrow 2Br(g); \quad \Delta H^\ominus = 193 \text{ kJ mol}^{-1}$$
$$H_2(g) + Br_2(g) \longrightarrow 2HBr(g); \quad \Delta H^\ominus = -104 \text{ kJ mol}^{-1}$$

calculate the standard enthalpy change for the reaction

$$H(g) + Br(g) \longrightarrow HBr(g)$$

---

**Hint :** Put the value of $\Delta H_F^\ominus$ under each substance in the equation.

---

**2** Standard enthalpy changes are given below for:

**a** the hydrogenation of cyclohexene $C_6H_{10}$ to cyclohexane $C_6H_{12}$

$$C_6H_{10}(l) + H_2(g) \longrightarrow C_6H_{12}(l); \quad \Delta H^\ominus = -120 \text{ kJ mol}^{-1}$$

**b** the hydrogenation of benzene $C_6H_6$ to cyclohexadiene, $C_6H_8$

$$C_6H_6(l) + H_2(g) \longrightarrow C_6H_8(l); \quad \Delta H^\ominus = +31 \text{ kJ mol}^{-1}$$

**c** the hydrogenation of benzene to cyclohexane $C_6H_{12}$

$$C_6H_6(l) + 3H_2(g) \longrightarrow +C_6H_{12}(l); \quad \Delta H^\ominus = -208 \text{ kJ mol}^{-1}$$

Comment on the values of $\Delta H^\ominus$.

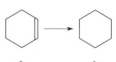

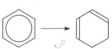

**3**  Ethanol is made by the catalytic hydration of ethene

$$C_2H_4(g) + H_2O(g) \longrightarrow C_2H_5OH(g); \quad \Delta H^\ominus = 144 \text{ kJ mol}^{-1}$$

Given that $\Delta H_F^\ominus(C_2H_4(g)) = +52 \text{ kJ mol}^{-1}$ and $\Delta H_F^\ominus(H_2O(g)) = -242 \text{ kJ mol}^{-1}$, calculate $\Delta H_F^\ominus$ of gaseous ethanol.

**Hint :** Again, it is a matter of putting the value of $\Delta H_F^\ominus$ under each substance in the equation.

**4**  Calculate the standard enthalpy change for the reaction

Anhydrous copper(II) sulphate + Water $\longrightarrow$ Copper(II) sulphate-5-water

Use the values for the standard enthalpy of solution: $CuSO_4(s)$ –66.5 kJ mol$^{-1}$; $CuSO_4 . 5H_2O(s)$ 11.7 kJ mol$^{-1}$.

**Hint :** Draw an enthalpy triangle

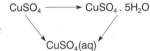

**5**  You are given the enthalpy changes for two processes.

$$KI(s) \longrightarrow K^+(g) + I^-(g); \quad \Delta H^\ominus = +629 \text{ kJ mol}^{-1}$$
$$K^+(g) + I^-(g) + aq \longrightarrow KI(aq); \quad \Delta H^\ominus = -615 \text{ kJ mol}^{-1}$$

Calculate the enthalpy change for $KI(s) + aq \longrightarrow KI(aq)$

**Hint :** Remember Hess's Law.

**6**  Calculate the standard enthalpy of formation of **a** sulphur dioxide **b** carbon dioxide. On burning in an excess of oxygen under standard conditions 1.00 g of sulphur evolves 9.28 kJ and 1.00 g of carbon evolves 32.8 kJ.

**Hint :** Write the equations and you can't go wrong.

**7**  Calculate the standard enthalpy change for the conversion:

Monoclinic sulphur $\longrightarrow$ Rhombic sulphur

Standard enthalpies of combustion are: rhombic sulphur –296.9 kJ mol$^{-1}$; monoclinic sulphur –297.2 kJ mol$^{-1}$.

**Hint :** Draw an enthalpy triangle

**8** Calculate $\Delta H^{\ominus}_{Formation}$ of ethane from the data:

$\Delta H^{\ominus}_{Combustion}$(ethane) = −1560 kJ mol$^{-1}$,   $\Delta H^{\ominus}_{Formation}$(CO$_2$) = −394 kJ mol$^{-1}$,

$\Delta H^{\ominus}_{Formation}$(H$_2$O) = −286 kJ mol$^{-1}$

---

**Hint :** Write the equation C$_2$H$_6$(g) + O$_2$(g) $\longrightarrow$

Then balance the equation and put $\Delta H^{\ominus}_F$ under each substance.

---

**9** The problem is to calculate the standard enthalpy of formation of ethene. Ethene is reduced to ethane:

$$CH_2\!=\!CH_2(g) + H_2(g) \longrightarrow CH_3CH_3(g); \quad \Delta H^{\ominus} = -138 \text{ kJ mol}^{-1}$$

Standard enthalpies of formation/kJ mol$^{-1}$ are: CO$_2$(g) – 394; H$_2$O(g) – 242.

Standard enthalpy of combustion of ethane  = −1560 kJ mol$^{-1}$.

**10** The standard enthalpies of combustion of graphite, hydrogen and ethanol are −393, −286 and −1367 kJ mol$^{-1}$ respectively. Calculate the standard enthalpy of formation of ethanol. Is it

**A** −139 kJ mol$^{-1}$   **B** −277 kJ mol$^{-1}$   **C** +277 kJ mol$^{-1}$   **D** +554 kJ mol$^{-1}$

---

**Hint :** Write the equation for the formation of ethanol, C$_2$H$_5$OH.

Then construct an enthalpy diagram for the oxidation to CO$_2$ + H$_2$O.

---

**11** Given the following values, calculate $\Delta H^{\ominus}$ for the formation of butane.

$\Delta H^{\ominus}_{Combustion}$/kJ mol$^{-1}$: butane, C$_4$H$_{10}$ – 2876.5; H$_2$ – 285.8; C – 393.5.

---

**Hint :** Start with the equation for the formation of butane. Then draw an enthalpy triangle for the combustion of the different substances.

---

**12** Calculate $\Delta H^{\ominus}_{Combustion}$ of ethanol from the following data.

$\Delta H^{\ominus}_{Formation}$/kJ mol$^{-1}$ C$_2$H$_5$OH(l) – 277; CO$_2$(g) – 394; H$_2$O(l) – 286.

**13** The following are standard enthalpies of combustion at 298 K, in kJ mol$^{-1}$:

| | | | | | |
|---|---|---|---|---|---|
| C(graphite) | −394 | C$_2$H$_6$(g) | −1561 | C$_4$H$_{10}$(l) | −3510 |
| H$_2$(g) | −286 | CH$_2$=CH$_2$(g) | −1393 | CH≡CH(g) | −1299 |
| CH$_3$CO$_2$H(l) | −876 | C$_2$H$_5$OH(l) | −1400 | CH$_3$OH(l) | −715 |
| C$_4$H$_6$(g) | −2542 | CH$_3$OCH$_3$(g) | −1455 | C$_2$H$_5$OH(g) | −1444 |
| CH$_4$(g) | −891 | C$_3$H$_8$(g) | −2220 | | |
| CH$_3$CO$_2$C$_2$H$_5$(l) | −2246 | C$_6$H$_{12}$(l) | – 3924 | | |

**a** Calculate the standard enthalpy change for the reaction:

$$2C(\text{graphite}) + 2H_2(g) + O_2(g) \longrightarrow CH_3CO_2H(l)$$

**b** Calculate the standard enthalpy change of formation of buta-1,3-diene, $C_4H_6(g)$.

**c** Calculate the standard enthalpy of formation of methane, $CH_4(g)$ and of ethene, $CH_2{=}CH_2(g)$.

**d** Calculate the standard enthalpy change in the hydrogenation of ethene(g) to ethane(g).

**e** Calculate the standard enthalpy change for the theoretical reaction:

$$CH_3OCH_3(g) \longrightarrow C_2H_5OH(g)$$

## 12.7 AVERAGE STANDARD BOND ENTHALPY

In the dissociation of methane,

$$CH_4(g) \longrightarrow C(g) + 4H(g); \quad \Delta H^{\ominus} = +1662 \text{ kJ mol}^{-1}$$

$\Delta H^{\ominus}$ is the standard enthalpy of atomisation of methane. Dividing the standard enthalpy of atomisation between four bonds gives an average value for the C—H bond of 415.5 kJ mol$^{-1}$. In other compounds the C—H bond has slightly different values. The average value in a range of compounds is 412 kJ mol$^{-1}$. This value is called **the average standard bond enthalpy** of the C—H bond. Since standard bond enthalpies vary somewhat from one compound to another, the use of standard bond enthalpies gives only approximate values for standard enthalpies of reaction calculated from them.

The average standard bond enthalpy is often called the **bond energy term**. We can say that the bond energy term for the C—H bond is 412 kJ mol$^{-1}$.

- The sum of all the bond energy terms for a compound is the standard enthalpy of atomisation of that compound *in the gaseous state*. It is not the same as the standard enthalpy of formation.

- The standard enthalpy of formation of a compound is made up of two term: the sum of the bond energy terms + the standard enthalpies of atomisation of all the atoms.

### Standard enthalpy of reaction from average standard bond enthalpies

Mean standard bond enthalpies can be used to give an approximate estimate of the standard enthalpy change which occurs in a reaction. During a reaction, energy is supplied to break the bonds in the reactants, and energy is given out when the bonds in the products form. The difference between the sum of the standard bond enthalpies of the products and the standard bond enthalpies of the reactants is the standard enthalpy of the reaction. The value obtained is less reliable than an experimental measurement.

**EXAMPLE 1**

Calculate the standard enthalpy of the reaction

$$CH_2{=}CH_2(g) + H_2(g) \longrightarrow CH_3CH_3(g)$$

Mean standard bond enthalpies are (in kJ mol$^{-1}$): C—H 416; C=C 612; C—C 348; H—H 436.

**METHOD**          Bonds broken are:

one C=C bond, of standard enthalpy = 612 kJ mol$^{-1}$

one H—H bond, of standard enthalpy = 436 kJ mol$^{-1}$

Total enthalpy absorbed = 1048 kJ mol$^{-1}$

Bonds created are:

one C—C bond, of standard enthalpy = 348 kJ mol$^{-1}$

two C—H bonds, of standard enthalpy = 832 kJ mol$^{-1}$

Total enthalpy released = −1180 kJ mol$^{-1}$

**ANSWER**          Standard enthalpy of reaction = −1180 + 1048 = −132 kJ mol$^{-1}$

**EXAMPLE 2**          Benzene has a standard enthalpy of formation of 83 kJ mol$^{-1}$. Calculate the standard enthalpy of formation from the following data:

Mean standard bond enthalpies are: (C—C) = 348; (C=C) = 615; (C—H) = 412 kJ mol$^{-1}$.

$\Delta H^{\ominus}$ for vaporisation of carbon = 715 kJ mol$^{-1}$

$\Delta H^{\ominus}$ for atomisation of hydrogen (per mole of H atoms) = 217.5 kJ mol$^{-1}$

Compare the experimental value and the theoretical value for $\Delta H_{\mathrm{F}}^{\ominus}$.

**METHOD**          Enthalpy is absorbed in atomising carbon and hydrogen.

Standard enthalpy absorbed = $(6 \times 715) + (6 \times 217.5)$
$$= 5595 \text{ kJ mol}^{-1}$$

Enthalpy is released when bonds are formed.

Standard enthalpy released = 6(C—H) = −2472 kJ mol$^{-1}$
+3(C—C) = −1044 kJ mol$^{-1}$
+3(C=C) = −1845 kJ mol$^{-1}$
Total = −5361 kJ mol$^{-1}$

**ANSWER**          $\Delta H_{\mathrm{F}}^{\ominus} = +5595 - 5361 = 234$ kJ mol$^{-1}$.

The calculated value for the standard enthalpy of formation is higher than the experimental value: the benzene molecule is more stable than it is calculated to be. The difference is the value of the energy of electron delocalisation, 151 kJ mol$^{-1}$.

**EXERCISE 12.4**  **Problems on bond energy terms**

**Reminder :** Rearranging equations § 1.1. Negative numbers § 1.4.

Estimate your answer § 1.8.

**1**  Some average standard bond enthalpies and standard enthalpies of atomisation are listed below.

| | | | | | |
|---|---|---|---|---|---|
| C—C | 348 | C=O | 743 | C(graphite) | 718 |
| C=C | 612 | H—Cl | 432 | $\frac{1}{2}H_2(g)$ | 218 |
| C≡C | 837 | C—Cl | 338 | $\frac{1}{2}O_2(g)$ | 248 |
| C—H | 412 | C—Br | 276 | $\frac{1}{2}Br_2(g)$ | 96.5 |
| C—O | 360 | H—Br | 366 | $\frac{1}{2}Cl_2(g)$ | 121 |
| H—O | 463 | | | | |

**a**  Calculate the enthalpy change for the reaction

$$CH_3CH{=}CH_2 + H_2 \longrightarrow CH_3CH_2CH_3$$

Is it

**A**  $-560\ kJ\ mol^{-1}$   **B**  $-133\ kJ\ mol^{-1}$   **C**  $+133\ kJ\ mol^{-1}$   **D**  $+289\ kJ\ mol^{-1}$

**Hint :** A: Sum of average standard bond enthalpies of bonds broken = positive value

B: Sum of standard bond enthalpies of bonds made = negative value

Sum of A + B = Standard enthalpy of reaction

**b**  Say which one of the following is the standard enthalpy change for the reaction

$$CH_4 + 2O_2(g) \longrightarrow 2CO_2(g) + 2H_2O(l)$$

**A**  $+2168\ kJ\ mol^{-1}$   **B**  $-636\ kJ\ mol^{-1}$   **C**  $-1242\ kJ\ mol^{-1}$   **D**  $-2168\ kJ\ mol^{-1}$

**c**  Calculate the standard enthalpy change for the reaction

$$C_2H_4 + HBr \longrightarrow C_2H_5Br$$

Is it

**A**  $+478\ kJ\ mol^{-1}$   **B**  $+62\ kJ\ mol^{-1}$   **C**  $-62\ kJ\ mol^{-1}$   **D**  $-338\ kJ\ mol^{-1}$

**d**  Calculate the standard enthalpy of formation of ethane and of ethene.

**e**  Find the standard enthalpy change for the reaction,

$$CH_2{=}CH{-}CH_3(g) + Br_2(g) \longrightarrow CH_2BrCHBrCH_3(g)$$

**f**  Find the standard enthalpy of formation of methoxymethane, $CH_3OCH_3(g)$.

**g**  Calculate the standard enthalpy of formation of gaseous ethyl ethanoate, $CH_3CO_2C_2H_5(g)$.

**h** Calculate the standard enthalpy of formation of benzene, assuming its structure is

Explain the difference between the value you have calculated and the value of $83 \text{ kJ mol}^{-1}$ obtained from measurements of the standard enthalpy of combustion.

**i** Find the standard enthalpy of formation of gaseous buta-1,3-diene, $CH_2{=}CH{-}CH{=}CH_2(g)$. How does this value compare with the value you obtained in part **b** from the standard enthalpy of combustion? How do you explain the difference?

**j** Estimate the standard enthalpy changes for the reactions:

(i) $Cl. + CH_4 \longrightarrow CH_3Cl + H.$

(ii) $Cl. + CH_4 \longrightarrow CH_3. + HCl$

Which of the two reactions will occur more readily?

## 12.8 THE BORN–HABER CYCLE

The Born–Haber cycle is a technique for applying Hess's law to the standard enthalpy changes which occur when an ionic compound is formed. Consider the reaction between sodium and chlorine to form sodium chloride. The steps which are involved in this reaction are:

**a** Vaporisation of sodium

$Na(s) \longrightarrow Na(g); \quad \Delta H_S^\ominus = $ standard enthalpy of sublimation

**b** Ionisation of sodium

$Na(g) \longrightarrow Na^+(g) + e^-; \quad \Delta H_I^\ominus = $ ionisation energy of sodium

**c** Dissociation of chlorine molecules

$\frac{1}{2}Cl_2(g) \longrightarrow Cl(g); \quad \Delta H_D^\ominus = $ standard enthalpy of atomisation of chlorine

**d** Ionisation of chlorine atoms

$Cl(g) + e^- \longrightarrow Cl^-(g); \quad \Delta H_E^\ominus = $ electron affinity of chlorine

**e** Reaction between ions

$Na^+(g) + Cl^-(g) \longrightarrow NaCl(s); \quad \Delta H_L^\ominus = $ standard lattice enthalpy

Definitions of the standard enthalpies used above are:

The **standard enthalpy of sublimation** is the enthalpy absorbed when one mole of sodium atoms are vaporised.

The **ionisation energy** of sodium is the enthalpy required to remove a mole of electrons from a mole of sodium atoms in the gas phase.

The **standard enthalpy of atomisation** of chlorine is the enthalpy required to dissociate $\frac{1}{2}$ mole of chlorine molecules into atoms.

The **electron affinity** of chlorine is the enthalpy absorbed when one mole of chlorine atoms form chloride ions. It has a negative value, showing that this reaction is exothermic.

The **standard lattice enthalpy** is the enthalpy absorbed when one mole of gaseous sodium ions and one mole of gaseous chloride ions form one mole of crystalline sodium chloride. It has a negative value.

The steps in the Born–Haber cycle are represented as going upwards if they absorb energy and downwards if they give out energy (see Figure 12.3).

According to Hess's law, the standard enthalpy of formation of sodium chloride is equal to the sum of the enthalpy changes in the various steps

$$\Delta H_F^{\ominus} = \Delta H_S^{\ominus} + \tfrac{1}{2}\Delta H_D^{\ominus} + \Delta H_I^{\ominus} + \Delta H_E^{\ominus} + \Delta H_L^{\ominus}$$
$$= +109 + 121 + 494 - 380 - 755 = -411 \text{ kJ mol}^{-1}$$

In practice, it is easier to measure standard enthalpies of formation than to measure some of the other steps. The electron affinity is the hardest term to measure experimentally, and the Born–Haber cycle is often used to calculate electron affinities.

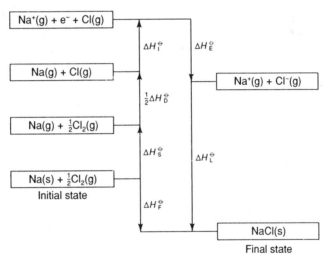

Fig. 12.3   *Born–Haber cycle for sodium chloride*

**EXAMPLE**

Use the data to calculate the electron affinity of chlorine.

Standard enthalpy of sublimation of potassium = 90 kJ mol$^{-1}$

Standard enthalpy of ionisation of potassium = 420 kJ mol$^{-1}$

Standard enthalpy of dissociation of chlorine = 244 kJ mol$^{-1}$

Standard lattice enthalpy of potassium chloride = –706 kJ mol$^{-1}$

Standard enthalpy of formation of potassium chloride = –436 kJ mol$^{-1}$

**METHOD**

Insert values/kJ mol$^{-1}$ for the quantities shown in Figure 12.3.

$$\Delta H_S^{\ominus} + \Delta H_I^{\ominus} + \tfrac{1}{2}\Delta H_D^{\ominus} + \Delta H_E^{\ominus} + \Delta H_L^{\ominus} = \Delta H_F^{\ominus}$$
$$+90 + 420 + 122 + \Delta H_E^{\ominus} - 706 = -436$$
$$\Delta H_E^{\ominus} = -362 \text{ kJ mol}^{-1}$$

**ANSWER**

Electron affinity of chlorine = –362 kJ mol$^{-1}$

## 12.9 STANDARD ENTHALPY OF SOLUTION

When an ionic compound dissolves, energy must be supplied to separate the ions. Under standard conditions, the amount of energy is equal to the standard lattice enthalpy. When the ions dissolve they are hydrated by water molecules, and energy is released. This energy is the standard enthalpy of hydration. If the standard enthalpy of hydration is greater than the standard lattice enthalpy there is a net release of energy, and this favours dissolution.

**EXAMPLE**

$NaF(s) \longrightarrow Na^+(g) + F^-(g); \quad \Delta H^{\ominus} = +925 \text{ kJ mol}^{-1}$

Values of $\Delta H^{\ominus}_{\text{Hydration}}$ are $Na^+(g) = -405 \text{ kJ mol}^{-1}$; $F^-(g) = -506 \text{ kJ mol}^{-1}$. Use these values and $\Delta H^{\ominus}$ to calculate $\Delta H^{\ominus}_{\text{Solution}}$ of NaF.

**METHOD**

Standard enthalpy required to dissociate the lattice = $+925 \text{ kJ mol}^{-1}$

Standard enthalpy released as ions are hydrated = $-405 - 506 \text{ kJ mol}^{-1} = -911 \text{ kJ mol}^{-1}$

Standard enthalpy of solution = $+925 - 911 \text{ kJ mol}^{-1} = +14 \text{ kJ mol}^{-1}$

**EXERCISE 12.5**  **Problems on the Born–Haber cycle and standard enthalpy of solution**

1  Use the data below to calculate the electron affinity of chlorine:

Standard enthalpy of formation of rubidium chloride = $-431 \text{ kJ mol}^{-1}$

Lattice energy of rubidium chloride = $-675 \text{ kJ mol}^{-1}$

First ionisation energy of rubidium = $+408 \text{ kJ mol}^{-1}$

Standard enthalpy of atomisation of rubidium = $+86 \text{ kJ mol}^{-1}$

Bond dissociation enthalpy of molecular chlorine = $+242 \text{ kJ mol}^{-1}$

2  From a Born–Haber cycle calculation, it can be estimated that the standard enthalpy of formation of magnesium(I) chloride, MgCl, would be $-130 \text{ kJ mol}^{-1}$. The standard enthalpy of formation of magnesium(II) chloride $MgCl_2$, is $-640 \text{ kJ mol}^{-1}$.

a  Why do you think that $MgCl_2$ is formed, and not MgCl, when magnesium reacts with chlorine?

b  Calculate the standard enthalpy change in the theoretical reaction
$$2MgCl(s) \longrightarrow Mg(s) + MgCl_2(s)$$

3  Calculate the lattice energy of sodium chloride from the following data:

| | $\Delta H^{\ominus}/\text{kJ mol}^{-1}$ |
|---|---|
| $Na(s) \longrightarrow Na(g)$ | +109 |
| $Na(g) \longrightarrow Na^+(g) + e^-$ | +494 |
| $Cl_2(g) \longrightarrow 2Cl(g)$ | +242 |
| $Cl(g) + e^- \longrightarrow Cl^-(g)$ | −360 |
| $Na(s) + \frac{1}{2}Cl_2(s) \longrightarrow NaCl(s)$ | −411 |

**4**  **a**  Data for the Born–Haber cycle for the formation of calcium chloride are:

$Ca(s) \longrightarrow Ca(g)$ $\quad\quad\quad\quad\quad$ $\Delta H^{\ominus} = +190 \text{ kJ mol}^{-1}$

$Ca(g) \longrightarrow Ca^{2+}(g) + 2e^-$ $\quad\quad$ $\Delta H^{\ominus} = +1730 \text{ kJ mol}^{-1}$

$\frac{1}{2}Cl_2(g) \longrightarrow Cl(g)$ $\quad\quad\quad\quad$ $\Delta H^{\ominus} = +121 \text{ kJ mol}^{-1}$

$Ca^{2+}(g) + 2Cl^-(g) \longrightarrow CaCl_2(s)$ $\quad$ $\Delta H^{\ominus} = -2184 \text{ kJ mol}^{-1}$

$Ca(s) + Cl_2(g) \longrightarrow CaCl_2(s)$ $\quad\quad$ $\Delta H^{\ominus} = -795 \text{ kJ mol}^{-1}$

Calculate the electron affinity of chlorine.

**b**  For the reactions

$Ca(g) \longrightarrow Ca^+(g) + e^-$ $\quad\quad\quad$ $\Delta H^{\ominus} = +590 \text{ kJ mol}^{-1}$

$Ca^+(g) + Cl^-(g) \longrightarrow CaCl(s)$ $\quad$ $\Delta H^{\ominus} = -760 \text{ kJ mol}^{-1}$

use these standard enthalpy changes and those given in **a** to calculate the standard enthalpy of formation of CaCl(s). Why do you think $CaCl_2$ is formed in preference to CaCl?

**5**  The lattice enthalpy of potassium iodide is $+645 \text{ kJ mol}^{-1}$. The enthalpy of dissolution of potassium iodide in water is $+22 \text{ kJ mol}^{-1}$. What is the enthalpy of solvation (in $\text{kJ mol}^{-1}$) of potassium iodide?

**A** +667    **B** +623    **C** −623    **D** −667

**6**  For sodium chloride:

Standard lattice dissociation enthalpy = $+788 \text{ kJ mol}^{-1}$

Standard enthalpy of solvation = $-784 \text{ kJ mol}^{-1}$.

Calculate the standard enthalpy of dissolution.

**7**  The values below (in $\text{kJ mol}^{-1}$) relate to the solubility of lithium chloride, sodium chloride and sodium fluoride:

| | LiCl | NaCl | NaF |
|---|---|---|---|
| Standard lattice enthalpy | −843 | −775 | −968 |
| Sum of standard hydration enthalpies of the separate ions | −883 | −778 | −965 |

What can you predict from these values for the standard enthalpy changes about the relative solubilities of **a** LiCl and NaCl **b** NaF and NaCl? Explain your answer.

## 12.10 FREE ENERGY AND ENTROPY

Some reactions which happen spontaneously are endothermic. The difference in enthalpy between the products and the reactants cannot be the only factor which decides whether a chemical reaction takes place. There must be an additional factor involved. It is often observed that reactions which occur spontaneously increase the randomness or disorder of the system. For example, when an ionic solid dissolves, it passes from the regular arrangement of a crystalline lattice to a random solution of ions. This is termed an increase in **entropy** of the system. The two factors combine to give the change in the **free energy** of the system:

Free energy     $G$ = Enthalpy $H$ − Temperature/K × Entropy $S$

$$G = H - TS$$

It follows that $\Delta G = \Delta H - T\Delta S$

For a physical or a chemical change to occur, $\Delta G$ for that change must be negative. The change is therefore assisted by a decrease in enthalpy ($\Delta H$ negative) and by an increase in entropy ($\Delta S$ positive).

### Calculation of change in standard entropy

The standard entropy change of a process is given by:

$$\begin{pmatrix} \text{Standard} \\ \text{entropy change} \end{pmatrix} = \begin{pmatrix} \text{Sum of standard} \\ \text{entropies of products} \end{pmatrix} - \begin{pmatrix} \text{Sum of standard} \\ \text{entropies of reactants} \end{pmatrix}$$

**EXAMPLE 1**

Calculate the standard entropy change for the reaction of chlorine and ethene, given the values (in $J\,K^{-1}\,mol^{-1}$):

$S^{\ominus}(Cl_2(g)) = 223$; $S^{\ominus}(CH_2{=}CH_2(g)) = 219$; $S^{\ominus}(CH_2ClCH_2Cl(l)) = 208$.

**METHOD**

The equation for the reaction is

$$CH_2{=}CH_2(g) + Cl_2(g) \longrightarrow CH_2ClCH_2Cl(l)$$

$$S^{\ominus}(\text{product}) = 208\ J\,K^{-1}\,mol^{-1}$$

$$S^{\ominus}(\text{reactants}) = 219 + 223 = 442\ J\,K^{-1}\,mol^{-1}$$

$$\Delta S^{\ominus} = 208 - 442 = -234\ J\,K^{-1}\,mol^{-1}$$

**ANSWER**

The standard entropy change for the reaction is $-234\ J\,K^{-1}\,mol^{-1}$.

The negative sign means a decrease in disorder. Since two moles of gas have formed one mole of liquid, this is no surprise.

### Calculation of change in standard free energy

The change in standard enthalpy, the change in standard entropy and the temperature must be known and inserted into the equation

$$\Delta G^{\ominus} = \Delta H^{\ominus} - T\Delta S^{\ominus}$$

**EXAMPLE 2**

Calculate the change in standard free energy and determine whether the reaction

$$Fe_2O_3(s) + 3H_2(g) \longrightarrow 2Fe(s) + 3H_2O(g)$$

will take place at **a** 20 °C, **b** 500 °C. Use the values (in kJ mol$^{-1}$):

| Substance | $Fe_2O_3$ | $H_2$ | Fe | $H_2O$ |
|---|---|---|---|---|
| Standard enthalpy | −822 | 0 | 0 | −242 |
| Standard entropy | 0.090 | 0.131 | 0.027 | 0.189 |

**METHOD**

$\Delta G^\ominus = \Delta H^\ominus - T\Delta S^\ominus$

$\Delta H^\ominus = (0 + 3(-242)) - (-822 + 0) = +96$ kJ mol$^{-1}$

$\Delta S^\ominus = (2 \times 0.027) + (3 \times 0.189) - 0.090 - (3 \times 0.131)$

$\quad = 0.054 + 0.567 - 0.090 - 0.393 = +0.138$ kJ mol$^{-1}$

**a** At 20 °C,

$\Delta G^\ominus = \Delta H^\ominus - T\Delta S^\ominus$

$\quad = +96 - (293 \times 0.138) = 96 - 40.43 = +55.57$ kJ mol$^{-1}$

**ANSWER**

$\Delta G^\ominus$ is 55.6 kJ mol$^{-1}$ which is positive, and the reaction will therefore not occur at 20 °C.

**b** At 500 °C,

$\Delta G^\ominus = +96 - (773 \times 0.138) = -10.7$ kJ mol$^{-1}$

**ANSWER**

$\Delta G^\ominus$ is −10.7 kJ mol$^{-1}$ which is negative, the reaction will occur at 500 °C.

**Note:** the assumption that $\Delta H^\ominus$ does not vary with temperature.

## EXERCISE 12.6   *Problems on standard entropy change and standard free energy change

**1** Refer to the following values of standard entropy (J mol$^{-1}$ K$^{-1}$) at 298 K:

| | | | | | |
|---|---|---|---|---|---|
| $H_2(g)$ | 131 | $H_2O(l)$ | 70 | $NH_4Cl(s)$ | 94.6 |
| $Cl_2(g)$ | 223 | $H_2O(g)$ | 189 | $N_2O_4(g)$ | 304 |
| $N_2(g)$ | 192 | $HCl(g)$ | 187 | $C_2H_4(g)$ | 220 |
| $O_2(g)$ | 205 | $NH_3(g)$ | 193 | $C_2H_6(g)$ | 230 |
| $Na(s)$ | 51 | $NO_2(g)$ | 240 | $HNO_3(l)$ | 156 |
| $NaCl(s)$ | 72.4 | | | | |

Calculate the standard entropy changes for the following reactions:

**a** $H_2(g) + Cl_2(g) \longrightarrow 2HCl(g)$

**b** $N_2(g) + 3H_2(g) \longrightarrow 2NH_3(g)$

**c** $H_2(g) + \frac{1}{2}O_2(g) \longrightarrow H_2O(l)$

**d** $H_2(g) + C_2H_4(g) \longrightarrow C_2H_6(g)$

**e** $N_2O_4(g) \longrightarrow 2NO_2(g)$

**f** $Na(s) + \frac{1}{2}Cl_2(g) \longrightarrow NaCl(s)$

**g** $NH_4Cl(s) \longrightarrow NH_3(g) + HCl(g)$

**h** $4HNO_3(l) \longrightarrow 4NO_2(g) + O_2(g) + 2H_2O(l)$

**2** Predict, without doing calculations, whether the following reactions will have a positive or negative value of $\Delta S^\ominus$:

**a** $NH_4NO_3(s) \longrightarrow N_2O(g) + 2H_2O(g)$

**b** $2H_2O_2(aq) \longrightarrow 2H_2O(l) + O_2(g)$

**c** $PH_3(g) + HI(g) \longrightarrow PH_4I(s)$

**d** $3O_2(g) \longrightarrow 2O_3(g)$

**e** $CO_2(g) + C(s) \longrightarrow 2CO(g)$

**f** $Ni(s) + 4CO(g) \longrightarrow Ni(CO)_4(g)$

**3** Use the following values of standard entropy content and standard enthalpy of formation to calculate standard free energy changes:

| Substance | $\Delta H_F^{\ominus}/kJ \, mol^{-1}$ | $S^{\ominus}/J \, K^{-1} \, mol^{-1}$ |
|---|---|---|
| HgO(s) (red) | −90.7 | 72.0 |
| HgO(s) (yellow) | −90.2 | 73.0 |
| HgS(s) (red) | −58.2 | 77.8 |
| HgS(s) (black) | −54.0 | 83.3 |

**a** Calculate the value of $\Delta G^{\ominus}$ for the change

$$HgO(s) \, (red) \longrightarrow HgO(s) \, (yellow)$$

at 25 °C and at 100 °C. At what temperature will the change take place? That is, at what temperature is $\Delta G^{\ominus} = \Delta H^{\ominus} - T\Delta S^{\ominus}$?

**b** Calculate the value of $\Delta G^{\ominus}$ for the change

$$HgS(s) \, (red) \longrightarrow HgS(s) \, (black)$$

at 25 °C. Will the change occur at this temperature?

**4** *Cis*-but-2-ene has $\Delta H_F^{\ominus} = -5.7 \, kJ \, mol^{-1}$ and $S^{\ominus} = 301 \, J \, K^{-1} \, mol^{-1}$; *trans*-but-2-ene has $\Delta H_F^{\ominus} = -10.1 \, kJ \, mol^{-1}$ and $S^{\ominus} = 296 \, J \, K^{-1} \, mol^{-1}$.
Calculate

**a** $\Delta G^{\ominus}$ for the transition *cis*-but-2-ene $\longrightarrow$ *trans*-but-2-ene and

**b** for the transition *trans*-but-2-ene $\longrightarrow$ *cis*-but-2-ene
Which is the more stable isomer?

## EXERCISE 12.7   Questions from examination papers

**1** The table below includes some values of standard enthalpies of formation ($\Delta H_F^{\ominus}$).

| Substance | $H_2O(l)$ | LiOH(s) | Li(s) |
|---|---|---|---|
| $\Delta H_F^{\ominus}/kJ \, mol^{-1}$ | −286 | −487 | 0 |

The standard enthalpy of solution of lithium hydroxide is given below.

$$LiOH(s) \longrightarrow Li^+(aq) + OH^-(aq) \qquad \Delta H^{\ominus} = -21 \, kJ \, mol^{-1}$$

**a** State why the standard enthalpy of formation of lithium is quoted as zero. (1)

**b** Write an equation for the chemical reaction which represents the formation of lithium hydroxide from its elements, in which the enthalpy change is equal to its standard enthalpy of formation. (2)

**c** Write an equation, including state symbols, for the reaction of lithium with water in which lithium ions are formed. (2)

**d** Use the data given above to calculate a value for the enthalpy change for the reaction of lithium with water. (3)

**e** State the observed trend in the reactivity of the Group I elements with water, from lithium to caesium. (1)

**f** When caesium reacts with water, the heat energy released ($\Delta H^{\ominus}$) is less than that for lithium reacting with water. State how this fact relates, if at all, to the observe difference in reactivity of lithium and caesium with water. Give a reason for your answer. (2)

*NEAB 99* (11)

**2 a** Some mean bond enthalpy values are shown below.

| Bond | Mean bond enthalpy/kJ mol$^{-1}$ |
|------|------|
| H—H | 436 |
| C—H | 413 |
| C—C | 348 |
| C=C | 612 |

**(i)** The hydrogenation of 1-ethylcyclohexene occurs according to the equation

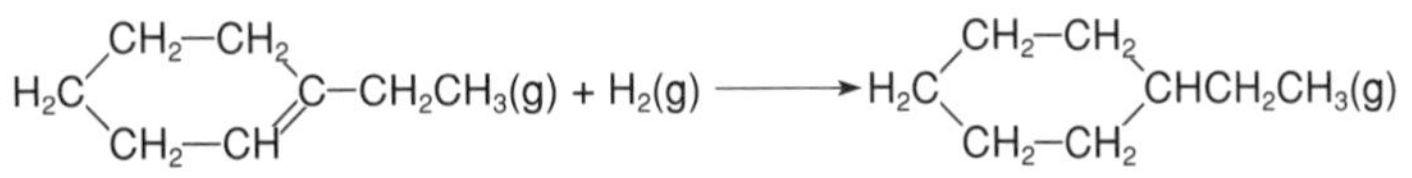

Use mean bond enthalpy values to determine the standard enthalpy of hydrogenation of this compound to form ethylcyclohexane.

**(ii)** Deduce the enthalpy change when the hypothetical compound

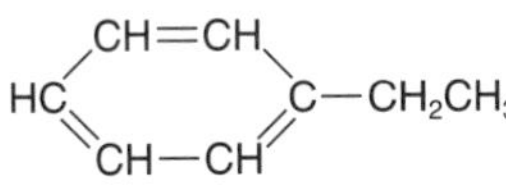

is fully hydrogenated to form ethylcyclohexane.

**(iii)** The standard enthalpy change for the hydrogenation of ethylbenzene, $C_6H_5CH_2CH_3$, is −210 kJ mol$^{-1}$. Which of the two compounds, ethylbenzene or 1-ethylcyclohexa-1,3,5-triene, is the more stable? Explain your answer. (6)

**b** At 298 K, the enthalpy of solution of calcium chloride is −123 kJ mol$^{-1}$ and the enthalpy of lattice formation of this salt is −2255 kJ mol$^{-1}$. The enthalpy of hydration of the calcium ion is −1650 kJ mol$^{-1}$.

**(i)** Write equations using calcium chloride or its ions to illustrate the terms *enthalpy of solution*, *enthalpy of lattice formation* and *enthalpy of hydration*.

**(ii)** Use the data above to determine the enthalpy of hydration of the chloride ion. (7)

*NEAB 99* (13)

**3** The first ionisation energies for the Group 1 elements are listed below.

**a**

| Element | Ionisation energy/kJ mol$^{-1}$ |
|---------|--------------------------------|
| Li | +519 |
| Na | +494 |
| K | +418 |
| Rb | +402 |
| Cs | +376 |

Explain the trend in ionisation energies with increasing atomic number within the group. (3)

**b** A Born–Haber cycle for the formation of lithium chloride is shown below.

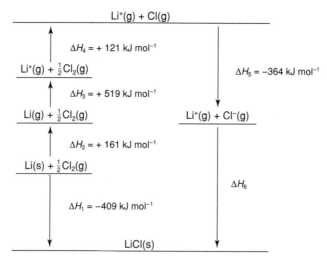

(i) Calculate $\Delta H_6$, the lattice enthalpy of lithium chloride. (2)

(ii) What enthalpy change does the value of $\Delta H_2$ represent? (1)

(iii) Suggest, with a reason, whether the value of $\Delta H_2$ would be larger or smaller for rubidium than it is for lithium. (2)

**c** (i) What are the factors which affect the magnitude of the lattice enthalpy? (2)

(ii) The calculated value for the lattice enthalpy of lithium chloride is less exothermic than the experimental value found in **b (i)**. Account for this difference. (2)

(iii) The calculated value for the lattice enthalpy of rubidium chloride is $-674$ kJ mol$^{-1}$; it is very similar to the experimental value of $-675$ kJ mol$^{-1}$. Give a reason for the similarity of these two values. (1)

**d** (i) When LiCl is heated in a bunsen flame a characteristic red flame is seen. Explain why lithium compounds produce a coloured flame in these circumstances. (2)

(ii) Lithium chloride can be made by burning lithium in chlorine.
Give a reason why rubidium chloride is not normally prepared in the laboratory in a similar way. (1)

$L$ (16)

4 Calcium fluoride and calcium chloride are typical ionic compounds which can be made from their respective elements.

$$Ca(s) + F_2(g) \longrightarrow CaF_2(s)$$
$$Ca(s) + Cl_2(g) \longrightarrow CaCl_2(s)$$

a (i) Explain why the two compounds have the general formula $CaX_2$. (1)

(ii) Describe, using trends in electronegativities, why both compounds consist of structures containing ions. (1)

(iii) Explain, in terms of the ions present, why calcium chloride has a melting temperature of 772 °C and calcium fluoride has a melting temperature of 1423 °C. (1)

b Calculate, using the data given below, the enthalpy of lattice **formation** at 298 K of calcium fluoride, $CaF_2$, in kJ mol$^{-1}$. (4)

| Process | | Enthalpy/kJ mol$^{-1}$ |
|---|---|---|
| $\Delta H^{\ominus}$ atomisation of calcium | $Ca(s) \longrightarrow Ca(g)$ | 193 |
| $\Delta H^{\ominus}$ atomisation of fluorine | $\frac{1}{2}F_2(g) \longrightarrow F(g)$ | 79 |
| $\Delta H^{\ominus}$ adding electron to fluorine | $F(g) + e^- \longrightarrow F^-(g)$ | −348 |
| $\Delta H^{\ominus}$ formation of $CaF_2(s)$ | $Ca(s) + F_2(g) \longrightarrow CaF_2(s)$ | −1214 |

| Process ionisation energies of calcium | | Ionisation energy/kJ mol$^{-1}$ |
|---|---|---|
| First | $Ca(g) \longrightarrow Ca^+(g) + e^-$ | 590 |
| Second | $Ca^+(g) \longrightarrow Ca^{2+}(g) + e^-$ | 1150 |

*WJEC(part)* (7)

5 The table below lists a number of mean bond enthalpy values.

| Bond | Mean bond enthalpy/kJ mol$^{-1}$ |
|---|---|
| C—C | 348 |
| C=C | 612 |
| C—H | 413 |
| O—H | 463 |

a Explain the meaning of the term *mean bond enthalpy*. (3)

b Given that the enthalpy of combustion to form carbon dioxide and steam is −2102 kJ mol$^{-1}$ for propane and −1977 kJ mol$^{-1}$ for propene, determine the enthalpy change for the oxidation of 1 mol of propane to propene and steam

$$C_3H_8(g) + \tfrac{1}{2}O_2(g) \longrightarrow C_3H_6(g) + H_2O(g)$$

using equations or a cycle to support your answer. (3)

c State the number and type of bonds made and broken in the oxidation of propane to propene and steam. Use the mean bond enthalpies in the table above, together with your answer to part **b**, to calculate the bond enthalpy of the O=O bond in the oxygen molecule. (4)

*NEAB 99* (10)

**6**  **a**  Define *lattice energy*. (2)

**b**  Lattice energy is often found using a Born–Haber cycle. The Born–Haber cycle for rubidium chloride, RbCl, is shown below.

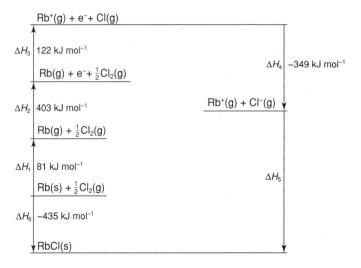

$$Rb^+(g) + e^- + Cl(g)$$

$\Delta H_3$  122 kJ mol$^{-1}$

$$Rb(g) + e^- + \tfrac{1}{2}Cl_2(g)$$

$\Delta H_4$  −349 kJ mol$^{-1}$

$$Rb^+(g) + Cl^-(g)$$

$\Delta H_2$  403 kJ mol$^{-1}$

$$Rb(g) + \tfrac{1}{2}Cl_2(g)$$

$\Delta H_1$  81 kJ mol$^{-1}$

$\Delta H_5$

$$Rb(s) + \tfrac{1}{2}Cl_2(g)$$

$\Delta H_6$  −435 kJ mol$^{-1}$

$$RbCl(s)$$

**(i)**  Write down the name for each of the energy terms listed below (and shown on the diagram): $\Delta H_2$; $\Delta H_4$; $\Delta H_6$.

**(ii)**  Use the cycle to calculate the lattice energy of rubidium chloride.

**(iii)**  Would you expect the value of the lattice energy of lithium chloride, LiCl, to be more or less exothermic than that for rubidium chloride? Give a reason for your answer. (7)

**c**  Suggest why a Born–Haber cycle is not generally used to consider the energy changes involved in the formation of solid silicon(IV) oxide, $SiO_2$. (1)

*OCR*  (10)

**7**  Below are some standard enthalpy changes including the standard enthalpy of combustion of nitroglycerine, $C_3H_5N_3O_9$

$$\tfrac{1}{2}N_2(g) + O_2(g) \longrightarrow NO_2(g) \qquad\qquad \Delta H_F^\ominus = +34\text{ kJ mol}^{-1}$$

$$C(s) + O_2(g) \longrightarrow CO_2(g) \qquad\qquad \Delta H_F^\ominus = -394\text{ kJ mol}^{-1}$$

$$H_2(g) + \tfrac{1}{2}O_2(g) \longrightarrow H_2O(g) \qquad\qquad \Delta H_F^\ominus = -242\text{ kJ mol}^{-1}$$

$$C_3H_5N_3O_9(l) + \tfrac{11}{4}O_2(g) \longrightarrow 3CO_2(g) + \tfrac{5}{2}H_2O(g) + 3NO_2(g) \qquad \Delta H_C^\ominus = -1540\text{ kJ mol}^{-1}$$

**a**  Standard enthalpy of formation is defined using the term *standard state*.

What does the term *standard state* mean? (2)

**b**  Use the standard enthalpy changes given above to calculate the standard enthalpy of formation of nitroglycerine. (4)

**c**  Calculate the enthalpy change for the following decomposition of nitroglycerine.

$$C_3H_5N_3O_9(l) \longrightarrow 3CO_2(g) + \tfrac{5}{2}H_2O(g) + \tfrac{3}{2}N_2(g) + \tfrac{1}{4}O_2(g)$$ (3)

**d**  Suggest one reason why the reaction in part **c** occurs rather than combustion when a bomb containing nitroglycerine explodes on impact. (1)

**e** An alternative reaction for the combustion of hydrogen, leading to liquid water, is given below.

$$H_2(g) + \tfrac{1}{2}O_2(g) \longrightarrow H_2O(l) \qquad \Delta H^\ominus = -286 \text{ kJ mol}^{-1}$$

Calculate the enthalpy change for the process $H_2O(l) \longrightarrow H_2O(g)$ and explain the sign of $\Delta H$ in your answer. (2)

*NEAB 99* (12)

**8** **a** Construct a Born–Haber cycle for the formation of the hypothetical crystalline solid magnesium(I) chloride, MgCl(s).

**b** The table below shows values of standard enthalpies for some processes involving magnesium and chlorine. Use these values to calculate the standard enthalpy of formation of the hypothetical MgCl(s).

**c** Use your answer to calculate the standard enthalpy change for the reaction

$$2MgCl(s) \rightleftharpoons MgCl_2(s) + Mg(s)$$

given that the standard enthalpy of formation of $MgCl_2(s)$, $\Delta H_F^\ominus$, is $-653$ kJ mol$^{-1}$.

**d** Explain why the standard entropy change in this reaction is likely to be negligibly small.

**e** Comment on the stability of MgCl(s) relative to that of $MgCl_2(s)$.

| | $\Delta H^\ominus$/kJ mol$^{-1}$ |
|---|---|
| $Mg^+(g) + Cl^-(g) \longrightarrow MgCl(s)$ | $-753$ |
| $Cl(g) + e^- \longrightarrow Cl^-(g)$ | $-364$ |
| $\tfrac{1}{2}Cl_2(g) \longrightarrow Cl(g)$ | $+121$ |
| $Mg(s) \longrightarrow Mg(g)$ | $+146$ |
| $Mg(g) \longrightarrow Mg^+(g) + e^-$ | $+736$ |

*NEAB 99* (10)

**9** **a** Sodium chloride is a compound which is almost completely ionic.
   **(i)** Draw a Born–Haber cycle and use it to explain why the formation of solid sodium chloride is energetically favourable.

   **(ii)** Explain why sodium chloride, unlike some metal chlorides, is predominantly ionic. (10)

**b** The extent of solubility in water of some ionic compounds, though not all, can be rationalised by a consideration of the lattice enthalpy of the compound and the hydration enthalpies of the ions. Draw a Hess's law cycle to show how this is done, and use the following data to explain the trend in solubility of the hydroxides of Group 2.

| Cation: | $Mg^{2+}$ | $Ca^{2+}$ | $Sr^{2+}$ | $Ba^{2+}$ |
|---|---|---|---|---|
| Enthalpy of hydration /kJ mol$^{-1}$ | $-1920$ | $-1650$ | $-1480$ | $-1360$ |
| Lattice energy of hydroxide /kJ mol$^{-1}$ | $-2843$ | $-2554$ | $-2354$ | $-2228$ |

(6)

**c** Saturated solutions of potassium iodate(VII), which is not very soluble in water, contain the following equilibrium when the solution is in contact with excess solid:

$$KIO_4(s) \rightleftharpoons K^+(aq) + IO_4^-(aq)$$

Potassium iodate(VII) is an oxidising agent, which will oxidise iodide ions to iodine in acidic solution, being reduced to iodine in the process.

**(i)** Derive the equation for the reaction between iodate(VII) ions and iodide ions in acidic solution. (2)

**(ii)** A saturated solution of $KIO_4$ at 10 °C was analysed by placing a 25.0 cm$^3$ sample into a conical flask, acidifying with dilute sulphuric acid, and adding an excess of potassium iodide. The liberated iodine was titrated with 0.100 mol dm$^{-3}$ sodium thiosulphate solution, 30.0 cm$^3$ being required.
Determine the concentration in mol dm$^{-3}$ of the potassium iodate(VII) solution. (4)

**(iii)** If a potassium salt is added to a saturated solution of potassium iodate(VII) which is then treated as above, the titre with sodium thiosulphate solution is much lower. Explain why this is so, and suggest what you would see as the potassium salt is added. (3)

*L* (25)

**10 a** Entropy is often linked with the idea of order and disorder. Write an equation for a chemical reaction in which the amount of *disorder* increases and state the sign of the entropy change which accompanies this reaction. (2)

**b** The combustion of graphite involves an increase in entropy from reactants to products of +3 J K$^{-1}$ mol$^{-1}$. The standard entropy of oxygen is 205 J K$^{-1}$ mol$^{-1}$ and that of carbon dioxide is 214 J K$^{-1}$ mol$^{-1}$ Calculate the standard entropy of graphite and suggest why it has a relatively low value. (4)

**c** **(i)** Write an equation that relates $\Delta G$ to $\Delta H$ and $\Delta S$ in a reaction.

**(ii)** Derive an equation which relates $\Delta S$ to $\Delta H$ when $\Delta G = 0$. (2)

**d** For certain reversible processes such as boiling and freezing, the Gibbs free energy change is zero. When water freezes, 6.0 kJ mol$^{-1}$ of heat energy are evolved. Use the equation you derived in part **c (ii)** to calculate the entropy change in 54 g of water when it freezes at 0 °C. (4)

*NEAB 96* (12)

**11 a** In terms of free energy, state what is meant by the term *spontaneous change*. (1)

**b** Explain why some spontaneous chemical reactions never seem to happen. (1)

**c** On heating, sodium hydrogencarbonate ($NaHCO_3$) decomposes into sodium carbonate, carbon dioxide and steam. At 298 K, the standard enthalpy change for the decomposition of one mole of $NaHCO_3$ is +65 kJ mol$^{-1}$ and the corresponding standard entropy change is +168 J K$^{-1}$ mol$^{-1}$.

**(i)** Write an equation for the decomposition of $NaHCO_3$.

**(ii)** Determine the standard Gibbs free energy change for this decomposition at 298 K.

**(iii)** Explain why heating is needed to make the reaction spontaneous. (6)

*NEAB 96* (8)

**12** **a** Define *standard enthalpy of reaction*. (1)

  **b** Zinc displaces copper from its compounds in aqueous solution.

  **(i)** Give the ionic equation for this reaction. (1)

  **(ii)** Explain how the enthalpy change for this reaction could be measured experimentally. (3)

  **(iii)** Show how your results could be used to calculate the enthalpy of reaction. (2)

  **c** Calculate the enthalpy change for the following reaction

$$P_4O_{10}(s) + 6H_2O(l) \longrightarrow 4H_3PO_4(s)$$

given the following enthalpies of formation, $\Delta H_F$. Include an enthalpy cycle in your answer.

| Compound | $\Delta H_F$/kJ mol$^{-1}$ |
|---|---|
| $P_4O_{10}(s)$ | −2984 |
| $H_2O(l)$ | −286 |
| $H_3PO_4(s)$ | −1279 |

(3)

*OCR* (10)

**13** Natural gas is converted into hydrogen by reaction with steam in the presence of a nickel catalyst.

$$CH_4(g) + H_2O(g) \longrightarrow CO(g) + 3H_2(g)$$

  **a** The standard enthalpies of formation, $\Delta H_F^{\ominus}$, and the standard entropies, $S^{\ominus}$, of the substances involved are shown in the table below.

| Substance | $\Delta H_F^{\ominus}$/kJ mol$^{-1}$ | $S^{\ominus}$/J K$^{-1}$ mol$^{-1}$ |
|---|---|---|
| $CH_4(g)$ | −74.9 | 186 |
| $H_2O(g)$ | −242 | 189 |
| $CO(g)$ | −111 | 198 |
| $H_2(g)$ | | 131 |

  **(i)** State the value of $\Delta H_F^{\ominus}$ for $H_2(g)$. (1)

  **(ii)** Calculate the entropy change, $\Delta S^{\ominus}$, for the reaction of steam with methane. (1)

  **(iii)** Calculate the enthalpy change, $\Delta H^{\ominus}$, for the reaction, showing your method clearly. (2)

  **(iv)** Use your answers to **(ii)** and **(iii)** to calculate the temperature at which the reaction will be feasible. (2)

  **(v)** Discuss the advantages and disadvantages of carrying out the reaction at a higher temperature than that calculated in **(iv)**. (3)

**b** **(i)** The nickel catalyst is deactivated by carbon which can be produced by a side reaction between carbon monoxide and hydrogen.

$$CO(g) + H_2(g) \rightleftharpoons C(s) + H_2O(g)$$

Explain how the nickel catalyst increases the rate of reaction, and suggest a reason why it is deactivated by carbon. (3)

**(ii)** Write an expression for the *equilibrium constant in terms of partial pressure*, $K_p$, for the reaction between carbon monoxide and hydrogen above. (1)

**(iii)** When the reaction was carried out under a pressure of $3.00 \times 10^6$ Pa and a temperature of 1100 K the mixture of gases present at equilibrium was found to have the percentage composition by volume shown below.

| Compound | % by volume |
|----------|-------------|
| CO       | 12.8        |
| $H_2O$   | 16.3        |
| $H_2$    | 70.9        |

Find the value of $K_p$ at this temperature. (3)

*OCR* (16)

# Reaction kinetics

# 13

Reaction kinetics is the study of the factors which affect the rates of chemical reactions.

## 13.1 REACTION RATE

The rate of a chemical reaction is the rate of change of concentration. Consider a reaction of the type $A \longrightarrow B$, where one molecule of the reactant forms one molecule of the product, Figure 13.1 shows how the concentration of product, $x$, increases as the time, $t$, which has passed since the start of the reaction increases. The initial concentration of reactant (the concentration at the start of the reaction) is $a$, and at any time after the start of the reaction, the concentration of reactant is $(a - x)$.

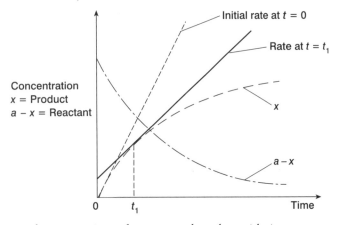

Fig. 13.1   *Variation of concentrations of reactant and product with time*

You can see that the rate of reaction is decreasing as the reaction proceeds and the reactant is being used up. We can only state the rate of reaction between certain times.

We can calculate the average rate of reaction over a certain interval of time in this way.

To 1 dm$^3$ of solution containing 0.300 mol methyl ethanoate is added a small amount of mineral acid. This catalyses the hydrolysis reaction

$$CH_3CO_2CH_3(aq) + H_2O(l) \rightleftharpoons CH_3CO_2H(aq) + CH_3OH(aq)$$

After 100 seconds, the concentration has decreased to 0.292 mol dm$^{-3}$. This means that 0.008 mol dm$^{-3}$ of methyl ethanoate has reacted, and 0.008 mol dm$^{-3}$ of methanol and ethanoic acid have been formed.

$$\left(\begin{array}{l}\text{Average rate of reaction} \\ \text{over this time interval}\end{array}\right) = \frac{\text{Change in concentration}}{\text{Time}}$$

$$= \frac{0.008}{100} = 8 \times 10^{-5} \text{ mol dm}^{-3} \text{ s}^{-1}$$

Since the rate of reaction varies with time, it is usual to state the **initial rate** of the reaction. This is the rate at the start of the reaction when an infinitesimally small amount of the reactant has been used up. In Figure 13.1, the gradient of the tangent to the curve at $t = 0$ gives the initial rate of the reaction.

## 13.2 THE EFFECT OF CONCENTRATION ON RATE OF REACTION

Consider a reaction between **A** and **B** to form **C**:

$$\textbf{A} + \textbf{B} \longrightarrow \textbf{C}$$

The rate of formation of **C** depends on the concentrations of **A** and **B**, but one cannot simply say that the rate of formation of **C** is proportional to the concentration of **A** and proportional to the concentration of **B**. The relationship is

$$\text{Rate of formation of } \textbf{C} \propto [\textbf{A}]^m[\textbf{B}]^n$$

where $m$ and $n$ are usually integers, often 0, 1 or 2, and are characteristic of the reaction. We say that the reaction is of order $m$ with respect to **A** and of order $n$ with respect to **B**. The order of reaction is $(m + n)$. We cannot tell the order simply by looking at the chemical equation for the reaction. For example, the reaction between bromate(V) ions and bromide ions and acid to give bromine

$$\text{BrO}_3^-(\text{aq}) + 5\text{Br}^-(\text{aq}) + 6\text{H}^+(\text{aq}) \longrightarrow 3\text{Br}_2(\text{aq}) + 3\text{H}_2\text{O}(\text{l})$$

The rate of disappearance of $\text{BrO}_3^-$ fits the expression

$$\text{Rate} \propto [\text{BrO}_3^-][\text{Br}^-][\text{H}^+]^2$$

It is first order with respect to bromate(V), first order with respect to bromide, second order with respect to hydrogen ion and fourth order overall.

If $\qquad\qquad$ Reaction rate $\propto [\textbf{A}]^m[\textbf{B}]^n \qquad$ it follows that

$$\text{Reaction rate} = k[\textbf{A}]^m[\textbf{B}]^n$$

The proportionality constant $k$ is called the **rate constant** for the reaction or the **rate coefficient** for the reaction.

## 13.3 ORDER OF REACTION

As a reaction proceeds, the concentrations of the reactants decrease, and the rate of reaction decreases, as shown in Figure 13.1. The shape of the curve depends on the order of the reaction (see Figure 13.2).

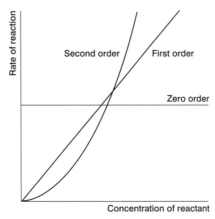

Fig. 13.2    *Graphs of rate against concentration*

## 13.4 *FIRST-ORDER REACTIONS

If the reaction

$$A \longrightarrow Products$$

is a first-order reaction, the rate equation will be

$$Rate = k[A]$$

The unit of $k$ is given by

$$\frac{Rate/mol\,dm^{-3}\,s^{-1}}{Concentration/mol\,dm^{-3}} = s^{-1}$$

The unit of $k$, the first-order rate constant, is $s^{-1}$.

### Half-life

The time taken for half the amount of **A** to react, $t_{1/2}$ is called the **half-life** of the reaction.

The half-life of a first-order reaction is independent of the initial concentration of the reactant. Radioactive decay is an example of first-order kinetics.

### Pseudo-first-order reactions

The acid-catalysed hydrolysis of an ester, e.g. ethyl ethanoate,

$$CH_3CO_2C_2H_5(aq) + H_2O(l) \longrightarrow CH_3CO_2H(aq) + C_2H_5OH(aq)$$

is first order with respect to ester and first order with respect to water. If water is present in excess, so that the fraction of the water which is used up in the reaction is small, the concentration of water is practically constant, and, since the acid catalyst is not used up, the rate depends only on the concentration of ester:

$$Rate = k'[CH_3CO_2C_2H_5]$$

$k'$ is constant for a certain concentration of acid, and the reaction obeys a first-order rate equation.

| | |
|---|---|
| **EXAMPLE 1** | The rate constant of a first-order reaction is $2.0 \times 10^{-6}$ s$^{-1}$. The initial concentration of the reactant is 0.10 mol dm$^{-3}$. What is the value of the initial rate in mol dm$^{-3}$ s$^{-1}$? |
| **METHOD** | The rate equation has the form |

$$\text{Rate} = k[\text{A}]$$

Putting the values of [A] and $k$ into this equation gives

**ANSWER** $\qquad$ Rate $= 2.0 \times 10^{-6}$ s$^{-1} \times 0.10$ mol dm$^{-3} = 2.0 \times 10^{-7}$ mol dm$^{-3}$ s$^{-1}$.

| | |
|---|---|
| **EXAMPLE 2** | Uranium-238 decays to form lead. The half-life of the decay is $4.5 \times 10^9$ years. In a sample of rock the molar ratio of uranium to lead is 1 : 3. How old is the rock? |
| **METHOD** | One mole of uranium has decayed to $\frac{1}{4}$ mol of uranium and $\frac{3}{4}$ mol of lead. This has taken two half-lives (from 1 mol to $\frac{1}{2}$ mol to $\frac{1}{4}$ mol) $= 9.0 \times 10^9$ years. |
| **ANSWER** | The rock is $9.0 \times 10^9$ years old. |

## 13.5 SECOND-ORDER REACTIONS WITH REACTANTS OF EQUAL CONCENTRATION

In the simplest case, the reaction

$$\text{A} + \text{B} \longrightarrow \text{Products}$$

has a rate equation

$$\text{Rate} = k[\text{A}][\text{B}]$$

| | |
|---|---|
| **EXAMPLE 1** | The initial rate of a second-order reaction is $8.0 \times 10^{-3}$ mol dm$^{-3}$ s$^{-1}$. The initial concentrations of the two reactants, **A** and **B**, are 0.20 mol dm$^{-3}$. What is the rate constant in dm$^3$ mol$^{-1}$ s$^{-1}$? |
| **METHOD** | Since |

$$\text{Rate} = k[\text{A}][\text{B}]$$

putting

$$\text{Rate} = 8.0 \times 10^{-3} \text{ mol dm}^{-3} \text{ s}^{-1} \text{ and } [\text{A}] = [\text{B}] = 0.20 \text{ mol dm}^{-3}$$

gives $\qquad 8.0 \times 10^{-3}$ mol dm$^{-3}$ s$^{-1} = k \times (0.20$ mol dm$^{-3})^2$

**ANSWER** $\qquad$ Rate constant, $k = 0.20$ dm$^3$ mol$^{-1}$ s$^{-1}$

| | |
|---|---|
| **EXAMPLE 2** | In the alkaline hydrolysis of an ester, both ester and alkali had the initial concentration of 0.0500 mol dm$^{-3}$. The following results were obtained. |

| Time/s | 100 | 200 | 300 | 400 | 600 |
|---|---|---|---|---|---|
| % of ester remaining | 70.5 | 55.8 | 44.5 | 37.8 | 29.7 |

The order of reaction is 2. Plot a graph of the percentage of ester remaining against time. By drawing a tangent to the graph, find the initial rate of reaction. Calculate the rate constant of this second order reaction.

**METHOD**

The graph of percentage of ester remaining against time is shown in Figure 13.3. The tangent to the graph at time $t = 0$ has a gradient

$$42\% \times 0.05 \text{ mol dm}^{-3}/100 \text{ s} = 2.1 \times 10^{-4} \text{ mol dm}^{-3} \text{ s}^{-1}$$

Since the reaction is second order, the rate constant, $k$, is given by

$$k \times (0.05 \text{ mol dm}^{-3})^2 = 2.1 \times 10^{-4} \text{ mol dm}^{-3} \text{ s}^{-1}$$

$$k = 8.4 \times 10^{-2} \text{ dm}^3 \text{ mol}^{-1} \text{ s}^{-1}$$

**ANSWER**

Note: In more advanced work, you will learn a more accurate graphical method of finding the rate constant of a second order reaction.

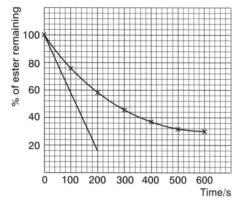

Fig. 13.3 *Percentage of ester remaining against time*

## 13.6 ZERO-ORDER REACTIONS

In a zero-order reaction, the rate is independent of the concentration of the reactant. In the reaction between propanone and iodine

$$CH_3COCH_3(aq) + I_2(aq) \longrightarrow CH_3COCH_2I(aq) + HI(aq)$$

the reaction rate does not change if the concentration of iodine is changed. The rate of reaction is independent of the iodine concentration, and the reaction is said to be zero order with respect to iodine.

**EXERCISE 13.1** **Problems on finding the order of reaction**

1 X and Y react together. For a three-fold increase in the concentration of X, there is a nine-fold increase in the rate of reaction. What is the order of reaction with respect to X?

Hint : What is the relationship between rate and [X]?

2 A and B react to form C. In one run, the concentration of A is doubled, while B is kept constant, and the initial rate is doubled. In a second run, the concentration of B is doubled while that of A is kept constant, and the initial rate is quadrupled. What can you deduce about the order of the reaction?

Hint : Spot the relationships: rate and [A], rate and [B].

**3**   Figure 13.4 shows that the rate of reaction is:

**A**   proportional to $[I_2]$

**B**   proportional to $[I_2]^2$

**C**   proportional to $1/[I_2]$

**D**   independent of $[I_2]$

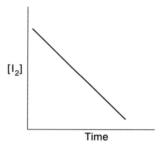

Fig. 13.4

**4**   **X** decomposes to form **Y** + **Z**. The following results were obtained in a study of the reaction:

| Initial [**X**]/mol dm$^{-3}$ | $2.0 \times 10^{-3}$ | $4.0 \times 10^{-3}$ | $5.0 \times 10^{-3}$ |
|---|---|---|---|
| Initial rate/mol dm$^{-3}$ s$^{-1}$ | $1.3 \times 10^{-6}$ | $1.3 \times 10^{-6}$ | $1.3 \times 10^{-6}$ |

What is the rate expression? What is the order of the reaction?

**Hint :** Find the relationship between rate and [**X**].

**5**   The reaction **A** + **B** $\longrightarrow$ **C** is first order with respect to **A** and to **B**. When the initial concentrations are [**A**] = $1.5 \times 10^{-2}$ mol dm$^{-3}$ and [**B**] = $2.5 \times 10^{-3}$ mol dm$^{-3}$, the initial rate of reaction is found to be $3.75 \times 10^{-4}$ mol dm$^{-3}$ s$^{-1}$. Calculate the rate constant for the reaction.

**Hint :** Start with the rate equation.

**6**   In the reaction

$$\text{A} + \text{B} \longrightarrow \text{P} + \text{Q}$$

the following results were obtained for the initial rates of reaction for different initial concentrations:

| [**A**]/mol dm$^{-3}$ | [**B**]/mol dm$^{-3}$ | Initial rate/mol dm$^{-3}$ s$^{-1}$ |
|---|---|---|
| 1.0 | 1.0 | $2.0 \times 10^{-3}$ |
| 2.0 | 1.0 | $4.0 \times 10^{-3}$ |
| 4.0 | 2.0 | $16 \times 10^{-3}$ |

Deduce the rate equation and calculate the rate constant.

**7** The rate of a reaction depends on the concentrations of the reactants. In the reaction between **X** and **Y**, the following results were obtained for runs at the same temperature.

| Initial concentration of **X**/mol dm$^{-3}$ | Initial concentration of **Y**/mol dm$^{-3}$ | Initial rate/ mol dm$^{-3}$ h$^{-1}$ |
|---|---|---|
| $2 \times 10^{-3}$ | $3 \times 10^{-3}$ | $3.0 \times 10^{-3}$ |
| $2 \times 10^{-3}$ | $6 \times 10^{-3}$ | $1.2 \times 10^{-2}$ |
| $4 \times 10^{-3}$ | $6 \times 10^{-3}$ | $2.4 \times 10^{-2}$ |

Deduce the order of the reaction with respect to: **a X**, **b Y**. Calculate the rate constant for the reaction.

**8** The following results were obtained for the decomposition of nitrogen(V) oxide

$$2N_2O_5(g) \longrightarrow 4NO_2(g) + O_2(g)$$

| Concentration of $N_2O_5$/mol dm$^{-3}$ | Initial rate/mol dm$^{-3}$ s$^{-1}$ |
|---|---|
| $1.6 \times 10^{-3}$ | 0.12 |
| $2.4 \times 10^{-3}$ | 0.18 |
| $3.2 \times 10^{-3}$ | 0.24 |

What is the rate expression for the reaction? What is the order of reaction? What is the initial rate of reaction when the concentration of $N_2O_5$ is:

**a** $2.0 \times 10^{-3}$ mol dm$^{-3}$   **b** $2.4 \times 10^{-2}$ mol dm$^{-3}$?

## EXERCISE 13.2   *Problems on first-order reactions

**1** A isomerises to form **B**. The reaction is first order. If 75% of **A** is converted to **B** in 2.5 hours, what is the half-life for the isomerisation?

**2** The reaction

$$2N_2O_5 \longrightarrow 4NO_2 + O_2$$

is first order. If the initial concentration of $N_2O_5$ is 1.25 mol dm$^{-3}$ and the initial rate is $1.38 \times 10^{-5}$ mol dm$^{-3}$ s$^{-1}$, what is the rate constant for the decomposition?

**Hint :** Rate = $k[N_2O_5]$

**3**

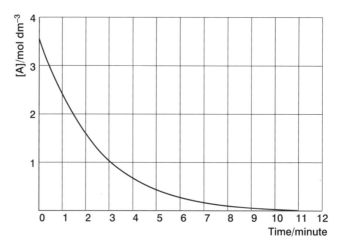

Fig. 13.5 *Graph of concentration of [A] plotted against time*

The curve shown in Figure 13.5 represents the decomposition of **A** at a certain temperature.

Calculate the gradients of the curve at: **a** 1 minute, **b** 3 minutes, **c** 5 minutes, and **d** 11 minutes. Plot a graph of the gradient against the concentration of **A** at 1, 3, 5 and 11 minutes. Calculate the rate constant for the reaction from the slope of the graph. What is the order of the reaction?

**Hint :** Rate = gradient of a tangent to the curve at that time. Find the relationship between rate and concentration.

**4**   Figure 13.6 shows the results of a study of the reaction

$$A + B \longrightarrow C$$

The experimental conditions were:

|          | Initial concentrations/mol dm$^{-3}$ | |
|          | [A] | [B] |
|----------|-----|-----|
| Curve 1  | 0.10 | 0.10 |
| Curve 2  | 0.10 | 0.20 |
| Curve 3  | 0.10 | 0.30 |

**a**  Find the initial rates of curves **1**, **2** and **3**.

**b**  What is the order of reaction with respect to **B**?

**c**  Inspect curve 3. What is the time required for completion of $\frac{1}{2}$ reaction, and of $\frac{3}{4}$ reaction? What is the order with respect to **A**?

**d**  Write an overall rate equation for the reaction.

**e**  Find the rate constant for the reaction.

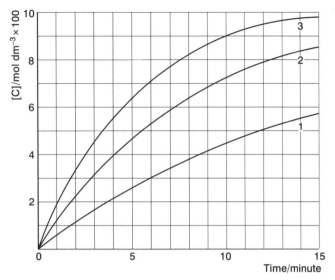

Fig. 13.6   *Graph of [C] plotted against time*

## EXERCISE 13.3   Problems on second-order reactions

**Reminder :** Graphs § 1.10. Units § 1.11. Check your answer § 1.12.

**1**   The following results were obtained from a study of the reaction between **P** and **Q**.

| Concentrations/mol dm$^{-3}$ | | Initial rate/mol dm$^{-3}$ s$^{-1}$ |
|---|---|---|
| [P] | [Q] | |
| $2.00 \times 10^{-3}$ | $2.00 \times 10^{-3}$ | $2.00 \times 10^{-4}$ |
| $1.80 \times 10^{-3}$ | $1.80 \times 10^{-3}$ | $1.62 \times 10^{-4}$ |
| $1.40 \times 10^{-3}$ | $1.40 \times 10^{-3}$ | $9.80 \times 10^{-5}$ |
| $1.10 \times 10^{-3}$ | $1.10 \times 10^{-3}$ | $6.05 \times 10^{-5}$ |
| $0.80 \times 10^{-3}$ | $0.80 \times 10^{-3}$ | $3.20 \times 10^{-5}$ |

Prove that the reaction is second order. Calculate the rate constant.

**2**   The following results were obtained for a reaction between **A** and **B**.

| Concentrations/mol dm$^{-3}$ | | Initial rate/mol dm$^{-3}$ s$^{-1}$ |
|---|---|---|
| [A] | [B] | |
| 0.5 | 1.0 | 2 |
| 0.5 | 2.0 | 8 |
| 0.5 | 3.0 | 18 |
| 1.0 | 3.0 | 36 |
| 2.0 | 3.0 | 72 |

What is the order of reaction with respect to **A** and with respect to **B**? What is the rate equation for the reaction? Calculate the rate constant. State the units in which it is expressed.

**3** The table shows the results of experiments on the reaction

$$ClO^-(aq) + I^-(aq) \xrightarrow{OH^-(aq)} IO^-(aq) + Cl^-(aq)$$

| Initial concentration/mol dm$^{-3}$ | | | Initial rate of formation of IO$^-$/ mol dm$^{-3}$ s$^{-1}$ × $10^6$ |
|---|---|---|---|
| ClO$^-$ | I$^-$ | OH$^-$ | |
| 0.015 | 0.015 | 1.00 | 1.65 |
| 0.030 | 0.015 | 1.00 | 3.30 |
| 0.015 | 0.030 | 1.00 | 3.30 |
| 0.015 | 0.015 | 0.50 | 1.65 |

**a** Give the order of reaction with respect to **(i)** ClO$^-$ **(ii)** I$^-$ **(iii)** OH$^-$.

**b** Write the rate equation for the reaction.

**4** Shown below are the results of some experiments on the reaction

$$A + B \longrightarrow C$$

| Experiment | Initial concentration/mol dm$^{-3}$ | | Initial rate/mol dm$^{-3}$ s$^{-1}$ |
|---|---|---|---|
| | [A] | [B] | |
| 1 | $1.0 \times 10^{-3}$ | $1.0 \times 10^{-3}$ | $5.0 \times 10^{-8}$ |
| 2 | $2.0 \times 10^{-3}$ | $1.0 \times 10^{-3}$ | $1.0 \times 10^{-7}$ |
| 3 | $2.0 \times 10^{-3}$ | $2.0 \times 10^{-3}$ | $2.0 \times 10^{-7}$ |

**a** State the order of reaction with respect to **(i)** A **(ii)** B, explaining your reasoning.

**b** Write the rate equation for the reaction, and give the units of the rate constant.

## EXERCISE 13.4 *Problems on radioactive decay

**Reminder :** Graphs § 1.10.

**1** Plot a graph, using the following figures, to show the radioactive decay of krypton. From the graph, find the half-life.

| Time/minute | 0 | 20 | 40 | 60 | 80 | 100 | 120 |
|---|---|---|---|---|---|---|---|
| Activity/count per second | 100 | 92 | 85 | 78 | 72 | 66 | 61 |

**2** A sample of gold was irradiated in a nuclear reactor. It gave the following results when its radioactivity was measured at various intervals. Plot the results, and deduce the half-life of the radioactive isotope of gold formed.

| Time/hour | 0 | 1 | 5 | 10 | 25 | 50 | 75 | 100 |
|---|---|---|---|---|---|---|---|---|
| Radioactivity/ count per minute | 300 | 296 | 285 | 270 | 228 | 175 | 133 | 103 |

**3**   A dose of $1.00 \times 10^{-4}$ g of astatine-211 is given to a patient for treatment of cancer of the thyroid gland. How much of this radioactive isotope ($t_{1/2}$ = 7.0 h) will remain in the body 28 hours later?

**4**   A radioactive source, after storing for 42 days, is found to have $\frac{1}{8}$th of its original activity. What is the half-life of the radioactive isotope present in the source?

**5**   Actinium B has a half-life of 36.0 min. What fraction of the original quantity of actinium remains after: **a** 180.0 min, **b** 1080 min?

## 13.7 *THE EFFECT OF TEMPERATURE ON REACTION RATES

An increase in temperature increases the rate of a reaction by increasing the rate constant. A plot of the logarithm of the rate constant, $k$, against $1/T$ is a straight line, with a negative gradient (see Figure 13.7).

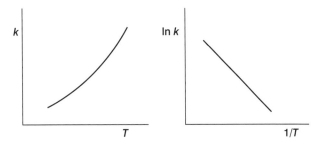

Fig. 13.7   *Dependence of rate constant on temperature*

---

**Reminder :** For logarithms see § 1.9.

---

The variation of rate constant with temperature obeys the **Arrhenius equation**

$$k = Ae^{-E/RT}$$

$A$ and $E$ are constant for a given reaction; $R$ is the gas constant. In order to react, two molecules must collide with a minimum amount of energy $E$, which is called the **activation energy**. The fraction of molecules possessing energy $E$ is given by $e^{-E/RT}$. The constant $A$, called the **pre-exponential factor**, represents the maximum rate which the reaction can reach when all the molecules have energy equal to or greater than $E$.

The Arrhenius equation can be written as

$$\ln k = \ln A - E/RT$$

(ln = logarithm to the base e; see § 1.9.)

A plot of $\ln k$ against $1/T$ is linear with a gradient of $-E/R$ and intercept $\ln A$. The values of $A$ and $E$ can thus be found from a plot of $\ln k$ against $1/T$.

**EXAMPLE**

The reaction

$$2N_2O_5(g) \longrightarrow 2N_2O_4(g) + O_2(g)$$

was studied at a number of temperatures, and the following values for the rate constant were obtained:

| Temperature/K | 293 | 308 | 318 | 338 |
|---|---|---|---|---|
| Rate constant/s$^{-1}$ | $1.76 \times 10^{-5}$ | $1.35 \times 10^{-4}$ | $4.98 \times 10^{-4}$ | $4.87 \times 10^{-3}$ |

Calculate the activation energy and the pre-exponential factor for the reaction.

**METHOD**

Since

$$\ln k = \ln A - \frac{E}{RT}$$

a plot of $\ln k$ against $1/T$ gives a straight line with a gradient of $-E/R$ and an intercept on the $\ln k$ axis of $\ln A$.

The table below gives the values which must be plotted:

| $T$/K | 293 | 308 | 318 | 338 |
|---|---|---|---|---|
| $k$/s$^{-1}$ | $1.76 \times 10^{-5}$ | $1.35 \times 10^{-4}$ | $4.98 \times 10^{-4}$ | $4.87 \times 10^{-3}$ |
| $1/T$/K$^{-1}$ | $3.41 \times 10^{-3}$ | $3.25 \times 10^{-3}$ | $3.14 \times 10^{-3}$ | $2.96 \times 10^{-3}$ |
| $\ln k$ | $-10.95$ | $-8.91$ | $-7.60$ | $-5.32$ |

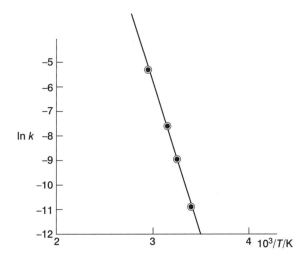

Fig. 13.8 *Plot of ln k against $10^3/T$*

The gradient $= \dfrac{5.63}{0.45} \times 10^{-3}\ \text{K}^{-1} = 1.25 \times 10^4\ \text{K}$

Then

$$-1.25 \times 10^4\ \text{K} = -\frac{E}{8.31}\ \text{J K}^{-1}\,\text{mol}^{-1}$$

$$E = 104\ 000\ \text{J mol}^{-1} = 104\ \text{kJ mol}^{-1}$$

Since $\ln A = \ln k + E/RT$

and since, from the graph, $\ln k = -9.0$ at $1/T = 3.25 \times 10^{-3} \, K^{-1}$

$$\ln A = -9.0 + \frac{104 \times 10^3 \, J\,mol^{-1} \times 3.25 \times 10^{-3} \, K^{-1}}{8.314 \, J\,K^{-1}\,mol^{-1}}$$

$$A = 5.6 \times 10^{13} \, s^{-1}$$

**ANSWER**     Energy of activation = 104 kJ mol$^{-1}$; pre-exponential factor = $5.6 \times 10^{13} \, s^{-1}$.

## EXERCISE 13.5   *Problems on activation energy

**Reminder :** Logarithms § 1.9. Graphs § 1.10.

**1**   The reaction

$$2A(g) + B(g) \longrightarrow C(g)$$

was studied at a number of temperatures, and the following results were obtained:

| Temperature/K | 285 | 333 | 385 | 476 | 565 |
|---|---|---|---|---|---|
| Rate constant/dm$^6$ mol$^{-2}$ s$^{-1}$ | 2.34 | 13.2 | 52.5 | 316 | 1000 |

Calculate the activation energy from a graph of $\ln k$ against $1/T$.

**2**   The following results were obtained in the study of the effect of temperature on the rate of a chemical reaction.

| Temperature/K | 2000 | 1136 | 909 | 676 | 540 |
|---|---|---|---|---|---|
| Rate constant/dm$^3$ mol$^{-1}$ s$^{-1}$ | 1.00 | 0.316 | 0.158 | 0.050 | 0.0158 |

From a plot of $\ln k$ against $1/T$, find the activation energy for the reaction.

**3**   Use the following results to calculate the energy of activation of the reaction:

| Temperature/K | 800 | 625 | 472 | 377 | 333 |
|---|---|---|---|---|---|
| Rate constant/dm$^3$ mol$^{-1}$ s$^{-1}$ | 0.562 | 0.316 | 0.132 | 0.0562 | 0.0316 |

**EXERCISE 13.6**  **Problems on reaction kinetics**

---

**Reminder :** Graphs § 1.10. Units § 1.11.

---

**1**  Nitrogen monoxide and oxygen react to form nitrogen dioxide:

$$2NO(g) + O_2(g) \longrightarrow 2NO_2(g)$$

The rate equation is

$$\text{Rate} = k[NO]^2[O_2]$$

The rate of disappearance of nitrogen monoxide is $4.8 \times 10^{-5}$ mol dm$^{-3}$ s$^{-1}$. What is the rate of disappearance of oxygen?

**2**  Benzene is nitrated by a mixture of concentrated nitric acid and concentrated sulphuric acid. When the reaction was carried out in an organic solvent the results shown below were obtained.

|  | [Benzene/mol dm$^{-3}$] | [Nitric acid/mol dm$^{-3}$] | Initial rate $\times 10^4$/ mol dm$^{-3}$ s$^{-1}$ |
|---|---|---|---|
| Experiment 1 | 0.30 | 0.30 | 1.64 |
| Experiment 2 | 0.30 | 0.60 | 3.28 |
| Experiment 3 | 0.60 | 0.90 | 4.92 |

**a**  State the order of reaction with respect to **(i)** benzene **(ii)** nitric acid.

**b**  Write the rate law for the reaction.

**3**  Hydrogen and iodine combine to form hydrogen iodide. The reaction is first order with respect to hydrogen and first order with respect to iodine. The rate constant is $2.78 \times 10^{-4}$ mol dm$^{-3}$ s$^{-1}$. If the concentrations are $[H_2] = 0.85 \times 10^{-2}$ mol dm$^{-3}$, and $[I_2] = 1.25 \times 10^{-2}$ mol dm$^{-3}$, what is the initial rate of reaction?

**4**  The reaction

$$2NO(g) + Cl_2(g) \longrightarrow 2NOCl(g)$$

is third order. The rate constant is $1.7 \times 10^{-5}$ dm$^6$ mol$^{-2}$ s$^{-1}$. If the concentrations of the reactants are each 0.20 mol dm$^{-3}$, what is the initial rate of reaction?

**5**  Hydrogen peroxide decomposes in aqueous solution:

$$2H_2O_2(aq) \longrightarrow 2H_2O(l) + O_2(g)$$

The following results show how the rate of decomposition varies with the initial hydrogen peroxide concentration:

| Rate/mol dm$^{-3}$ s$^{-1}$ | [H$_2$O$_2$]/mol dm$^{-3}$ |
|---|---|
| $3.64 \times 10^{-5}$ | 0.05 |
| $7.41 \times 10^{-5}$ | 0.10 |
| $1.51 \times 10^{-4}$ | 0.20 |
| $2.21 \times 10^{-4}$ | 0.30 |

Plot the rate of decomposition against the concentration. Deduce: **a** the order of the reaction, and **b** the rate constant under the conditions of the experiment.

**6**   P and Q reacted in solution to give **R**. At constant temperature the results shown below were obtained.

| [P]/mol dm$^{-3}$ | [Q]/mol dm$^{-3}$ | Initial rate of formation of C/mol dm$^{-3}$ s$^{-1}$ |
|---|---|---|
| 0.10 | 0.01 | $1.19 \times 10^{-3}$ |
| 0.10 | 0.03 | $1.07 \times 10^{-2}$ |
| 0.05 | 0.03 | $5.35 \times 10^{-3}$ |

**a**  Find the order of reaction with respect to  **(i)** P  **(ii)** Q.

**b**  Calculate the rate constant, $k$.

**7**   Results of measurements on the course of the reaction **A** $\longrightarrow$ **B** are shown below.

| [A]/mol dm$^{-3}$ | 0.150 | 0.119 | 0.0630 | 0.0750 | 0.0600 | 0.0460 | 0.0375 |
|---|---|---|---|---|---|---|---|
| Time/minutes | 0 | 10 | 20 | 30 | 40 | 50 | 60 |

Use the results to show that the reaction is first order with respect to **A**.

**8**   Iodine reacts with propanone in acidic solution:

$$CH_3COCH_3(aq) + I_2(aq) \longrightarrow CH_3COCH_2I(aq) + HI(aq)$$

In a study of the reaction, 0.10 mol of propanone and 0.010 mol of hydrochloric acid were placed in a flask at constant temperature. $4.0 \times 10^{-4}$ mol of iodine in 10 cm$^3$ of water was added, and timing was begun. 10 cm$^3$ samples were withdrawn periodically at 10 minute intervals. They were neutralised with an excess of sodium hydrogencarbonate and titrated with 0.010 mol dm$^{-3}$ sodium thiosulphate. The results are shown below.

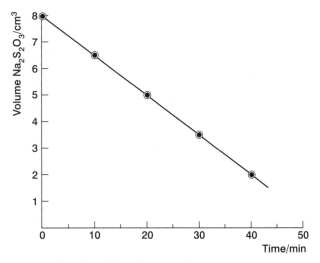

Fig. 13.9   *Graph of volume of sodium thiosulphate against time*

**a**  State the order with respect to iodine.

**b**  If the initial concentration of iodine were doubled, what would be the effect on the gradient of the graph?

**c**  If the concentration of HCl is doubled, the rate doubles. What can you deduce from this?

**9** The reaction $A \longrightarrow B$ was studied in the gas phase. The results obtained at a certain temperature are shown below.

| Time/hour | 0 | 2 | 5 | 10 | 20 | 30 |
|---|---|---|---|---|---|---|
| % of A reacted | 0 | 9 | 21 | 37 | 60 | 75 |

**a** Plot a graph of % of A reacted against time.

**b** Use the graph to find **(i)** the order of reaction with respect to A **(ii)** the half-life and **(iii)** the initial rate (in % A hour$^{-1}$) of the reaction.

**c** If the initial concentration of A were multiplied by 4, what would be **(i)** the half-life and **(ii)** the initial rate of reaction?

**10** **a** For an atom of the isotope $^{47}_{20}Ca$ state the number of protons and the number of neutrons.

**b** The isotope decays by emitting a $\beta$ particle. Complete the equation

$$^{47}_{20}Ca \longrightarrow {}^{0}_{-1}e +$$

**c** The half-life of $^{47}_{20}Ca$ is 6.5 hours. What fraction will remain after 13 hours? Is this a suitable half-life to allow the isotope to be used as a tracer for diagnostic purposes in hospital patients?

## EXERCISE 13.7 Questions from examination papers

**1** **a** The order of reaction for the vapour phase decomposition of ethanal (CH$_3$CHO) into methane and carbon monoxide can be determined experimentally.

$$CH_3CHO(g) \longrightarrow CH_4(g) + CO(g)$$

The reaction is catalysed by the oxides of transition elements.

   **(i)** Explain what you understand by *order of reaction*. (1)

   **(ii)** Suggest how the rate of the above reaction at constant temperature could be monitored. (1)

   **(iii)** On a copy of the axes below, draw a curve to show the distribution of molecular energies in a gas at temperature, *T*. (1)

   **(iv)** Explain how a catalyst increases the rate of the reaction. (2)

**b** In an experiment where the initial concentration of ethanal was steadily increased the following results for the initial rate of the reaction were obtained.

| $[CH_3CHO(g)]$ $/mol\ dm^{-3}$ | Initial rate $/mol\ dm^{-3}\ s^{-1}$ |
|---|---|
| 0.100 | $3.74 \times 10^{-4}$ |
| 0.200 | $1.50 \times 10^{-3}$ |
| 0.300 | $3.37 \times 10^{-3}$ |
| 0.400 | $5.98 \times 10^{-3}$ |
| 0.500 | $9.35 \times 10^{-3}$ |

   **(i)** On graph paper, plot a suitable graph, and hence deduce the order of this
       reaction. (3)

   **(ii)** Give a rate equation for this reaction. (1)

   **(iii)** Calculate the rate constant for this reaction, including its units. (2)

                                                                      OCR  (11)

**2**   The following information refers to a procedure to determine the *order of reaction* with
     respect to iodide ions for the reaction represented by the equation

$$2I^-(aq) + H_2O_2(aq) + 2H^+(aq) \longrightarrow 2H_2O(l) + I_2(aq)$$

Rate is measured by the time taken for the iodine produced to react with a small *fixed
amount* of sodium thiosulphate added to the constant volume system. The faster the iodine
is produced, the shorter the time taken for the sodium thiosulphate to be used up. The
reciprocal of this time can be used as a measure of the initial rate of reaction. The results
are given below.

| Experiment | $[KI(aq)]/mol\ dm^{-3}$ | Time $(t)/s$ | Reciprocal of time $(1/t)/s^{-1}$ |
|---|---|---|---|
| I | 0.004 | 74 | 0.0135 |
| II | 0.006 | 49.4 | 0.0202 |
| III | 0.008 | 37 | 0.0270 |
| IV | 0.010 | 30 | 0.0333 |
| V | 0.012 | 25 | 0.0400 |

**a**   **(i)** In the experiments the concentrations of acid and hydrogen peroxide were far
         more concentrated than that of potassium iodide. Explain why this was
         necessary. (1)

   **(ii)** In each of the experiments the aqueous hydrogen peroxide was always added last.
         State why this was necessary. (1)

   **(iii)** Explain why the volume of the system was kept constant for all the
         experiments. (1)

**b**   **(i)** Plot a graph of the initial rate $(1/t)$ on the vertical axis against the concentration of
         potassium iodide used on the horizontal axis. (2)

   **(ii)** Use your graph to determine the order, **n**, of the reaction with respect to iodide
         ions. Carefully state your reasoning. (3)

**c**   Further studies show that the rate equation for the reaction is

$$\text{Rate}(mol\ dm^{-3}\ s^{-1}) = k[I^-(aq)]^n[H_2O_2(aq)]$$

    **(i)** From this overall rate equation, state what can be deduced about the role of aqueous hydrogen ions. (1)

    **(ii)** State the units of $k$ in the above rate equation. (1)

    *WJEC* (10)

**3**   **a**  Catalysts are widely used in industry to alter the rate of a chemical reaction. Describe how catalysts carry out this function. Your answer should contain reference to activation energy, homogeneous catalysis and heterogeneous catalysis. Include diagrams and appropriate examples. (12)

    **b**  The hydrolysis of the ester ethyl ethanoate is represented by the following equation.

$$CH_3COOC_2H_5 + H_2O \rightleftharpoons CH_3COOH + C_2H_5OH$$

A student investigating the rate of this reaction obtained the results shown below.

| Concentration of ester/mol dm$^{-3}$ | Time/$10^4$ s |
|---|---|
| 0.50 | 0.00 |
| 0.39 | 0.20 |
| 0.31 | 0.40 |
| 0.17 | 0.90 |
| 0.13 | 1.10 |
| 0.09 | 1.40 |

    **(i)** Plot a graph of these results. Deduce, with a reason, the order of the reaction with respect to ethyl ethanoate.

    **(ii)** Write the rate equation for this reaction, assuming that it is zero order with respect to water. Explain any symbols you use.

    **(iii)** Use your graph to find the rate of reaction after $0.60 \times 10^4$ seconds. (9)

    *OCR* (21)

**4**   Iodine reacts with propanone in the presence of an acid catalyst according to the equation:

$$CH_3COCH_3 + I_2 \xrightarrow{\text{H}^+} CH_3COCH_2I + HI$$

Data concerning an experiment to determine the rate equation for this reaction are given in the following table:

| Relative concentrations | | | Relative rate |
|---|---|---|---|
| [CH$_3$COCH$_3$] | [I$_2$] | [H$^+$] | |
| 1 | 1 | 1 | 2 |
| 1 | 2 | 1 | 2 |
| 2 | 1 | 1 | 4 |
| 1 | 1 | 2 | 4 |

  **a**   **(i)** State why a rate equation cannot be written from a knowledge only of the chemical equation representing the reaction. (2)

    **(ii)** Use the data to deduce the order of reaction with respect to propanone, iodine, hydrogen ions. (3)

**(iii)** What does the value for iodine tell you about the part that iodine plays in the rate determining step of the reaction? (1)

**(iv)** Write the rate equation for the reaction. (1)

**b** **(i)** Use the bond enthalpies given, in kJ mol$^{-1}$, to calculate the enthalpy change for the iodination of propanone: (2)

| C—H | I—I | C—I | H—I |
|------|------|------|------|
| +413 | +151 | +228 | +298 |

**(ii)** Draw an enthalpy level diagram for the reaction, showing on it both catalysed and uncatalysed pathways. (3)

*L* (12)

**5** It is often important that industrial chemical processes are carried out reasonably quickly. Chemists therefore need to understand the factors which affect rates of reaction.

**a** Explain what is meant by the terms

**(i)** activation energy, (include a diagram in your answer)

**(ii)** order of reaction,

**(iii)** half-life of a reaction,

**(iv)** rate-determining step. (11)

**b** The concentration of hydrochloric acid affects the rate of its reaction with magnesium. This can be investigated by measuring the time taken for a fixed mass of magnesium to dissolve in acid solutions of different concentrations. Using this method, the following results were obtained.

| Concentration of HCl/ mol dm$^{-3}$ | 1.0 | 0.8 | 0.3 |
|-------------------------------------|-----|-----|-----|
| Time/s | 12 | 19 | 125 |

**(i)** Plot a graph of

1. 1/time against concentration,
2. 1/time against (concentration)$^2$.

**(ii)** Given that 1/time is a measure of the rate of reaction, explain why these graphs show that the reaction is neither zero order nor first order with respect to hydrochloric acid.

**(iii)** Suggest why these graphs show that the reaction is second order with respect to hydrochloric acid.

**(iv)** Assuming that the reaction is second order, write a rate equation for this reaction and hence calculate a value for the rate constant. What are the units of the rate constant? (10)

*OCR* (21)

6 The rate equation for a reaction between substances **A**, **B** and **C** is of the form:

$$\text{Rate} = k[\textbf{A}]^x[\textbf{B}]^y[\textbf{C}]^z \text{ where } x + y + z = 4$$

The following data were obtained in a series of experiments at a constant temperature.

| Experiment | Initial concentration of **A**/mol dm$^{-3}$ | Initial concentration of **B**/mol dm$^{-3}$ | Initial concentration of **C**/mol dm$^{-3}$ | Initial rate/ mol dm$^{-3}$ s$^{-1}$ |
|---|---|---|---|---|
| 1 | 0.10 | 0.20 | 0.20 | $8.0 \times 10^{-5}$ |
| 2 | 0.10 | 0.05 | 0.20 | $2.0 \times 10^{-5}$ |
| 3 | 0.05 | 0.10 | 0.20 | $2.0 \times 10^{-5}$ |
| 4 | 0.10 | 0.10 | 0.10 | to be calculated |

a Use the data in the table to deduce the order of reaction with respect to **A** and the order of reaction with respect to **B**. Hence deduce the order of reaction with respect to **C**. (3)

b Calculate the value of the rate constant, $k$, stating its units and also the value of the initial rate in experiment 4. (4)

c How does the value of $k$ change when the temperature of the reaction is increased? (1)
*NEAB 99* (8)

7 a A fixed mass of marble is reacted with dilute hydrochloric acid at a constant temperature. Explain why the rate of the reaction is increased if the lumps of marble are reduced in size. (2)

b The initial rate of the reaction between substances **A** and **B** was measured in a series of experiments and the following rate equation was deduced.

$$\text{Rate} = k[\textbf{A}][\textbf{B}]^2$$

(i) Complete a copy of the table of data below for the reaction between **A** and **B**.

| Expt | Initial [**A**] /mol dm$^{-3}$ | Initial [**B**] /mol dm$^{-3}$ | Initial rate /mol dm$^{-3}$ s$^{-1}$ |
|---|---|---|---|
| 1 | 0.020 | 0.020 | $1.2 \times 10^{-4}$ |
| 2 | 0.040 | 0.040 | |
| 3 | | 0.040 | $2.4 \times 10^{-4}$ |
| 4 | 0.060 | 0.030 | |
| 5 | 0.040 | | $7.2 \times 10^{-4}$ |

(ii) Using the data for Experiment 1, calculate a value for the rate constant, $k$, and state its units. (7)
*NEAB 99* (9)

8 Oxygen reacts with nitrogen monoxide to make nitrogen dioxide.

$$O_2(g) + 2NO(g) \rightleftharpoons 2NO_2(g)$$

In an experiment to investigate the effects of changing concentrations on the rate of reaction, the following results were obtained.

| Experiment number | Initial concentration $O_2/10^{-2}$ mol dm$^{-3}$ | Initial concentration $NO/10^{-2}$ mol dm$^{-3}$ | Initial rate of disappearance of $NO/10^{-4}$ mol dm$^{-3}$ s$^{-1}$ |
|---|---|---|---|
| 1 | 1.0 | 1.0 | 0.7 |
| 2 | 1.0 | 2.0 | 2.8 |
| 3 | 1.0 | 3.0 | 6.3 |
| 4 | 2.0 | 2.0 | 5.6 |
| 5 | 3.0 | 3.0 | 18.9 |

**a** Deduce the order of reaction with respect to $O_2$ and NO, explaining your reasoning. (4)

**b** **(i)** Write the rate equation for this reaction.

**(ii)** Calculate the value of the rate constant, $k$. State its units. (3)

**c** When the rate of disappearance of NO is $2.8 \times 10^{-4}$ mol dm$^{-3}$ s$^{-1}$, determine the rate of disappearance of $O_2$. (1)

OCR   (8)

# Equilibria

# 14

## 14.1 CHEMICAL EQUILIBRIUM

An example of a reversible reaction between gases is the reaction between hydrogen and iodine to form hydrogen iodide:

$$H_2(g) + I_2(g) \rightleftharpoons 2HI(g)$$

If the reaction takes place in a closed vessel, the combination of hydrogen and iodine gradually slows down as the concentrations of these gases decrease. At first, there is very little decomposition of hydrogen iodide into hydrogen and iodine, but, as the concentration of hydrogen iodide increases, the rate of decomposition of hydrogen iodide into hydrogen and iodine increases until the rates of the forward and reverse reactions are equal, and the concentration of each species is constant.

An example of a reversible reaction which takes place in solution is the reaction between ethanoic acid and ethanol to form ethyl ethanoate:

$$CH_3CO_2H(l) + C_2H_5OH(l) \rightleftharpoons CH_3CO_2C_2H_5(l) + H_2O(l)$$

As the concentrations of ester and water increase, the reverse reaction – hydrolysis of the ester to form the acid and alcohol – speeds up. At equilibrium, the rate of the forward reaction is equal to the rate of the reverse reaction. Esterification is catalysed by inorganic acids. The presence of a catalyst speeds up the rate at which equilibrium is established.

## 14.2 THE EQUILIBRIUM LAW

### Reactions in solution

If a reversible reaction is allowed to reach equilibrium, it is found that the product of the concentrations of the products divided by the product of the concentrations of the reactants has a constant value at a particular temperature. In the esterification reaction,

$$CH_3CO_2H(l) + C_2H_5OH(l) \rightleftharpoons CH_3CO_2C_2H_5(l) + H_2O(l)$$

it is found that

$$\frac{[CH_3CO_2C_2H_5][H_2O]}{[CH_3CO_2H][C_2H_5OH]} = K_c$$

where $K_c$ is the equilibrium constant for the reaction in terms of concentration.

EXAMPLE 1    Calculate the equilibrium constant for the esterification

$$CH_3CO_2H(l) + C_2H_5OH(l) \rightleftharpoons CH_3CO_2C_2H_5(l) + H_2O(l)$$

1.00 mol of ethanoic acid was allowed to react with 4.00 mol of ethanol. At equilibrium, the amount of ethanoic acid remaining was 0.07 mol.

**METHOD**

Since 0.07 mol ethanoic acid remains, 0.93 mol has reacted and 0.93 mol of ethyl ethanoate and 0.93 mol of water have been formed.

The amount of ethanol that has reacted is the same as the amount of acid, 0.93 mol, so the amount of ethanol remaining is 4.00 mol – 0.93 mol = 3.07 mol.

Let $V$ be the volume of the mixture.

$$K_c = \frac{[CH_3CO_2C_2H_5][H_2O]}{[CH_3CO_2H][C_2H_5OH]}$$

$$= \frac{(0.93 \text{ mol}/V) \times (0.93 \text{ mol}/V)}{(0.07 \text{ mol}/V)(3.07 \text{ mol}/V)} = 4.0$$

Note that the concentration units cancel out so $K_c$ has no unit.

**ANSWER**

The value of the equilibrium constant is 4.0.

## Reactions in the gas phase

**Reminder :** Partial pressure of a gas § 10.4.

In the reaction between hydrogen and iodine,

$$H_2(g) + I_2(g) \rightleftharpoons 2HI(g)$$

and

$$\frac{[HI]^2}{[H_2][I_2]} = K_c$$

Since this is a reaction between gases, the concentration of each gas can be expressed as a partial pressure. Then,

$$\frac{p_{HI}^2}{p_{H_2} \times p_{I_2}} = K_p$$

$K_p$ is the equilibrium constant in terms of partial pressures.

**EXAMPLE 2**

One mole of hydrogen and one mole of iodine are mixed and allowed to reach equilibrium at 300 K and a pressure of 1 atm. The amount of hydrogen iodide present at equilibrium is 1.5 mol. Calculate the equilibrium constant $K_p$ at 300 K.

**METHOD**

The equation,

$$H_2(g) + I_2(g) \rightleftharpoons 2HI(g)$$

shows that, if 1.5 mol of HI has been formed, then 0.75 mol of $H_2$ has reacted and 0.25 mol of $H_2$ remains. The amount of $I_2$ that remains is the same.

Mole fractions are: $H_2$ : 0.25/2.0, $I_2$ : 0.25/2.0, HI 1.5/2.0

Partial pressures are:

$$p_{H_2} = p_{I_2} = 1 \text{ atm} \times 0.25/2.0$$
$$p_{HI} = 1 \text{ atm} \times 1.5/2.0$$
$$K_p = \frac{p_{HI}^2}{p_{H_2} \times p_{I_2}}$$
$$= \frac{(1 \text{ atm} \times 1.5/2.0)^2}{(1 \text{ atm} \times 0.25/2.0)^2}$$
$$= 36$$

Notice that the pressure units cancel out and $K_p$ for this reaction has no unit.

**ANSWER**     At 300 K, the value of $K_p$ is 36.

**EXAMPLE 3**     Phosphorus pentachloride dissociates at high temperatures:

$$PCl_5(g) \rightleftharpoons PCl_3(g) + Cl_2(g)$$

The system reaches equilibrium at a certain temperature when 39% of a sample of $PCl_5$ has dissociated. The total equilibrium pressure is 2.00 atm. Calculate the value of the equilibrium constant.

**METHOD**

$$K_p = \frac{p_{PCl_3} \times p_{Cl_2}}{p_{PCl_5}}$$

| Since | $PCl_5(g) \rightleftharpoons$ | $PCl_3(g)$ + | $Cl_2(g)$ | Total |
|---|---|---|---|---|
| Original amount/mol | 1.00 | 0 | 0 | 1.00 |
| Equilibrium amount/mol | 0.61 | 0.39 | 0.39 | 1.39 |

Mole fractions are    $PCl_5 : 0.61 \text{ mol}/1.39 \text{ mol}$,
$PCl_3 : 0.39 \text{ mol}/1.39 \text{ mol}$
$Cl_2 : 0.39 \text{ mol}/1.39 \text{ mol}$

Partial pressures are:    $p_{PCl_5} = 2.00 \text{ atm} \times 0.61/1.39$
$p_{PCl_3} = 2.00 \text{ atm} \times 0.39/1.39$
$p_{Cl_2} = 200 \text{ atm} \times 0.39/1.39$
$$K_p \frac{(2.00 \text{ atm} \times 0.39/1.39)^2}{(2.00 \text{ atm} \times 0.61/1.39)} = 0.359 \text{ atm}$$

**ANSWER**     The value of $K_p$ is 0.36 atm.

## Reactions between solids and gases

In the reaction between iron and steam,

$$3Fe(s) + 4H_2O(g) \rightleftharpoons Fe_3O_4(s) + 4H_2(g)$$

The equilibrium constant is given by

$$\frac{p_{H_2}^4}{p_{H_2O}^4} = K_p$$

The solids do not appear in the expression. Their vapour pressures are constant (at a constant temperature) as long as there is some of each solid present. These constant vapour pressures are incorporated into the value of the constant $K_p$.

**EXAMPLE 4**

A mixture of iron and steam is allowed to come to equilibrium at 600 °C. The equilibrium pressures of hydrogen and steam are 3.2 kPa and 2.4 kPa. Calculate the equilibrium constant $K_p$ for the reaction.

**METHOD**

The reaction is

$$3Fe(s) + 4H_2O(g) \rightleftharpoons 4H_2(g) + Fe_3O_4(s)$$

The equilibrium constant is given by

$$K_p = \frac{p_{H_2}^4}{p_{H_2O}^4}$$

Substituting in this equation gives

$$K_p = \left(\frac{3.2}{2.4}\right)^4$$

**ANSWER**

$$K_p = 3.1$$

**EXERCISE 14.1** **Problems on equilibria**

> **Reminder** : Mole fraction of **A** in mixture $= \dfrac{\text{Number of moles of } \mathbf{A} \text{ in mixture}}{\text{Total number of moles in mixture}}$ § 10.3.
>
> Partial pressure = Mole fraction × Total pressure § 10.4.

**1** Write the expression for the equilibrium constant for the reaction,

$$A + B \rightleftharpoons C + D$$

**a** when **A**, **B**, **C** and **D** are gases

**b** when **A**, **B**, **C** and **D** are solutions

**c** when **A** and **C** are solids, **B** and **D** are gases

**d** when **C** is a solid and **A**, **B** and **D** are solutions.

**2** Hydrogen and iodine combine to give hydrogen iodide in a reversible reaction:

$$H_2(g) + I_2(g) \rightleftharpoons 2HI(g)$$

At a certain temperature, the equilibrium partial pressures of the gases are:

| Gas | Partial pressure/kPa |
|-----|---------------------|
| $H_2$ | 2.5 |
| $I_2$ | 22.5 |
| HI | 90.0 |

Calculate the value of the equilibrium constant $K_p$ at this temperature.

**STEPS TO TAKE**

First write the expression for $K_p$. Then put in the partial pressures of the gases.

**3** One mole of ethanol and 1 mol of ethanoic acid were mixed and allowed to reach equilibrium at 25 °C.

$$C_2H_5OH(l) + CH_3CO_2H(l) \rightleftharpoons CH_3CO_2C_2H_5(l) + H_2O(l)$$

The amount of ethyl ethanoate in the equilibrium mixture was found to be twice the amount of ethanol. Calculate the equilibrium constant $K_c$ for the reaction at 25 °C.

**STEPS TO TAKE**    Write the expression for $K_c$. Let amount of ethanol used = $x$ mol; then amount of ethanol remaining = $(1 - x)$ mol. Also amount of ethanoic acid remaining = $(1 - x)$ mol. You know that the amount of ethyl ethanoate is twice this amount and is equal to the amount of water. Concentration = amount/volume. Now put the concentrations into the expression for $K_c$.

**4** Equal amounts of hydrogen and iodine are allowed to reach equilibrium:

$$H_2(g) + I_2(g) \rightleftharpoons 2HI(g)$$

If 80% of the hydrogen is converted into hydrogen iodide, what is the value of $K_p$ at this temperature?

**STEPS TO TAKE**    Say that 1 mol of $H_2$ and 1 mol of $I_2$ are used. Then when 80% of $H_2$ has reacted, amount of $H_2$ left = 0.20 mol = amount of $I_2$ left, and amount of HI = 1.60 mol

$$H_2(g) + I_2(g) \rightleftharpoons 2HI(g)$$

| | | | |
|---|---|---|---|
| Initial amounts/ mol | 1 | 1 | 0 |
| Equilibrium amounts | 0.20 | 0.20 | 1.60 |

Now work out the mole fraction of each component and the partial pressure of each component. Then find $K_p$

**5** Equal amounts of ethanol and ethanoic acid were mixed and allowed to reach equilibrium at room temperature. 60% of the acid esterified. Calculate $K_c$ for the reaction.

**STEPS TO TAKE**    The equation is

$$C_2H_5OH(l) + CH_3CO_2H(l) \rightleftharpoons CH_3CO_2C_2H_5(l) + H_2O(l)$$
$$\text{Ethanol} \qquad \text{Ethanoic acid} \qquad \text{Ethyl ethanoate}$$

Say that 1 mol of ethanol and 1 mol of acid were used. Then the equilibrium amounts are: 0.40 mol ethanol, 0.40 mol acid, 0.60 mol ester. Concentration = amount/volume. Now put the values of concentration into the expression for $K_c$.

**6** When ethane is 'cracked',

$$C_2H_6(g) \rightleftharpoons C_2H_4(g) + H_2(g)$$

When 1.00 mol ethane is cracked at 180 kPa and 1000 K the yield of ethene is 0.36 mol. Calculate the value of $K_p$ at 1000 K.

**Hint :** Amount of ethane left ... amount of hydrogen ... partial pressures ... $K_p$.

**7** A mixture of 3 mol of hydrogen and 1 mol of nitrogen is allowed to reach equilibrium at a pressure of $5 \times 10^6 \text{ N m}^{-2}$. The composition of the gaseous mixture is then 8% $NH_3$, 23% $N_2$, 69% $H_2$ by volume. Calculate $K_p$.

---

**Hint :** Mole fractions of $NH_3$, $N_2$, $H_2$ ... partial pressures ... $K_p$.

---

**8** The equilibrium constant $K_p$ for the reaction

$$H_2(g) + I_2(g) \rightleftharpoons 2HI(g)$$

is 60 at 450 °C. The number of moles of hydrogen iodide in equilibrium with 2 mol of hydrogen and 0.3 mol of iodine at 450 °C is

**A** 1/100 **B** 1/10 **C** 6 **D** 36 **E** 3

**STEPS TO TAKE** Let amount of HI = $x$ mol. Then work out the mole fractions of $H_2$, $I_2$, HI ... and their partial pressures ... and insert these into the expression for $K_p$.

**9** A mixture of 1 mol of nitrogen and 3 mol of hydrogen is allowed to reach equilibrium at $1.0 \times 10^7$ Pa and 500 °C. The equilibrium mixture contains 20% of ammonia by volume. Calculate the value of $K_p$ at 500 °C for the reaction

$$N_2(g) + 3H_2(g) \rightleftharpoons 2NH_3(g)$$

**10** The oxidation of sulphur dioxide is a reversible process:

$$2SO_2(g) + O_2(g) \rightleftharpoons 2SO_3(g)$$

**a** Calculate the value of the equilibrium constant $K_p$ at 1000 K from the following data on partial pressures at 1000 K.

| Partial pressures/N m$^{-2}$ | | |
|---|---|---|
| $p(SO_2)$ | $p(O_2)$ | $p(SO_3)$ |
| 10 000 | 68 800 | 80 100 |

**b** Sulphur dioxide and oxygen in the ratio 2 mol : 1 mol were allowed to reach equilibrium at a pressure of 5 atm. At equilibrium, $\frac{1}{3}$ of the $SO_2$ was converted into $SO_3$. Calculate the equilibrium constant $K_p$ for the reaction.

**11** Ethanol is manufactured by the catalytic hydration of ethene.

$$C_2H_4(g) + H_2O(g) \rightleftharpoons C_2H_5OH(g); \quad \Delta H^\ominus = -45 \text{ kJ mol}^{-1}$$

**a** When ethene and water were mixed in a ratio of 1.0 mol $C_2H_4$ : 0.6 mol $H_2O$ at 50 atm, what was the partial pressure of ethene?

**b** In the equilibrium mixture formed at 570 K, the partial pressures of the three gases were: ethene 36.77 atm, water 21.29 atm, ethanol 1.94 atm. Calculate the total pressure of the system.

**c** What percentage of ethene has been converted into ethanol?

**d** **(i)** Write an expression for $K_p$ for the reaction.

**(ii)** Calculate the value of $K_p$ at 570 K.

**EXERCISE 14.2    Questions from examination papers**

**1**    At a temperature of 107 °C the reaction

$$CO(g) + 2H_2(g) \rightleftharpoons CH_3OH(g)$$

reaches equilibrium under a pressure of 1.59 MPa with 0.122 mol of carbon monoxide and 0.298 mol of hydrogen present at equilibrium in a vessel of volume 1.04 dm$^3$.

Use these data to answer the questions that follow.

**a**  Assuming ideal gas behaviour, determine the total number of moles of gas present. Hence calculate the number of moles of methanol in the equilibrium mixture.    (3)

**b**  Calculate the value of the equilibrium constant, $K_c$, for this reaction and state its units.    (3)

**c**    **(i)** Write an expression for the equilibrium constant, $K_p$, for this equilibrium.

**(ii)** Calculate the mole fraction of each of the three gases present in the equilibrium mixture.

**(iii)** Calculate the partial pressure of hydrogen present in the equilibrium mixture.

**(iv)** Calculate the value of the equilibrium constant, $K_p$, and state its units.    (8)

*NEAB 99*  (14)

**2**  **a**  **(i)** Give balanced equations for the **three** main stages in the Contact Process for the manufacture of sulphuric acid.    (3)

**(ii)** Name the catalyst used in the Contact Process, and state its function.    (2)

**b**  Sulphuric acid vapour dissociates when heated according to the equation

$$H_2SO_4(g) \rightleftharpoons H_2O(g) + SO_3(g)$$

**(i)** Write an expression for the equilibrium constant, $K_p$, for the reaction, stating the units.    (2)

**(ii)** The following data show the equilibrium partial pressures of $H_2SO_4$, $H_2O$ and $SO_3$ at two temperatures:

| $T/K$ | Partial pressures/Pa | | |
|-------|-------------|----------|---------|
|       | $H_2SO_4(g)$ | $H_2O(g)$ | $SO_3(g)$ |
| 400   | 4.5         | 3.2      | 2.9     |
| 493   | 470         | 300      | 270     |

I.  Calculate the equilibrium constant, $K_p$, at each temperature. Hence deduce, with an explanation, whether the dissociation of sulphuric acid vapour is exothermic or endothermic.    (3)

II.  Sulphuric acid vapour was placed in a flask at 400 K, and left to decompose for a time, after which the partial pressures (Pa) in the mixture were found to be.

$$H_2O(g) = 1.0 \quad SO_3(g) = 1.0 \quad H_2SO_4(g) = 5.0$$

Deduce whether or not the mixture has achieved equilibrium, explaining your answer.    (2)

**c** **(i)** State how *dilute* sulphuric acid may be safely prepared from the concentrated acid. (1)

**(ii)** The sulphuric acid molecule, $H_2SO_4$, contains two S=O bonds. Use a sketch to show the bonds present in the $H_2SO_4$ molecule, and name its shape. (2)

**(iii)** A solution of a sodium salt, NaX, produces a yellow precipitate with lead(II) nitrate solution. Solid NaX reacts with hot *concentrated* sulphuric acid producing several products, including a purple vapour. Identify NaX (no explanation is required), and write an **ionic** equation for the reaction of NaX(aq) with $Pb(NO_3)_2$(aq). (2)

**d** Ethanol is dehydrated with excess concentrated $H_2SO_4$ at 170 °C producing compound **A**. Reaction of **A** with $H_2O$ at high pressure and temperature produces compound **B** (a volatile liquid). Reaction of **A** with hot $H_2$ and a Ni catalyst produces compound **C**. State the names of **A**, **B** and **C**. (3)

*WJEC* (20)

**3** The dissociation of hydrogen iodide can be represented by the equation below.

$$2HI(g) \rightleftharpoons H_2(g) + I_2(g) \qquad \Delta H^{\ominus} = -53 \text{ kJ mol}^{-1}$$

**a** Using this reaction as an example, explain what is meant by

**(i)** a reversible reaction,

**(ii)** a dynamic equilibrium. (2)

**b** Explain, giving a reason, the effect on the equilibrium above of

**(i)** increasing the temperature whilst keeping the pressure constant,

**(ii)** increasing the pressure whilst keeping the temperature constant. (4)

**c** State the effect on the equilibrium above of adding a catalyst whilst keeping pressure and temperature constant. (1)

**d** At a given temperature, in a vessel of volume 10 $dm^3$, one mole of hydrogen iodide was found to be 20% dissociated.

**(i)** Write an expression for the equilibrium constant, $K$, for the following equilibrium.

$$2HI(g) \rightleftharpoons H_2(g) + I_2(g)$$

**(ii)** Calculate how many moles of each constituent are present in the mixture at equilibrium.

**(iii)** Calculate the value of the equilibrium constant, $K$.

**(iv)** Determine the units of the equilibrium constant, $K$.

**(v)** If the reaction were carried out in a vessel of volume 20 $dm^3$, assuming all other conditions remain the same, the value of $K$ is unchanged. Suggest a reason why the value of $K$ is unchanged. (6)

*OCR* (13)

**4  a** The reaction between nitrogen and hydrogen is given by the following equation:

$$N_2 + 3H_2 \rightleftharpoons 2NH_3$$

   **(i)** Write an expression for the equilibrium constant, $K_c$, for the reaction. (1)

   **(ii)** Nitrogen and hydrogen were mixed at 450 °C and the system allowed to reach equilibrium. The equilibrium concentrations of nitrogen, hydrogen and ammonia were 0.600, 1.80 and 0.800 mol dm$^{-3}$ respectively. Calculate the value of the equilibrium constant, $K_c$, at 450 °C and state its units. (2)

   **b** When ammonia is dissolved in water the following reaction occurs:

$$NH_3(g) + H_2O(l) \rightleftharpoons NH_4^+(aq) + OH^-(aq)$$

   The pH of a 0.110 mol dm$^{-3}$ solution of ammonia is 11.1 at 25 °C. The value of the ionic product for water, $K_w$, is $1.00 \times 10^{-14}$ mol$^2$ dm$^{-6}$ at 25 °C. Calculate

   **(i)** the hydrogen ion concentration, (1)

   **(ii)** the concentration of OH$^-$ ions. (2)

   **c** **(i)** Explain why it is that a solution containing ethanoic acid and sodium ethanoate acts as a buffer solution. (3)

   **(ii)** 25.00 cm$^3$ of a solution of ethanoic acid required 21.20 cm$^3$ of aqueous sodium hydroxide for neutralisation. State how you would use this information to prepare a buffer solution containing equal concentrations (mol dm$^{-3}$) of ethanoic acid and sodium ethanoate. Explain your method. (3)

   **d** State the reagents and conditions used to carry out the conversions **(i)** and **(ii)** below. Give the type of reaction in each case.

   **(i)** Propan-1-ol into propanoic acid. (2)

   **(ii)** Sodium propanoate into ethane. (2)

   *WJEC* (16)

**5** Phosphorus(V) chloride dissociates at high temperatures according to the equation

$$PCl_5(g) \rightleftharpoons PCl_3(g) + Cl_2(g)$$

83.4 g of phosphorus(V) chloride are placed in a vessel of volume 9.23 dm$^3$. At equilibrium at a certain temperature, 11.1 g of chlorine are produced at a total pressure of 250 kPa.

Use these data, where relevant, to answer the questions that follow.

   **a** Calculate the number of moles of each of the gases in the vessel at equilibrium. (3)

   **b** **(i)** Write an expression for the equilibrium constant, $K_c$, for the above equilibrium.

   **(ii)** Calculate the value of the equilibrium constant, $K_c$, and state its units. (4)

   **c** **(i)** Write an expression for the equilibrium constant, $K_p$, for the above equilibrium.

   **(ii)** Calculate the mole fraction of chlorine present in the equilibrium mixture.

   **(iii)** Calculate the partial pressure of PCl$_5$ present in the equilibrium mixture.

   **(iv)** Calculate the value of the equilibrium constant, $K_p$, and state its units. (7)

   *NEAB 98* (14)

**6** Water gas is a mixture of hydrogen and carbon monoxide. It is made by passing steam over heated coke.

$$H_2O(g) + C(s) \rightleftharpoons H_2(g) + CO(g)$$

It was used for many years as a commercial fuel.

**a** This reaction was carried out at normal atmospheric pressure. Suggest, and explain, how increasing the pressure affects:

  **(i)** the time taken to reach equilibrium

  **(ii)** the equilibrium yield of water gas. (4)

**b** Write the expression for the equilibrium constant, $K_p$, for this reaction. (1)

**c** During the conversion of steam into water gas, it was found that 30% of the steam had been converted.

  **(i)** Calculate the partial pressure of each gas in the equilibrium mixture if the total pressure was 100 kPa.

  **(ii)** Calculate the value of $K_p$ under these conditions. Include the units of $K_p$ in your answer. (5)

*OCR* (10)

**7 a (i)** The reaction of sulphur dioxide and oxygen to produce sulphur trioxide,

$$2SO_2(g) + O_2(g) \longrightarrow 2SO_3(g) \qquad [1]$$

(the 'forward reaction') obeys the rate equation:

$$\text{Rate of reaction} = k\,[SO_2(g)][SO_3(g)]^{-1/2} \qquad [2]$$

  I. State the *overall* order of this reaction. (1)

  II. State what is unusual about the form of the rate equation. (1)

  III. Discuss, giving reasons, whether or not the rate equation is consistent with a mechanism in which $SO_2$ and $O_2$ react by a simple collision. (2)

  IV. Outline how the rate of consumption of $SO_2$, and the individual order of the reaction with respect to $SO_2$, may be determined experimentally. (5)

  V. State the effect on the rate of reaction [1] if the $SO_3$ concentration were to be increased by a factor of nine. (Assume that the $SO_2$ concentration and temperature remain fixed.) (1)

  **(ii)** The equilibrium constant for any reaction is related to the rate constants of the forward ($k_f$) and back ($k_b$) reactions by the expression

$$K_c = \frac{k_f}{k_b} \qquad [3]$$

It is also true, that at equilibrium

  *rate of forward reaction = rate of back reaction.* [4]

  I. Use expressions [3] and [4] in conjunction with equation [2], to determine the rate equation for the back reaction, i.e. $2SO_3(g) \longrightarrow 2SO_2(g) + O_2(g)$. (4)

II. State and explain the effect of a catalyst upon $K_c$ (Link your answer with expression [3] (2)

**b** 2-nitrophenol shows both intermolecular and intramolecular ('within the molecule') hydrogen-bonding. Draw molecular structures showing the two types of bonding present in pure 2-nitrophenol. State and explain which type of hydrogen-bonding would influence the boiling temperature of 2-nitrophenol. (4)

**c** **(i)** Water was added to an aluminium cooking pot, and the concentration of aluminium ions, $[Al^{3+}(aq)]$, released into the water measured. Known amounts of acid and base were added to the water, and $[Al^{3+}(aq)]$ measured at each pH, giving the following plot:

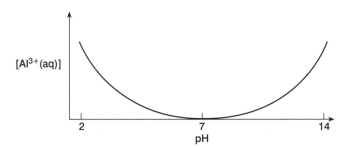

I. What does this plot show about the behaviour of aluminium *metal* in acids and bases?

II. From your own knowledge, briefly explain what evidence exists that aluminium *compounds* behave in a similar way to aluminium metal in acids and bases. (*Equations are not required.*)

III. Predict the general shape of a plot of concentration of ions released against pH, if the aluminium pot were replaced by an *iron* pot. (3)

**(ii)** When $Al_2Cl_6(s)$ is heated in a closed container, it is found that the gas pressure within the container increases faster than expected. Explain this observation, briefly discussing the chemistry involved. (2)

*WJEC(s)* (25)

# Organic chemistry

15

All the techniques which you need to tackle problems in organic chemistry have been covered in earlier chapters. You know methods of obtaining empirical and molecular formulae and methods of finding the molar masses of substances, including mass spectrometry. You can do calculations based on chemical equations. You have already calculated standard enthalpy changes, equilibrium constants and rates of reaction for organic reactions.

Numerical problems in organic chemistry give you some quantitative data and ask you to use it in conjunction with your knowledge of the reactions of organic compounds. There is no set pattern for tackling such problems. They are solved by a combination of calculation, familiarity with the reactions of the compounds involved and logic. The following examples and problems will show you what to expect.

**EXAMPLE 1**

When 0.2500 g of a hydrocarbon **X** burns in a stream of oxygen, it forms 0.7860 g of carbon dioxide and 0.3210 g of water. When 0.2500 g of **X** is vaporised, the volume which it occupies (corrected to s.t.p.) is 80.0 cm$^3$. Deduce the molecular formula of **X**.

Reminder : Empirical formula § 6.1. Molecular formula § 6.2.

**METHOD**

**X** burns to form carbon dioxide and water.

Mass of C in 0.7860 g of $CO_2 = \dfrac{12.0}{44.0} \times 0.7860$ g $=$ 0.2143 g

Mass of H in 0.3210 g of $H_2O = \dfrac{2.02}{18.0} \times 0.3210$ g $=$ 0.0360 g

Therefore 0.2500 g of **X** contains 0.2143 g of C and 0.0360 g of H

These masses give the molar ratio for C : H of $\dfrac{0.2143}{12.0}$ to $\dfrac{0.0360}{1.01}$

$$= 0.0178 \text{ to } 0.0360 = 1 \text{ to } 2$$

Thus, the empirical formula is $CH_2$.

Since 80.0 cm$^3$ is the volume occupied by 0.2500 g of **X**,

22.4 dm$^3$ is occupied by $\dfrac{22.4 \text{ dm}^3}{80.0 \times 10^{-3} \text{ dm}^3} \times 0.2500$ g of **X** $= 70.0$ g of **X**

The formula mass of $CH_2$ is 14. To give a molar mass of 70.0 g mol$^{-1}$, the empirical formula must be multiplied by 5. Therefore:

**ANSWER**

The molecular formula is $C_5H_{10}$.

**EXAMPLE 2**    An organic liquid, **P**, contains 52.2% carbon, 13.0% hydrogen and 34.8% oxygen by mass. Mild oxidation converts **P** into **Q**, and, on further oxidation, **R** is formed. **P** and **R** react to give **S**, which has a molecular formula of $C_4H_8O_2$. Identify compounds **P** to **S**, and explain the reactions involved.

**METHOD**    First, calculate the empirical formula of **P**. This comes to $C_2H_6O$. This must be the molecular formula also as **P** and **R** combine to form **S** which has 4C in the molecule. Since **P** contains one oxygen atom, it is an alcohol, an aldehyde, a ketone or possibly an ether. Other classes of compounds are ruled out by the absence of nitrogen and halogens. Oxidation proceeds in two stages, evidence that **P** is probably an alcohol, being oxidised first to an aldehyde and then to an acid.

According to the formulae, **P** would be $C_2H_5OH$, and **Q** would be $CH_3CHO$, and **R** would be $CH_3CO_2H$.

**ANSWER**    The reaction between **P** and **R** to form **S** is

$$CH_3CO_2H(l) + C_2H_5OH(l) \longrightarrow CH_3CO_2C_2H_5(l) + H_2O(l)$$

The molecular formula $C_4H_8O_2$ for **S** fits $CH_3CO_2C_2H_5$.

**P** = $C_2H_5OH$, ethanol; **Q** = $CH_3CHO$, ethanal; **R** = $CH_3CO_2H$, ethanoic acid; **S** = $CH_3CO_2C_2H_5$, ethyl ethanoate.

**EXAMPLE 3**    **X** is a liquid containing 31.5% by mass of C, 5.3% H and 63.2% O. An aqueous solution of **X** liberates carbon dioxide from sodium carbonate, with the formation of a solution from which a substance **Y** of formula $C_2H_3O_3Na$ can be obtained. **X** reacts with phosphorus(V) chloride to give hydrogen chloride and a compound, **Z**, of molecular formula $C_2H_2OCl_2$. Identify **X**, **Y** and **Z**. Write equations for the two reactions, **X** $\longrightarrow$ **Y** and **X** $\longrightarrow$ **Z**.

**METHOD**    The empirical formula of **X** is easily shown to be $C_2H_4O_3$. As **X** liberates carbon dioxide from a carbonate, it must be an acid. Taking $CO_2H$ from $C_2H_4O_3$ leaves $CH_3O$ as the formula for the rest of the molecule. This could be $CH_2OH$, making **X** $HOCH_2CO_2H$.

In the reaction with $PCl_5$, $C_2H_4O_3$ is converted into $C_2H_2OCl_2$. This would fit in with two hydroxyl groups being replaced by two chlorine atoms. This would be the case for the reaction

**ANSWER**    $$HOCH_2CO_2H(l) + 2PCl_5(l) \longrightarrow ClCH_2COCl(l) + 2POCl_3(l) + 2HCl(g)$$

If **X** is $HOCH_2CO_2H$, its reaction with sodium carbonate has an equation

$$2HOCH_2CO_2H(aq) + Na_2CO_3(s) \longrightarrow 2HOCH_2CO_2Na(aq) + CO_2(g) + H_2O(l)$$

and **Y** is $HOCH_2CO_2Na$.

All the information agrees with **X** = $HOCH_2CO_2H$, hydroxyethanoic acid, **Y** = $HOCH_2CO_2Na$, sodium hydroxyethanoate; **Z** = $ClCH_2COCl$, chloroethanoyl chloride.

**EXAMPLE 4**    1.220 g of a dicarboxylic aliphatic acid is dissolved in water and made up to 250 $cm^3$. A 25.0 $cm^3$ portion of the solution requires 21.0 $cm^3$ of 0.100 mol $dm^{-3}$ sodium hydroxide solution for neutralisation. Deduce the molecular formula of the acid, and write a structural formula for it.

**METHOD**  The equation for the neutralisation is

$$R(CO_2H)_2(aq) + 2NaOH(aq) \longrightarrow R(CO_2Na)_2(aq) + 2H_2O(l)$$

Amount of NaOH $= 21.0 \times 10^{-3}$ dm$^3 \times 0.100$ mol dm$^{-3} = 2.10 \times 10^{-3}$ mol

From the equation, No. of moles of acid $= \frac{1}{2} \times$ No. moles of NaOH
$$= 1.05 \times 10^{-3}$$ mol

Mass of acid $= \frac{1}{10} \times 1.220$ g $= 0.1220$ g

∴ Mass of $1.05 \times 10^{-3}$ mol $= 0.1220$ g

Mass of 1 mol $= 116$ g

The molar mass is 116 g mol$^{-1}$. Subtracting 90 for $(CO_2H)_2$ leaves 26 g mol$^{-1}$ for the rest of the molecule. This is the mass of $(CH_2)_2$. The molecular formula is $C_4H_4O_4$, and the structural formulae are

**ANSWER**

$$\begin{array}{ccc} HCCO_2H & \text{and} & HO_2CCH \\ \| & & \| \\ HCCO_2H & & HCCO_2H \end{array}$$

for *cis*- and *trans*-butenedioic acid.

**EXERCISE 15.1**   **Problems on organic chemistry**

> **Reminder :** Empirical formula § 6.1. Empirical formula by combustion § 7.4.
>
> Molecular formula § 6.2. Molar mass of a gas § 10.5. Units § 1.11.
>
> Amount = Mass/Molar mass § 3.1. Amount = Volume × Concentration § 8.1.
>
> GMV = 24.0 dm$^3$ mol$^{-1}$ at r.t.p. and 22.4 dm$^3$ mol$^{-1}$ at s.t.p.

**1**   A compound **X** contains carbon and hydrogen only. 0.135 g of **X**, on combustion in a stream of oxygen, gave 0.410 g of carbon dioxide and 0.209 g of water. Calculate the empirical formula of **X**.

**X** is a gas at room temperature, and 0.29 g of the gas occupy 120 cm$^3$ at room temperature and 1 atm. What is the molecular formula of **X**?

**2**   A monobasic organic acid **C** is dissolved in water and titrated with sodium hydroxide solution. 0.388 g of **C** require 46.5 cm$^3$ of 0.095 mol dm$^{-3}$ sodium hydroxide for neutralisation. Calculate the molar mass of **C** and deduce its formula.

**3**   An organic compound has a composition by mass of 83.5% C; 6.4% H; 10.1% O. The molar mass is 158 g mol$^{-1}$. Find the molecular formula.

**4**   A compound **A** contains 5.20% by mass of nitrogen. The other elements present are carbon, hydrogen and oxygen. Combustion of 0.0850 g of **A** in a stream of oxygen gave 0.224 g of carbon dioxide and 0.0372 g of water. Calculate the empirical formula of **A**.

5   An alkene **A** contains one double bond per molecule. 0.560 g of bromine is required to react completely with 0.294 g of **A**. When **A** is treated with ozone and the product is hydrolysed and then oxidised, **B**, which is a monobasic carboxylic acid is formed. 0.740 g of this acid require 100 cm$^3$ of 0.100 mol dm$^{-3}$ sodium hydroxide for neutralisation. Deduce the formulae of **A** and **B** and the structural formula of **A**.

6   An organic compound, **A**, contains 70.6% carbon, 5.88% hydrogen and 25.5% oxygen, by mass. It has a molar mass of 136 g mol$^{-1}$. When **A** is refluxed with sodium hydroxide solution, and the resulting liquid is distilled, a liquid **B** distils over, and a solution of **C** remains. On addition of dilute hydrochloric acid to **C**, a white precipitate **D** forms. When this precipitate is mixed with soda lime and heated, the vapour burns with a smoky flame. **B** reacts with ethanoyl chloride to give a product with a fruity smell. **B** does not give a positive result in the iodoform test. Identify **A**, **B** and **C**.

7   A compound contains C, H, N and O. When 0.225 g of the compound was heated with aqueous sodium hydroxide, the ammonia evolved neutralised 15.45 cm$^3$ of 0.100 mol dm$^{-3}$ sulphuric acid. The mass spectrum showed a molecular peak at $m/z = 73$. A 0.195 g sample of the compound gave on complete oxidation 0.352 g of carbon dioxide and 0.168 g of water.

   a   Find the molecular formula of the compound.

   b   Suggest its identity.

8   Phenol, C$_6$H$_5$OH, reacts with bromine to form a crystalline product. 25.0 cm$^3$ of a solution of phenol of concentration 0.100 mol dm$^{-3}$ were added to 30.0 cm$^3$ of aqueous potassium bromate(V). An excess of potassium bromide and hydrochloric acid was added to liberate bromine.

$$BrO_3^-(aq) + 5Br^-(aq) + 6H^+(aq) \longrightarrow 3Br_2(aq) + 3H_2O(l) \quad [1]$$

The excess bromine was estimated by adding potassium iodide and titrating the iodine displaced with aqueous sodium thiosulphate. 30.0 cm$^3$ of a 0.100 mol dm$^{-3}$ solution were required.

$$I_2(aq) + 2S_2O_3^{2-}(aq) \longrightarrow I_2(aq) + S_2O_4^{2-}(aq) \quad [2]$$

Deduce the equation for the reaction between phenol and bromine.

**STEPS TO TAKE**   a   Find the amount (mol) of phenol.

   b   Referring to equation [1], find the amount of bromine added.

   c   Referring to equation [2], find the amount of iodine titrated by thiosulphate. This is equal to the amount of bromine remaining after the reaction. Subtraction gives the amount of bromine used.

   d   From the ratio moles of Br$_2$ : moles of phenol, deduce the equation for the reaction.

9   Measurements on an alkene showed that 100 cm$^3$ of the gas weighed 0.228 g at 25 °C and $1.01 \times 10^5$ Pa. 25.0 cm$^3$ of the alkene reacted with 25.0 cm$^3$ of hydrogen.

   a   Find the molar mass of the alkene and its molecular formula.

   b   Give the names and structural formulae of two alkenes with this molecular formula.

**10** An organic acid has the percentage composition by mass: C 41.4%, H 3.4%, O 55.2%. A solution containing 0.250 g of the acid, which is dibasic, required 21.6 cm$^3$ of 0.200 mol dm$^{-3}$ aqueous sodium hydroxide for neutralisation.

Calculate **a** the empirical formula and **b** the molecular formula of the acid.
**c** Give its name and write its structural formula or formulae.

**11** An organic liquid contains carbon, hydrogen and oxygen. On oxidation 0.250 g of the liquid gave 0.595 g of carbon dioxide and 0.304 g of water. When vaporised 0.250 g of the liquid occupied 81 cm$^3$ at r.t.p. When the liquid was treated with phosphorus(V) chloride a pungent acidic gas was evolved. Find:

**a** the empirical formula and

**b** the molecular formula of the liquid.

**c** Write the structural formulae of the compounds which fit the facts.

**12** **A** is an organic compound with the percentage composition by mass C 71.1%, N 10.4%, O 11.8%, H 6.7% and a molar mass of 135 g mol$^{-1}$. On hydrolysis with aqueous sodium hydroxide, **A** gives an oily liquid **B**. **B** has the percentage composition by mass C 77.1%, N 15.1%, H 7.5% and a molar mass of 93 g mol$^{-1}$. **B** is basic and gives a precipitate with bromine water.

Find the molecular formulae for **A** and **B**. From their reactions deduce the identity of **A** and **B**.

## EXERCISE 15.2 Questions from examination papers

**1** **a** A compound, **X**, is a colourless, sweet smelling, neutral liquid of elemental percentage composition by mass of C 54.53%; H 9.15%; O 36.32%. **X** has a molar mass 88.1 g mol$^{-1}$.

The simplified infrared spectrum of **X** is shown below.

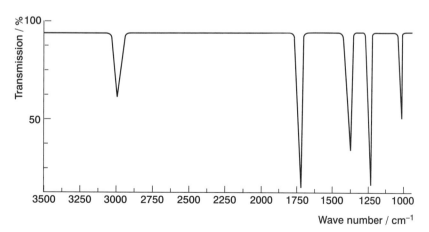

Some infrared data are given in the table below.

| Molecule or group | Vibration type | Vibration frequency/cm$^{-1}$ |
|---|---|---|
| Alkyl group | C—H stretch | 2960–2850 |
| Alcohol (ROH) | C—O stretch<br>O—H stretch<br>(hydrogen bonded) | 1200–1050<br>3200–3400 |
| Carboxylic acid (RCOOH) | C=O stretch | 1725–1700 |
| Ester (RCO$_2$R') | C=O stretch<br>C—O stretch | 1750–1730<br>1300–1050 |

By **carefully** considering **all** the data provided and **giving full reasoning** deduce the possible structure(s) for **X**.　　　　　　　　　　　　　　　　　(6)

**b** Use the information given below to determine structures for the compounds **A** to **E**.

An organic compound, **A**, of molecular formula C$_3$H$_6$O$_2$ reacts with aqueous sodium carbonate to give a colourless gas.

When **A** is heated with sodalime a second colourless gas, **B**, of formula C$_2$H$_6$ is formed.

On treatment with phosphorus pentachloride **A** gives a compound **C** of molecular formula C$_3$H$_5$ClO.

On reduction with lithium tetrahydridoaluminate(III), LiAlH$_4$ **A** gives compound **D** of molecular formula C$_3$H$_8$O.

**A** and **D** react together in the presence of concentrated sulphuric acid to give **E**, C$_6$H$_{12}$O$_2$.　　　　　　　　　　　　　　　　　(5)

**c** In a preparation of aspirin from 2-hydroxybenzenecarboxylic acid, C$_6$H$_4$(OH)COOH, 30.0 g of pure aspirin, C$_6$H$_4$(OCOCH$_3$)COOH, was obtained from 45.0 g of 2-hydroxybenzenecarboxylic acid. Calculate the percentage yield in this preparation.　(4)
*WJEC* (15)

**2** Benzene is the starting material for making a number of important organic compounds.

**a** It can be used to make cyclohexane which is needed for the manufacture of nylon.

**(i)** Draw the **displayed** formula of cyclohexane.　　　　　　　　　　(1)

**(ii)** Name the reagent and catalyst used to make cyclohexane from benzene.　(2)

**(iii)** Severe conditions of 200 °C and 30 atmospheres pressure are used. Explain why high pressure is needed when operating at the high temperature required for this reaction.　　　　　　　　　　　　　　　　　(1)

**b** Benzene can also be used to make dodecylbenzenesulphonic acid.

   **(i)** Give the reagent and catalyst for **Step 1**. (2)

  **(ii)** Both Step 1 and Step 2 involve attack by electrophiles on the benzene ring.

     Give the formula of the electrophile in **Step 2** and explain what is meant by the term **electrophile**? (2)

 **(iii)** State a use of substances like dodecylbenzenesulphonic acid. (1)

**c** Another use for benzene is to make diazo compounds using a sequence of reactions.

   **(i)** Name substances C and D. (2)

  **(ii)** State the type of reaction that occurs when D is converted into phenylamine. (1)

 **(iii)** Name a useful class of compounds that can be made from one of the products in this reaction sequence. (1)

**d** Benzenediazonium chloride decomposes in aqueous solution to form phenol, $C_6H_5OH$, and nitrogen. The rate of this reaction can be measured readily.

   **(i)** Suggest a method for measuring the rate of decomposition of benzenediazonium chloride solution. (2)

  **(ii)** The rate of the reaction was measured at different temperatures. In the table below the data have been converted into values of l/temperature and ln rate.

| l/temperature/$K^{-1}$ | ln rate |
|---|---|
| $3.0 \times 10^{-3}$ | $-1.4$ |
| $3.1 \times 10^{-3}$ | $-2.6$ |
| $3.2 \times 10^{-3}$ | $-3.9$ |
| $3.3 \times 10^{-3}$ | $-5.1$ |
| $3.4 \times 10^{-3}$ | $-6.3$ |

     Plot a graph of ln rate, on the vertical axis, against l/temperature, on the horizontal axis. (3)

 **(iii)** Calculate the activation energy, $E_A$, for the reaction, using your graph and the relationship:

$$\ln \text{rate} = \text{constant} - \frac{E_A}{R}(1/T)$$

     where $R = 8.3 \, J \, K^{-1} \, mol^{-1}$.

     Remember to include a sign and units in your answer which should be given to two significant figures. (3)

L(N) (21)

**3**  **a**  An organic compound, **Z**, contains 66.7% carbon, 11.1% hydrogen and 22.2% oxygen by mass. The mass spectrum of **Z** shows a molecular ion at $m/e = 72$.

Show that the molecular formula of **Z** is $C_4H_8O$. (3)

**b**  An orange precipitate is formed when a few drops of **Z** are added to 2,4-dinitrophenylhydrazine solution.

**(i)**  Give the structure of the group in **Z** that is responsible for this reaction. (1)

**(ii)**  Draw the structural formulae for three possible isomers of **Z** containing this group. (3)

**c**  Compound **Z** can be prepared from compound **Y**, which is a secondary alcohol.

**(i)**  Indicate with a letter **Z** the isomer drawn in **b(ii)** which corresponds to **Z** and give its name. (2)

**(ii)**  Give the names of the reagents that could be used to prepare **Z** from **Y** and state the type of reaction that occurs. (2)

**d**  Describe a simple chemical test, including what you would see, that could be used to distinguish between **Z** and the other isomers drawn for $C_4H_8O$. (2)

**e**  In addition to the molecular ion, the mass spectrum of **Z** shows fragment ions at $m/e$ values of 57, 43, and 29. Suggest possible structures for these fragment ions. (3)

**f**  **Z** boils at 79.6 °C while pentane, $C_5H_{12}$, which has the same relative molecular mass, boils at 36.3 °C. Account for the difference in boiling temperatures. (3)

**g**  **(i)**  Draw the structure of the secondary alcohol **Y**. (1)

**(ii)**  State the type of isomerism, other than structural isomerism, displayed by **Y**. (1)

**(iii)**  Describe the feature of the molecule that is responsible for this isomerism. (1)

$L$  (22)

**4**  **a**  **(i)**  Describe briefly the mechanism for the nitration of benzene. (3)

**(ii)**  The reaction between concentrated sulphuric and nitric(V) acids in a nitrating mixture is shown below:

$$HNO_3 + H_2SO_4 \longrightarrow NO_2^+ + H_2O + HSO_4^-.$$

The rate expression for the above reaction when there is a large excess of sulphuric acid is:

$$Rate = k[HNO_3]$$

with $k = 1 \times 10^3$ s$^{-1}$ at 298 K.

I.  1000 cm$^3$ of nitrating mixture initially contains 0.10 mol of $HNO_3$. Calculate the initial rate of the above reaction at 298 K, stating the units. (2)

II.  Explain briefly why the rate of production of $NO_2^+$ and $H_2O$ in a mixture of acids falls continuously with time. (1)

**b** Consider the following reaction sequence and further information about compounds **P–S**.

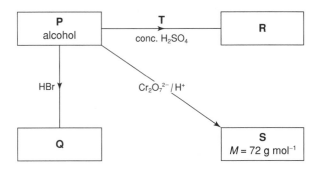

Mass spectrum of **T** gives an intense peak due to the parent ion at $m/e = 122$.

**(i)** 0.010 mol of **T** neutralises 0.010 mol of sodium hydrogencarbonate, $NaHCO_3$.
State the number of carboxylic acid groups that are present in one molecule
of **T**. (1)

**(ii)** **S** does **not** react with Fehling's solution but **S** does react with
2,4-dinitrophenylhydrazine, producing a yellow-orange precipitate. To which class
of organic compounds does **S** belong? (1)

**(iii)** Write down shortened **structural formulae** for compounds **P–T**, clearly showing
any functional groups present. (5)

**(iv)** State the reagent(s) required to convert

I. **Q** to **P**    II. **S** to **P** (2)

**(v)** Give the general formulae for the molecular groups identified by the *iodoform
reaction*. Identify **two** compounds, in the above scheme, which would give a
positive iodoform reaction. (4)

**(vi)** Draw diagrams showing the **two** optical isomers of **R**, showing clearly in what way
they are different. (1)

*WJEC* (20)

**5** An organic acid **X** contains the elements carbon, hydrogen and oxygen only. It was found
to contain 40.7% carbon and 5.10% hydrogen by mass. A solution of the acid was made
of concentration 0.0500 mol dm$^{-3}$. 25.0 cm$^3$ of this solution was neutralized completely by
31.25 cm$^3$ of 0.0800 mol dm$^{-3}$ sodium hydroxide solution. The mass spectrum shows that
the molecular ion peak has a mass/charge ratio of 118.

**a** **(i)** Which peak of a mass spectrum corresponds to the molecular ion? (1)

**(ii)** Find the molecular and structural formulae of acid **X** showing all your reasoning
clearly. (4)

**b** State and explain how many peaks would be seen in the high resolution n.m.r. spectrum
of **X**, and give their relative areas. (3)

c Acid **X** dissolves in ethoxyethane as well as water although it is more soluble in water. The partition coefficient of acid **X**, when distributed between ethoxyethane and water, is 0.200. 100 cm$^3$ of a second aqueous solution of **X** were shaken with 50.0 cm$^3$ of ethoxyethane. At equilibrium the concentration of the acid in the water layer was found to be 0.150 mol dm$^{-3}$.

    **(i)** Define the term *partition coefficient*. (1)

    **(ii)** Find the mass of **X** which was dissolved in the ethoxyethane layer assuming that the two solvents are completely immiscible. (2)

OCR (11)

**6** Furan is an organic compound containing only carbon, hydrogen and oxygen. It is used to make solvents and some forms of nylon. Furan consists of 70.6% of carbon and 5.8% of hydrogen, by mass.

**a** Showing each stage of your working, calculate the empirical formula of furan. (2)

**b** By experiment, the relative molecular mass of furan was found to be approximately 70. Use this, and your answer to **a**, to show that its molecular formula is $C_4H_4O$. (1)

**c** **(i)** Write the equation, including state symbols, for the standard enthalpy change of formation of furan, $C_4H_4O(l)$.

    **(ii)** Use your equation in **c (i)**, together with the data below, to calculate the standard enthalpy change for the following reaction.

$$C_4H_4O(l) \longrightarrow 4C(g) + 4H(g) + O(g)$$

$$C(s) \longrightarrow C(g) \qquad \Delta H^\ominus = +717 \text{ kJ mol}^{-1}$$

$$\tfrac{1}{2}H_2(g) \longrightarrow H(g) \qquad \Delta H^\ominus = +218 \text{ kJ mol}^{-1}$$

$$\tfrac{1}{2}O_2(g) \longrightarrow O(g) \qquad \Delta H^\ominus = +248 \text{ kJ mol}^{-1}$$

$$\Delta H_f^\ominus(\text{furan}) = -63 \text{ kJ mol}^{-1} \qquad (4)$$

**d** The structure of furan is shown below.

Use this structure, and bond energies from the table in Exercise 12.4, question 1, to calculate a value for the enthalpy change for the following process.

$$C_4H_4O(g) \longrightarrow 4C(g) + 4H(g) + O(g) \qquad (2)$$

**e** Use your answers to **c (ii)** and **d** to calculate a value for the enthalpy change for the following process.

$$C_4H_4O(l) \longrightarrow C_4H_4O(g) \qquad (1)$$

OCR (10)

**7**   Study the scheme below and the mass spectrum shown and then answer the questions
which follow.

**Note :** Strong oxidation of a carbon-containing aromatic side-chain, e.g. $CH_3$, yields a
carboxyl group.

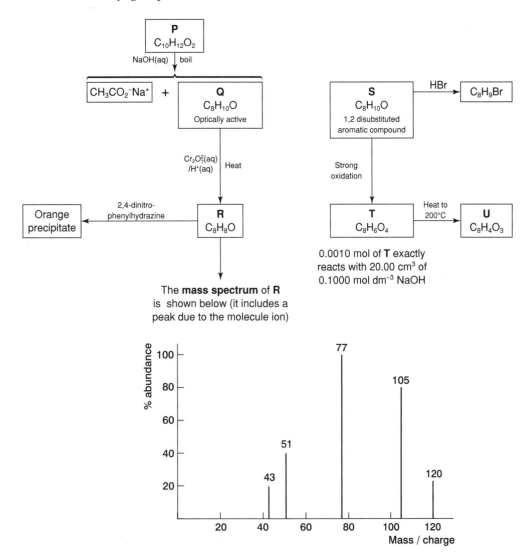

The **mass spectrum of R**
is  shown below (it includes a
peak due to the molecule ion)

Deduce the structural formulae of compounds **P–U**. Explain the nature of the reactions
involved and also the numerical and mass spectrum information given.

*WJEC(S, part)*  (16)

# Synoptic questions

# 16

You will often find that examination questions cover more than one type of calculation. This chapter is a selection of questions from past papers which require more than one of the techniques which you have met in the previous chapters.

**EXERCISE 16.1   Synoptic questions from examination papers**

1   The conversion of sulphur dioxide to sulphur trioxide is a vital step in the industrial manufacture of sulphuric acid. The equation for this reaction is

$$2SO_2(g) + O_2(g) \rightleftharpoons 2SO_3(g) \qquad \Delta H = -197 \text{ kJ mol}^{-1}$$

To achieve the conversion, a mixture of sulphur dioxide and air is passed through a catalyst bed at 700 K. The emergent gases are cooled and passed through a second catalyst bed, again at 700 K, to increase the yield of sulphur trioxide. These operations are repeated twice more so that a minimum of 99.5% conversion is achieved.

a   **(i)** Write the full expression for the equilibrium constant, $K_c$, for the reaction, including its units. (2)

   **(ii)** Why are the gases cooled between successive passes through the catalyst beds? (2)

   **(iii)** Suggest TWO conditions, other than temperature control, which will help to maximise the amount of sulphur dioxide converted to sulphur trioxide. (2)

   **(iv)** For what reason must the escape to the environment of these oxides of sulphur be rigorously prevented? (1)

b   **(i)** Predict whether there will be an increase or a decrease in the entropy of the system during the formation of sulphur trioxide in this reaction. Give a reason for your answer. (1)

   **(ii)** Calculate the entropy change of the surroundings, assuming the reaction occurs at 700 K. Include the sign and units in your answer. (2)

   **(iii)** In the light of your answer to **b (ii)**, what can you deduce about the numerical value of the entropy change of the system? (1)

c   **(i)** Calculate the pH of a 0.075 mol dm$^{-3}$ solution of sulphuric acid, assuming it to be fully ionized according to the equation

$$H_2SO_4(aq) \longrightarrow 2H^+(aq) + SO_4^{2-}(aq)$$ (2)

   **(ii)** A total of 25.0 cm$^3$ of ammonia solution was added in small portions from a burette to 10.0 cm$^3$ of 0.075 M sulphuric acid. The pH of the solution was followed as the ammonia was added.

Sketch a graph showing how the pH changed assuming that 20.0 cm³ of the ammonia solution was sufficient to neutralize the fully ionized sulphuric acid. (2)

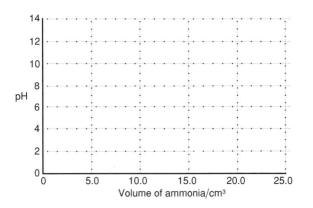

**(iii)** Calculate the concentration in mol dm⁻³ of the ammonia solution. (2)

**(iv)** Not all indicators are suitable for this titration. Explain why some indicators cannot be used. (1)

*L(N)* (18)

**2** **a** **(i)** Give **balanced** equations describing **two** reactions of calcium metal in which $Ca^{2+}$ ions are formed. Compare the reactions of zinc metal with the same reagents. (4)

**(ii)** State how the thermal stability of hydroxides and carbonates changes as Group II is descended. Describe **briefly** a simple laboratory experiment which would confirm whether or not a carbonate is thermally stable or unstable. (3)

**b** 14.78 g of a pure sample of carbonate of an element **Z**, which belongs to **either** Group I **or** Group II, was completely decomposed upon heating, producing exactly 4.48 dm³ of carbon dioxide at 0 °C and 1 atm pressure (101 kPa). Calculate the number of moles of $CO_2$ produced. Deduce the relative atomic mass of **Z**. Hence, using the Periodic Table, identify **Z**. (5)

[The molar volume of an ideal gas at 0 °C and 1 atm pressure (101 kPa) is 22.4 dm³.]

**c** Tests were carried out on a colourless solution which contains a soluble salt **MX₂** at a concentration of about 0.05 mol dm⁻³. The following table shows the reagents used and the observations made.

| Reagent added | Observation |
|---|---|
| Dilute sodium hydroxide | No precipitate |
| Dilute sulphuric acid | Dense white precipitate |
| Copper(II) sulphate | Off-white precipitate in a brown solution |

Identify the cation **M** and the anion **X**, carefully explaining your reasoning. Write an ionic equation for the reaction that occurs when copper(II) ions are added to the solution. (4)

**d** Use the standard enthalpy change of formation data given below to calculate the enthalpy change of reaction for the reduction of **0.10 mol** of iron(III) oxide by carbon monoxide. (4)

$$Fe_2O_3(s) + 3CO(g) \longrightarrow 2Fe(s) + 3CO_2(g)$$

$\Delta H_f^\ominus$ (298 K)/kJ mol$^{-1}$   $Fe_2O_3(s)$ –824;   $CO(g)$ –110;   $Fe(s)$ 0;   $CO_2(g)$ –394.

WJEC   (20)

**3  a  (i)** Give the ion/electron half-equation in which the $MnO_4^-$ ion is reduced to the $Mn^{2+}$ ion in acidic solution. (1)

**(ii)** The ion/electron half-equation for the oxidation of hydrogen peroxide is:

$$H_2O_2(aq) - 2e^- \longrightarrow O_2(g) + 2H^+(aq)$$

Give the **overall ionic** equation for the reaction of $H_2O_2$ with $MnO_4^-$ in acidic solution. (1)

**(iii)** 25.00 cm$^3$ of a solution, consisting of acidified hydrogen peroxide, required 19.00 cm$^3$ of 0.0200 mol dm$^{-3}$ $MnO_4^-$ for complete reaction. Calculate the concentration of hydrogen peroxide (mol dm$^{-3}$) in the solution. (3)

**(iv)** The oxidation number (state) of each hydrogen atom in the $H_2O_2$ molecule is +1. Give the oxidation number of each oxygen atom in this molecule. (1)

**b** The standard enthalpy change of formation of hydrogen peroxide gas, $\Delta H_f^\ominus(H_2O_2(g))$, is equal to the enthalpy change of the following reaction:

$$H_2(g) + O_2(g) \longrightarrow H_2O_2(g)$$

Use the following bond energies to calculate $\Delta H_f^\ominus(H_2O_2(g))$ at 298 K, stating the units. (The structure of $H_2O_2$ is H—O—O—H.) (3)

| Bond | H—H | O=O | O—H | O—O |
|------|-----|-----|-----|-----|
| Bond Energy (298 K)/kJ mol$^{-1}$ | 436 | 496 | 464 | 144 |

**c** The radium isotope $^{266}_{88}Ra$ decays by alpha emission into radon gas, Rn. The half-life of $^{226}_{88}Ra$ is 1600 years.

**(i)** Give a balanced equation representing the alpha emission. (1)

**(ii)** Explain the meaning of the term **half-life**. (1)

**(iii)** A sample of $^{226}_{88}Ra$, of initial mass 1.00 g, decays for 3200 years.

   I. Calculate the number of **moles** of $^{226}_{88}Ra$ left after this period.

   II. Calculate the number of moles of radon gas produced.

   III. Calculate the volume of radon gas produced at 0 °C and 1 atmosphere pressure. (4)

[The molar volume of an ideal gas at 0 °C and 1 atm (101 kPa) is 22.4 dm$^3$.]

WJEC   (15)

**4** **a** Using the bond energies given below calculate the enthalpy change for the direct hydration of propene.

$$CH_3CH{=}CH_2(g) + H_2O(g) \longrightarrow CH_3CHOHCH_3(g)$$ (3)

| Bond | Bond energy/kJ mol$^{-1}$ |
|------|---------------------------|
| C—C | 348 |
| C—O | 360 |
| C—H | 412 |
| O—H | 463 |
| C=C | 612 |

**b** The reaction in **a** is reversible. Explain, giving your reasoning, the effect upon the equilibrium yield of propan-2-ol by

**(i)** increasing the temperature at which equilibrium is attained (1)

**(ii)** increasing the total pressure at equilibrium. (1)

**c** For the equilibrium

$$CH_3CH{=}CH_2(g) + H_2O(g) \rightleftharpoons CH_3CHOHCH_3(g)$$

at a certain temperature, the equilibrium partial pressures of propene and steam in the system are $7.5 \times 10^6$ Pa and $9.5 \times 10^6$ Pa respectively, and the total pressure is $2.0 \times 10^7$ Pa.

Calculate the value for $K_p$ at this temperature, clearly stating its units. (2)

**d** The incomplete graph below shows the changes in pH when aqueous sodium hydroxide of concentration 0.1000 mol dm$^{-3}$ is added to 25.00 cm$^3$ of aqueous ethanoic acid of concentration 0.1000 mol dm$^{-3}$.

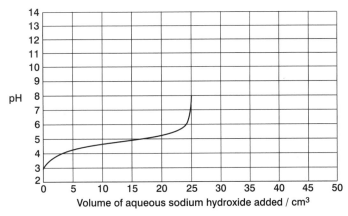

**(i)** Sketch on a copy of the graph a continuation of the curve until approximately 50 cm$^3$ of the alkali have been added. (1)

*You are not expected to determine accurate pH values in this region.*

**(ii)** State, by referring to the pH curve, the pH range of an indicator suitable for the titration of aqueous ethanoic acid with aqueous sodium hydroxide. (1)

**(iii)** When 25.00 cm$^3$ of sodium hydroxide of concentration 0.1000 mol dm$^{-3}$ have been added and reacted exactly with the 25.00 cm$^3$ of ethanoic acid of concentration 0.1000 mol dm$^{-3}$, according to

$$CH_3COOH(aq) + NaOH(aq) \longrightarrow CH_3COONa(aq) + H_2O(l)$$

the pH value of the resulting solution is greater than 7.

Give an explanation for this. No calculation is required (1)

*WJEC* (10)

**5** Methanol can be synthesised from carbon monoxide and hydrogen according to the equation

$$CO(g) + 2H_2(g) \rightleftharpoons CH_3OH(g)$$

Use the thermodynamic data below to answer the questions that follow.

| Substance | $\Delta H_F^\ominus$/kJ mol$^{-1}$ | $S^\ominus$/J K$^{-1}$ mol$^{-1}$ |
|---|---|---|
| CO(g) | −110 | 198 |
| H$_2$(g) | 0 | 131 |
| CH$_3$OH(g) | −201 | 240 |

**a** Determine the standard enthalpy change and the standard entropy change for the synthesis of methanol from carbon monoxide and hydrogen. (4)

**b** Explain what is meant by the term *feasible reaction* and determine the temperature at which the methanol synthesis reaction is no longer feasible. (3)

**c** Industrially, methanol can be manufactured from a 2 : 1 hydrogen : carbon monoxide mixture at high pressure and a temperature of 300 °C in the presence of a finely divided solid catalyst. Under these conditions, the equilibrium yield of methanol is only of the order of 1%, yet these are the conditions chosen for a continuous commercial process in which an overall conversion of 95% can be achieved.

**(i)** Explain why the relatively low temperature of 300 °C is chosen.

**(ii)** Why is a catalyst necessary?

**(iii)** Explain how a 95% conversion can be achieved in a reaction which has an equilibrium yield of only 1%. (3)

*NEAB* (10)

**6** **a** By considering the numbers of electrons in the 3d and 4s orbitals, give an explanation for the variation in the $E^\ominus(M^{3+}(aq)/M^{2+}(aq))$ values shown below. (6)

| Half-reaction | $E^\ominus$/V |
|---|---|
| $Cr^{3+}(aq) + e^- \rightleftharpoons Cr^{2+}(aq)$ | −0.41 |
| $Mn^{3+}(aq) + e^- \rightleftharpoons Mn^{2+}(aq)$ | +1.49 |
| $Fe^{3+}(aq) + e^- \rightleftharpoons Fe^{2+}(aq)$ | +0.77 |

**b** **(i)** I. Solution **A** is 0.0800 mol dm$^{-3}$ with respect to an oxidising ion **X**$^{4+}$ and is orange/yellow due to that ion.

25.00 cm$^3$ of a solution containing 5.490 g dm$^{-3}$ of iron(II) ions was titrated with solution **A** using ferroin indicator. 30.65 cm$^3$ of solution **A** was required for exact reaction. Deduce the formula of the simple cation formed as a result of the reduction of **X**$^{4+}$.

II. Given that 26.58 g of the anhydrous sulphate of **X** was used to make up 1 dm$^3$ of the solution containing **X**$^{4+}$, calculate the relative atomic mass of **X** and hence identify **X**. (8)

**(ii)** Ferroin indicator changes from orange-red to pale blue when it is oxidised. Ferroin in its reduced form contains an octahedral complex formed from iron(II) ions and 1,10-phenanthroline (C$_{12}$H$_8$N$_2$). The structural formula of 1,10-phenanthroline is

($C_{12}H_8N_2$)

The complex present after oxidation is also octahedral and is made up of iron(III) ions and 1,10-phenanthroline. In each case the ligand attaches to the metal ion via the lone pairs on the nitrogen atoms. Give the formula of the iron(II) complex and that of the iron(III) complex and write an equation for their interconversion. (3)

**c** Consider the following standard electrode potential values:

| Half-reaction | | $E^{\ominus}/V$ |
|---|---|---|
| $[Fe(CN)_6]^{3-}(aq) + e^- \rightleftharpoons$ | $[Fe(CN)_6]^{4-}(aq)$ | +0.36 |
| $\frac{1}{2}I_2(aq) + e^- \rightleftharpoons$ | $I^-(aq)$ | +0.54 |
| $Fe(H_2O)_6^{3+}(aq) + e^- \rightleftharpoons$ | $Fe(H_2O)_6^{2+}(aq)$ | +0.77 |

It is found that

1. on adding aqueous iron(III) ions to aqueous potassium iodide, the colour changes from colourless to red-brown

2. a solution of potassium hexacyanoferrate(II) will decolorise aqueous iodine.

Explain these observations and state whether they are consistent with the electrode potentials. What overall conclusion can be drawn from the changes? (8)

*WJEC* (25)

**7** Sulphur dioxide reacts with oxygen to form sulphur trioxide according to the equation

$$2SO_2(g) + O_2(g) \rightleftharpoons 2SO_3(g)$$

Data for this reaction are shown in the table below.

| | $\Delta H_f^{\ominus}/kJ\ mol^{-1}$ | $S^{\ominus}/J\ K^{-1}\ mol^{-1}$ |
|---|---|---|
| $SO_3(g)$ | −396 | +257 |
| $SO_2(g)$ | −297 | +248 |
| $O_2(g)$ | 0 | +204 |

**a** Determine the standard enthalpy, the standard entropy and standard free energy changes at 298 K for this reaction. (7)

**b** The reaction is said to be feasible. In terms of free energy change, explain the meaning of the term *feasible*. Calculate the temperature at which the reaction between sulphur dioxide and oxygen ceases to be feasible. (3)

**c** At a fixed temperature and a total pressure of 540 kPa, a vessel contains an equilibrium mixture of $SO_2(g)$ (0.050 mol), $O_2(g)$ (0.080 mol) and $SO_3(g)$ (0.070 mol). Determine the equilibrium partial pressure of each of the three gases in this mixture and hence the value of the equilibrium constant, $K_p$, for the reaction, stating its units. (7)

*NEAB 99* (17)

**8**   Hydrogen reacts with iodine to form hydrogen iodide.

$$H_2(g) + I_2(g) \rightleftharpoons 2HI(g) \qquad \Delta H = -10.4 \text{ kJ mol}^{-1}$$

The reaction is found to be first order with respect to both hydrogen and iodine.

**a** **(i)** Write a rate equation for the reaction. (1)

**(ii)** At 425 °C the rate of the reaction at equilibrium is $4.14 \times 10^{-7}$ mol dm$^{-3}$ s$^{-1}$.

Find the rate constant for the forward reaction at this temperature, given the composition of the equilibrium mixture below. (2)

| Substance | Equilibrium concentration/mol dm$^{-3}$ |
|-----------|------------------------------------------|
| $H_2(g)$  | $2.27 \times 10^{-3}$ |
| $I_2(g)$  | $2.84 \times 10^{-3}$ |
| $HI(g)$   | $1.72 \times 10^{-2}$ |

**(iii)** The Arrhenius equation can be used to show how the rate constant, $k$, varies with temperature, $T$.

$$\ln\left(\frac{k_1}{k_2}\right) = \frac{-E_A}{R}\left(\frac{1}{T_1} - \frac{1}{T_2}\right)$$

[The gas constant, $R$, may be taken as 8.31 J K$^{-1}$ mol$^{-1}$]

The rate constant at 300 °C is $1.37 \times 10^{-4}$ mol$^{-1}$ dm$^3$ s$^{-1}$. Use this and your answer to part **(ii)** to calculate the activation energy for the reaction. (3)

**(iv)** Draw a labelled energy profile diagram for the reaction. (2)

**b** **(i)** Calculate the standard entropy change, $\Delta S^{\ominus}$, for the reaction between hydrogen and iodine given the following standard molar entropies, $S^{\ominus}$ (2)

| Substance | $S^{\ominus}$/J K$^{-1}$ mol$^{-1}$ |
|-----------|--------------------------------------|
| $H_2(g)$  | 131 |
| $I_2(g)$  | 261 |
| $HI(g)$   | 206 |

(ii) Discuss how the value of the free energy change, $\Delta G^{\ominus}$, for the reaction will vary with temperature, and comment on the thermodynamic feasibility of the reaction. (3)

*OCR* (13)

**9** The percentage of nitrogen, N, in an organic compound may be determined by reaction of the compound with hot sodium metal, which converts any nitrogen present into sodium cyanide (NaCN).

The percentages of hydrogen and carbon in a compound are found by measuring the mass of $CO_2$ and $H_2O$ produced in the combustion of the compound in excess oxygen.

Study the following facts concerning a pure compound **T**, which contains only carbon, hydrogen, nitrogen and oxygen.

1. When 0.4576 g of compound **T** was reacted with excess sodium, 0.003 720 mol of sodium cyanide was produced.

2. When 0.2100 g of **T** was burned in excess oxygen, 0.2290 dm$^3$ of $CO_2$ gas was produced at 1 atm pressure and 273 K; 0.004 270 mol of water was also produced.
   [The molar volume of an ideal gas at 1 atm pressure and 273 K is 22.40 dm$^3$.]

**a** Calculate the simplest formula of **T**, showing your working. (12)

**b** The mass spectrum of **T** gives an ion signal at $m/e = 77$, due to the ion **P**. Identify **P** and explain briefly what this suggests about the structure of compound **T**. Write an equation showing the formation of the ion **P** in the spectrometer chamber. (3)

*WJEC (S, part)* (15)

**10 a (i)** Give a balanced equation which applies to the standard enthalpy change of formation ($\Delta H_f^{\ominus}$ (298 K)) of barium chloride. (1)

**(ii)** Calculate $\Delta H_f^{\ominus}$ ($BaCl_2(s)$) from the following data:

| Equation defining enthalpy change | $\Delta H^{\ominus}$ (298 K)/kJ mol$^{-1}$ |
|---|---|
| $Ba(s) \longrightarrow Ba(g)$ | 176 |
| $Ba(g) \longrightarrow Ba^+(g) + e^-$ | 502 |
| $Ba^+(g) \longrightarrow Ba^{2+}(g) + e^-$ | 966 |
| $Cl_2(g) \longrightarrow 2Cl(g)$ | 242 |
| $Cl(g) + e^- \longrightarrow Cl^-(g)$ | -364 |
| $BaCl_2(s) \longrightarrow Ba^{2+}(g) + 2Cl^-(g)$ | 2018 |

(4)

**(iii)** Given the following standard enthalpies of hydration, $\Delta H_{hyd}^{\ominus}$ (298 K)

$$Ba^{2+}(g) \quad -1360 \text{ kJ mol}^{-1} \qquad Cl^-(g) \quad -364 \text{ kJ mol}^{-1}$$

calculate the enthalpy change that occurs when one mole of $BaCl_2$ is hydrated. Comment upon the significance of this enthalpy change in determining the solubility, or otherwise, of $BaCl_2(s)$. (2)

**b** **(i)** I. Give a balanced equation for the reaction of BaO with water, naming the **product**. (2)

II. Name the reagent required to convert the product in **b (i)**, above, to barium chloride. (1)

**(ii)** Give brief practical details of the use of barium chloride solution as a confirmatory test for the presence of sulphate ions. State what would be **observed** if sulphate ions were present, and give an **ionic** equation for the reaction. (3)

**c** State the colour observed when barium compounds are strongly heated in a bunsen flame. Briefly explain the origin of this colour. (2)

**d** **(i)** State which element in the molecule **X**, $C_{19}H_{23}O_3N$, is responsible for its weakly basic properties. (1)

**(ii)** An aqueous solution of **X** contains $1.0 \times 10^{-5}$ mol dm$^{-3}$ of OH$^-$ ion. Calculate the H$^+$(aq) concentration, and hence the pH of the solution.

$(K_w(298 \text{ K}) = 1.0 \times 10^{-14} \text{ mol}^2 \text{ dm}^{-6})$ (2)

**(iii)** **X** is closely related to a second molecule **Y**, but only **Y** contains OH groups. Briefly explain how infrared spectroscopy would confirm the identity of a solid, known to be **either Y or X**, stating the approximate position (wavenumber) of any band used in the identification. (2)

*Table of infrared absorption*

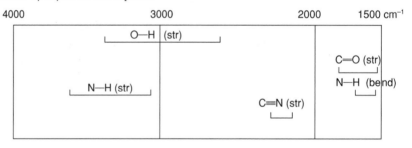

*WJEC* (20)

**11** **a** An excess of calcium carbonate was added to some hydrochloric acid in a flask on a balance which was immediately set at zero. The mass reading on the balance was then recorded over a period of time. Other similar experiments were performed at the same temperature and the results are shown below.

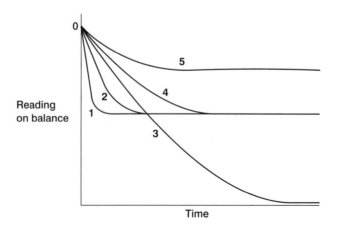

The reaction of lumps of calcium carbonate with 100 cm$^3$ of 0.1 M hydrochloric acid gave curve **4**. Which **one** of the curves, labelled **1–5**, could have been produced using the following mixtures? Each curve may be used **once** only.

  **(i)** 100 cm$^3$ of 0.1 M hydrochloric acid with finely powdered calcium carbonate?

  **(ii)** 50 cm$^3$ of 0.1 M hydrochloric acid with lumps of calcium carbonate?

  **(iii)** 50 cm$^3$ of 0.2 M hydrochloric acid with lumps of calcium carbonate? (3)

**b** Explain why the reaction of lumps of calcium carbonate is much slower with 0.1 M ethanoic acid than with 0.1 M hydrochloric acid at the same temperature. (3)

**c** Calcium ethanoate, $Ca(CH_3COO)_2$, is produced when ethanoic acid reacts with calcium carbonate as in the reaction in part **b**. One mole of calcium ethanoate decomposes on heating to form one mole of calcium carbonate and one mole of an organic compound, **W**, only.

  **(i)** Deduce the molecular formula of **W**.

  **(ii)** **W** has a strong absorption in its infra-red spectrum at 1720 cm$^{-1}$ but has no reaction with Tollens' or Fehling's solutions. State the two deductions which can be made about **W** from these observations and hence suggest a structure for **W**. A table of infra-red absorption data is given below. (4)

*Table of infra-red absorption data*

| Bond | Wavenumber/cm$^{-1}$ |
|---|---|
| C—H | 2850–3300 |
| C—C | 750–1100 |
| C=C | 1620–1680 |
| C=O | 1680–1750 |
| C—O | 1000–1300 |
| O—H (alcohols) | 3230–3550 |
| O—H (acids) | 2500–3000 |

*NEAB 99* (10)

# Answers

**Exercise 1.1** Equations

1. (a) $c = a - b$
   (b) (i) $x = y + z$   (ii) $y = x - z$
   (c) (i) $p = r + s + q$   (ii) $q = p - r - s$
2. (a) (i) Distance = Speed × Time   (ii) Time = Distance/Speed
   (b) (i) $[PCl_5] = K_c[PCl_3][Cl_2]$,   (ii) $[Cl_2] = [PCl_5]/K_c[PCl_3]$
   (c) (i) $[H_2] = [HI]^2/K_c[I_2]$   (ii) $[HI] = \sqrt{K_c[H_2][I_2]}$
3. (i) $T = PV/R$   (ii) $V = RT/P$
4. (a) (i) Amount of solute = Concentration × Volume
   (ii) Volume = Amount of solute/Concentration

   (b)

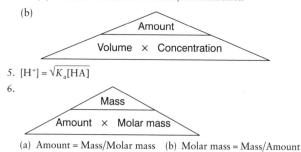

5. $[H^+] = \sqrt{K_a[HA]}$
6.

   (a) Amount = Mass/Molar mass   (b) Molar mass = Mass/Amount

**Exercise 1.2** Ratios

1. 165 mg = 0.165 g
2. 195 g
3. 225 dm$^3$
4. 11 g
5. 500 tonnes

**Exercise 1.3** Percentages

1. 40.0%, 66.7%, 66.7%, 20.0%
2. 25% $\frac{1}{4}$
3. 2.73 g
4. 94.5%
5. 36%

**Exercise 1.4** Negative numbers

1. $-30.6$ kJ mol$^{-1}$
2. (a) $-0.51$ V   (b) $+1.21$ V   (c) $+0.51$ V
   (d) $+0.93$ V   (e) $-0.62$ V

**Exercise 1.5** Standard form

1. (a) $6 \times 10^5$   (b) $4 \times 10^3$   (c) $4.5 \times 10^4$   (d) $1.23 \times 10^{-2}$
   (e) $6 \times 10^{-3}$ (f) $4.9 \times 10^{-4}$

**Exercise 1.6** Decimal places and significant figures

1. (a) 2   (b) 3   (c) 3   (d) 5   (e) 4   (f) 2   (g) 2   (h) 4
2. (a) 15.8 cm$^3$   (b) 206 tonnes   (c) 1120 cm$^3$
   (d) 49.5%   (e) 3.33 mol   (f) 0.145 mol dm$^{-3}$

(g) $-109$ kJ mol$^{-1}$   (h) 2.05 dm$^3$ mol$^{-1}$ s$^{-1}$
(i) 0.0821 dm$^3$ atm K$^{-1}$ mol$^{-1}$

3. (a) 83.4   (b) 2410   (c) 0.641
4. (a) 47.12   (b) 8.33   (c) 8.35   (d) 0.06   (e) 0.57

**Exercise 1.7** Graphs

1. $y = 3x + 4$
2. $x = \dfrac{56}{y}$
3. On the vertical axis plot (Rate/mol dm$^{-3}$ s$^{-1}$) × 10$^4$. The four values are 0.364, 0.741, 1.51, 2.21. On the horizontal axis plot [H$_2$O$_2$]. A straight line can be drawn through the origin and four other points, showing that

   Rate $\propto$ [H$_2$O$_2$]
   or rate = $k$ [H$_2$O$_2$]
   where $k$ is a constant.
4. Plot the count rate on the vertical axis against time on the horizontal axis. Spread the results from 100 cpm to 50 cpm over the whole of the graph paper; there is no need to go down to zero cpm. Draw a smooth curve through the points. The time at which the count rate has dropped to 50 cpm can be read from the curve. It is 170 minutes.

**Exercise 2.1** Practice with equations

1. (a) $H_2(g) + CuO(s) \longrightarrow Cu(s) + H_2O(g)$
   (b) $C(s) + CO_2(g) \longrightarrow 2CO(g)$
   (c) $C(s) + O_2(g) \longrightarrow CO_2(g)$
   (d) $Mg(s) + H_2SO_4(aq) \longrightarrow H_2(g) + MgSO_4(aq)$
   (e) $Cu(s) + Cl_2(g) \longrightarrow CuCl_2(s)$
2. (a) $Ca(s) + 2H_2O(l) \longrightarrow H_2(g) + Ca(OH)_2(aq)$
   (b) $2Cu(s) + O_2(g) \longrightarrow 2CuO(s)$
   (c) $4Na(s) + O_2(g) \longrightarrow 2Na_2O(s)$
   (d) $Fe(s) + 2HCl(aq) \longrightarrow FeCl_2(aq) + H_2(g)$
   (e) $2Fe(s) + 3Cl_2(g) \longrightarrow 2FeCl_3(s)$
3. (a) $Na_2O(s) + H_2O(l) \longrightarrow 2NaOH(aq)$
   (b) $2KClO_3(s) \longrightarrow 2KCl(s) + 3O_2(g)$
   (c) $2H_2O_2(aq) \longrightarrow 2H_2O(l) + O_2(g)$
   (d) $3Fe(s) + 2O_2(g) \longrightarrow Fe_3O_4(s)$
   (e) $3Mg(s) + N_2(g) \longrightarrow Mg_3N_2(s)$
   (f) $4NH_3(g) + 3O_2(g) \longrightarrow 2N_2(g) + 6H_2O(g)$
   (g) $3Fe(s) + 4H_2O(g) \longrightarrow Fe_3O_4(s) + 4H_2(g)$
   (h) $2H_2S(g) + 3O_2(g) \longrightarrow 2H_2O(g) + 2SO_2(g)$
   (i) $2H_2S(g) + SO_2(g) \longrightarrow 2H_2O(l) + 3S(s)$

**Exercise 3.1** Problems on relative formula mass

| | | |
|---|---|---|
| 64 | 40 | 101 |
| 84 | 278 | 95 |
| 148 | 99 | 161 |
| 98 | 63 | 246 |
| 136 | 685 | 142 |
| 106 | 74 | 123.5 |
| 159.5 | 162 | 249.5 |
| 400 | 286 | 278 |

**Exercise 3.2** Problems on percentage composition

**Section 1**

1. (a) C = 80%      H = 20%
   (b) Na = 57.5%   O = 40%      H = 2.5%
   (c) S = 40%      O = 60%
   (d) C = 90%      H = 10%

2. (a) C = 84%      H = 16%
   (b) Mg = 72%     N = 28%
   (c) Na = 15.3%   1 = 84.7%
   (d) Ca = 20%     Br = 80%

**Section 2**

1. (a) C = 85.7%    H = 14.3%
   (b) N = 35%      H = 5%       O = 60%
   (c) Fe = 62.2%   O = 35.6%    H = 2.2%
   (d) C = 26.7%    H = 2.2%     O = 71.1%

2. (a) C = 60%      H = 13%      O = 27%
   (b) C = 40.0%    H = 6.7%     O = 53.3%
   (c) C = 40.0%    H = 6.7%     O = 53.3%
   (d) Al = 36%     S = 64%

3. 34 000

4. 1.1 kg

**Exercise 4.1** Problems on the mole

**Section 1**

1. (a) 26 g  (b) 8 g  (c) 4 g  (d) 6 g
   (e) 4 g  (f) 8 g

2. (a) 2.0  (b) 2.0  (c) 0.25  (d) 0.10
   (e) 0.25 mol

3. (a) 2070 g  (b) 10.6 g  (c) 25.4 g  (d) 20.0 g
   (e) 10.0 g  (f) 40.0 g  (g) 42.0 g  (h) 13.0 g
   (i) 35.5 g  (j) 2.00 g

4. (a) 1.00  (b) 0.25  (c) 0.50  (d) 0.20
   (e) 0.20  (f) 3.0  (g) 0.10  (h) 2.0

5. (a) $6 \times 10^{23}$  (b) $6 \times 10^{23}$  (c) $6 \times 10^{22}$  (d) $3.6 \times 10^{24}$
   (e) $1.2 \times 10^{24}$  (f) $6 \times 10^{22}$  (g) $1.5 \times 10^{22}$  (h) $1.2 \times 10^{24}$

6. (a) 65 g  (b) 0.065 g

7. (a) 9.0 g  (b) 0.027 g

8. (a) 12 g  (b) 0.040 g

9. (a) 20.0 g  (b) 12.0 g  (c) 16.25 g  (d) 115 g

**Section 2**

1. £$1.25 \times 10^{8}$

2. (a) 0.2 mol  (b) 0.4 mol  (c) 1.2 mol  (d) 0.2 mol

3. $1.0 \times 10^{23}$

4. 55.5 mol

5. 2.92 mol

**Exercise 5.1** Problems on reacting masses of solids

**Section 1**

1. 1250 tonnes

2. 0.05 g

3. 0.25 g

4. (a) $2Al(OH)_3$, $3H_2SO_4$, $3H_2O$  (b) (i) 0.46 kg  (ii) 0.86 kg

5. Loss of 83p

6. (a) 72 g

7. 19 kg/year

8. Yes, 1 mol C (12 g) combines with 4 mol F ($4 \times 19$ g)

**Section 2**

1. 127

2. (a) 4.05 tonnes  (b) 8.33 tonnes

3. (a) (i) 1060 tonnes  (ii) 1030 tonnes
   (b) natural limestone; manufactured ammonia  (c) Ammonium sulphate is a fertiliser

4. 9.78 kg

5. 523 tonnes

6. 0.1435 g

7. 2.808 g

8. (a) 7.00 tonnes  (b) 7.24 tonnes

9. 36.5 g

10. 4.481 g

**Exercise 5.2** Problems on deriving equations

1. $Zn(s) + 2Ag^{+}(aq) \longrightarrow Zn^{2+}(aq) + 2Ag(s)$

2. $TiCl_4 + 2Mg \longrightarrow Ti + 2MgCl_2$

3. $S_2O_8^{2-} + 2I^{-} \longrightarrow 2SO_4^{2-} + I_2$

4. $H_2NSO_3^{-} + OH^{-} \longrightarrow NH_3 + SO_4^{2-}$

5. (a) Molar mass allows for 4C, and $54 - 48 = 6 \Rightarrow$ formula is $C_4H_6$. $CH_2{=}CH{-}CH{=}CH_2$
   (b) $CH_2{=}CH{-}CH{=}CH_2 + 2Br_2 \longrightarrow CH_2BrCHBrCHBrCH_2Br$

6. $C_6H_5NH_2 + 3Br_2 \longrightarrow Br_3C_6H_2NH_2 + 3HBr$

**Exercise 5.3** Problems on percentage yield

1. 91%

2. 93%

3. 62%

4. C

5. 50%

6. 89%

7. 91%

8. 99%

**Exercise 5.4** Problems on limiting reactant

1. 45%

2. 816 tonnes

3. (a) 20.4 mol  (b) 58.8 mol; $H_2SO_4$ is limiting; 2.7 kg

4. 304 kg

5. 94%

**Exercise 5.5** Questions from examination papers

1. Expt (1) indicates $Cl_2$ displacing $Br_2$ and $I_2$
   Expt (2): mass of AgI = 0.564 g $=\longrightarrow$ mass of AgBr = 0.338 g
   Then find the masses of NaI and NaBr
   % NaBr = 33.9%, % NaI = 66.1%

### Exercise 6.1 Problems on empirical formulae

1. C
2. $C_2H_4O$
3. $NiSO_4$
4. $C_2HCl_3$
5. $x = 4$
6. $Mg_3N_2$
7. $Fe_3O_4$
8. $Al_2O_3$
9. $BaCl_2.2H_2O$
10. $PbO_2$

### Exercise 6.2 Problems on formulae

1. $\mathbf{a} = C_2F_4$      $\mathbf{b} = C_4H_8O_2$
   $\mathbf{c} = C_2H_6$      $\mathbf{d} = C_6H_6$
   $\mathbf{e} = C_3H_6$      $\mathbf{f} = C_2H_6O_2$
   $\mathbf{g} = C_2H_4Cl_2$      $\mathbf{h} = C_6H_3N_3O_6$
2. D
3. $MO_2$
4. (a) $MgO$ (b) $CaCl_2$ (c) $FeCl_3$ (d) $CuS$
   (e) $LiH$
5. (a) $FeO$ (b) $Fe_2O_3$ (c) $Fe_3O_4$ (d) $K_2CrO_4$
   (e) $K_2Cr_2O_7$
6. (a) $a = 5$ (b) $b = 6$ (c) $c = 2$
7. (a) $C_2H_4O$ (b) $C_2H_4O$
8. (a) $C_5H_{10}O_2$ (b) $C_5H_{10}O_2$
9. $C_9H_{10}O_2$

### Exercise 7.1 Problems on reacting volumes of gases

#### Section 1

1. (a) $30 \text{ cm}^3$ (b) 40%
2. C
3. C
4. (b) (i) $579 \text{ m}^3$ (ii) $308 \text{ cm}^3$ (c) $44, CO_2$
5. (a) (i) $240 \text{ dm}^3$ (ii) $10 : 1$
   (b) to avoid incomplete combustion with the formation of CO
6. A
7. C
8. $C_xH_y(g) + 2O_2(g) \longrightarrow CO_2(g) + 2H_2O(g)$
   $\Rightarrow x = 1, y = 4; CH_4$
9. $C_aH_b(g) + O_2(g) \longrightarrow 3CO_2(g) + 4H_2O(g)$
   $a = 3, b = 8; C_3H_8$

#### Section 2

1. (a) $C_2H_4(g) + 3O_2(g) \longrightarrow 2CO_2(g) + 2H_2O(l)$
   (b) $20 \text{ cm}^3$ ethane $+ 10 \text{ cm}^3$ ethene
2. (a) $2 \text{ dm}^3$ (b) $750 \text{ cm}^3$ (c) $625 \text{ cm}^3$ (d) $938 \text{ cm}^3$ (e) $2 \text{ dm}^3$
3. $500 \text{ cm}^3 SO_2$
4. D

### Exercise 7.2 Problems on reactions of solids and gases

#### Section 1

1. $25 \text{ g}, 6 \text{ dm}^3$
2. $3250 \text{ g}, 1200 \text{ dm}^3$

3. $600 \text{ cm}^3, 1.20 \text{ dm}^3$
4. (a) $2H_2O$ on LHS (b) $1.33 \text{ g}$
5. (a) $2H_2O$ on RHS (b) (i) $16.0 \text{ g}$ (ii) $40 \text{ g}$

#### Section 2

1. $90 \text{ dm}^3$
2. $3.5 \text{ g}$
3. 50%
4. (a) $54 \text{ m}^3 Cl_2$, (b) $115.5 \text{ kg } CCl_4$
5. $360 \text{ g}$ (because $180 \text{ g} \longrightarrow 24 \text{ dm}^3$)
6. $384 \text{ m}^3$ (because $4 \times 4 \times 10^3 \text{ mol} = 16 \times 10^3 \times 24 \text{ dm}^3$)
7. £90 daily
8. $267 \text{ dm}^3$
9. $3.50 \text{ dm}^3$
10. $1.11 \text{ g}$
11. $2.39 \text{ g}$
12. $3.65 \text{ g}$
13. $11.5 \text{ dm}^3$
14. $2460 \text{ dm}^3$

### Exercise 7.3 Questions from examination papers

1. (c) (i) $\mathbf{T} = [CrCl_2(H_2O)_4]^+ \ Cl^-. 2H_2O, \mathbf{U} = [CrCl(H_2O)_5]^{2+} \ 2Cl^-. H_2O$
   (ii) I. $0.02 \text{ mol } N_2 + 0.02 \text{ mol } Cr_2O_3 + 0.08 \text{ mol } H_2O$
   $\mathbf{V} \longrightarrow N_2 + Cr_2O_3 + 4H_2O$ and $\mathbf{V}$ is $(NH_4)_2Cr_2O_7$
   II. $NH_4^+(s) + OH^-(aq) \longrightarrow NH_3(g) + H_2O(l)$
   1. $Cr_2O_7^{2-}(aq) + 14H^+(aq) + 6Fe^{2+}(aq) \longrightarrow$
   $\qquad\qquad\qquad 2Cr^{3+}(aq) + 6Fe^{3+}(aq) + 7H_2O(l)$
   2. $Cr_2O_7^{2-}(aq) + H_2O(l) \rightleftharpoons 2CrO_4^{2-}(aq) + 2H^+(aq)$
   $OH^-(aq)$ removes $H^+(aq)$ and drives the equilibrium from L to R. $CrO_4^{2-}$ is yellow.

### Exercise 8.1 Problems on concentration

1. (a) $0.0500 \text{ mol dm}^{-3}$ (e) $0.100 \text{ mol dm}^{-3}$
   (b) $1.00 \text{ mol dm}^{-3}$ (f) $0.125 \text{ mol dm}^{-3}$
   (c) $0.250 \text{ mol dm}^{-3}$ (g) $0.250 \text{ mol dm}^{-3}$
   (d) $2.00 \text{ mol dm}^{-3}$ (h) $0.200 \text{ mol dm}^{-3}$
2. (a) $0\ 250 \text{ mol}$ (b) $0.125 \text{ mol}$ (c) $0.005\ 00 \text{ mol}$ (d) $2.50 \text{ mol}$
   (e) $0.0500 \text{ mol}$ (f) $0.0250 \text{ mol}$ (g) $0.360 \text{ mol}$ (h) $0.300 \text{ mol}$
3. (a) $2.00 \text{ g}$ (b) $5.30 \text{ g}$ (c) $9.45 \text{ g}$ (d) $42.0 \text{ g}$ (e) $8.3 \text{ g}$
4. (a) $0.500 \text{ dm}^3$ (b) $0.0750 \text{ dm}^3$ or $75.0 \text{ cm}^3$ (c) $12.5 \text{ cm}^3$
   (d) $125 \text{ cm}^3$ (e) $200 \text{ cm}^3$

### Exercise 8.2 Problems on neutralisation

#### Section 1

1. (a) True (b) False (c) True (d) False
   (e) True (f) False
2. (a) True (b) False (c) True (d) False
   (e) True (f) False
3. (a) False (b) False (c) True (d) True
   (e) True (f) False
4. (a) True (b) False (c) False (d) False
   (e) True (f) True
5. 75%
6. $1.1 \text{ mol dm}^{-3}$
7. $0.145 \text{ mol dm}^{-3}$

8. 84%

9. (b) $3600 \, dm^3$

10. (a) 0.050%   (b) $2.0 \times 10^{-3} \, mol \, dm^{-3}$

11. $0.109 \, mol \, dm^{-3}$

12. (a) $1.20 \times 10^{-2} \, mol$   (b) $1.20 \times 10^{-2} \, mol$
    (c) $1.20 \times 10^{-2} \, mol$   (d) 0.276 g   (e) 9.2% Na

## Section 2

1. (b) (i) $4.44 \times 10^{-3} \, mol$   (ii) $8.88 \times 10^{-3} \, mol$
   (c) 0.475 g   (d) 95.0%

2. (b) $2.48 \times 10^{-3} \, mol$   (c) same
   (d) (i) $2.00 \times 10^{-2} \, mol$   (ii) $1.75 \times 10^{-2} \, mol$
   (e) (i) $8.76 \times 10^{-3} \, mol$   (ii) 0.876 g   (f) 87.6%

3. (b) $2.18 \times 10^{-3} \, mol$   (c) $2.18 \times 10^{-3} \, mol$   (d) $2.00 \times 10^{-2} \, mol$
   (e) $1.78 \times 10^{-2} \, mol$   (f) (i) $8.91 \times 10^{-3} \, mol$   (ii) 1.524 g
   (iii) 92.2%

4. (b) 0.0126 mol   (c) (i) 0.0252 mol   (ii) 0.428 g   (d) 18.0%

5. (a) $1.57 \times 10^{-3} \, mol$   (b) same
   (c) (i) $7.85 \times 10^{-4} \, mol$   (ii) 0.1036 g   (d) 46.1%

6. (a) $7.5 \times 10^{-3} \, mol$   (b) same   (c) $2.5 \times 10^{-3} \, mol$   (d) 3

7. (b) $4.88 \times 10^{-3} \, mol$   (c) $2.44 \times 10^{-3} \, mol$
   (d) (i) $9.76 \times 10^{-2} \, mol$   (ii) 10.35 g   (e) $n = 10$

8. NaOH $30.0 \times 10^{-3} \, mol$ added, $5.0 \times 10^{-3} \, mol$ unused,
   $25.0 \times 10^{-3} \, mol$ used; ethyl ethanoate $88 \times 25 \times 10^{-3} \, g = 2.2 \, g$;
   92.4%

9. 200

## Exercise 8.3 Problems on oxidation numbers

1. (a) +2   (f) +1   (k) 0    (o) +3   (s) +6   (w) +4
   (b) +2   (g) +3   (l) +7   (p) +6   (t) +4   (x) +1
   (c) 0    (h) +5   (m) +2   (q) +5   (u) −1   (y) +4
   (d) +1   (i) +5   (n) −1   (r) −1   (v) +6   (z) +5
   (e) +3   (j) +4

2. (a) Sn is oxidised from +2 to +4; Pb is reduced from +4 to +2.
   (b) Mn is oxidised from +2 to +7; Bi is reduced from +5 to +3.
   (c) As is oxidised from +3 to +7; Mn is reduced from +7 to +2.

3. (a) F is reduced from 0 to −1.
   (b) Cl disproportionates from 0 to +5 and −1.
   (c) N disproportionates from −3 and +5 to −1.
   (d) Cr is reduced from +6 to +3.
   (e) C is oxidised from +3 to +4.

4. (a) −2   (b) +2

## Exercise 8.4 Problems on redox reactions

### Section 1

1. (a) $2MnO_4^-(aq) + 5H_2O_2(aq) + 6H^+(aq) \longrightarrow$
   $$5O_2(g) + 2Mn^{2+}(aq) + 8H_2O(l)$$
   (b) $MnO_2(s) + 4H^+(aq) + 2Cl^-(aq) \longrightarrow$
   $$Mn^{2+}(aq) + Cl_2(g) + 2H_2O(l)$$
   (c) $2MnO_4^-(aq) + 5C_2O_4^{2-}(aq) + 16H^+(aq) \longrightarrow$
   $$2Mn^{2+}(aq) + 10CO_2(g) + 8H_2O(l)$$
   (d) $Cr_2O_7^{2-}(aq) + 3C_2O_4^{2-}(aq) + 14H^+(aq) \longrightarrow$
   $$2Cr^{3+}(aq) + 6CO_2(g) + 7H_2O(l)$$
   (e) $Cr_2O_7^{2-}(aq) + 6I^-(aq) + 14H^+(aq) \longrightarrow$
   $$2Cr^{3+}(aq) + 3I_2(aq) + 7H_2O(l)$$

2. (a) $5.0 \times 10^{-3} \, mol$   (b) $2.5 \times 10^{-3} \, mol$   (c) $2.5 \times 10^{-3} \, mol$
   (d) $2.5 \times 10^{-3} \, mol$   (e) $5.0 \times 10^{-2} \, mol$

3. (a) $6.0 \times 10^{-3} \, mol$   (b) $3.0 \times 10^{-3} \, mol$   (c) $6.0 \times 10^{-3} \, mol$
   (d) $3.0 \times 10^{-3} \, mol$   (e) $3.0 \times 10^{-2} \, mol$

4. (a) $2.0 \times 10^{-3} \, mol$   (b) $1.0 \times 10^{-3} \, mol$   (c) $1.0 \times 10^{-3} \, mol$
   (d) $2.0 \times 10^{-3} \, mol$   (e) $0.33 \times 10^{-3} \, mol$

5. 6.21%

6. **A**

7. **B**

8. **C**

9. (a) $3.83 \, mol \, m^{-3}$   (b) $245 \, g \, m^{-3}$

10. $0.0900 \, mol \, dm^{-3}$

11. $1.95 \times 10^{-2} \, mol \, dm^{-3}$

12. $8.94 \times 10^{-2} \, mol \, dm^{-3}$

13. 99.5%

### Section 2

1. $1.64 \times 10^{-2} \, mol \, dm^{-3}$

2. (a) $[Fe^{2+}] = 0.0600 \, mol \, dm^{-3}$   (b) $[Fe^{3+}] = 0.0160 \, mol \, dm^{-3}$

3. (a) $20.0 \, cm^3$   (b) $22.4 \, cm^3$

4. 95%

5. 63.5%

6. $7.63 \times 10^{-2} \, mol$

7. +4

8. 82.3%

9. (b) $0.74 \, mol \, dm^{-3}$

## Exercise 8.5 Problems on complexometric titrations

1. 11.2%

2. $9.37 \times 10^{-3} \, mol \, dm^{-3}$

3. 78 ppm

4. 60.0% CaO    40.0% MgO

5. 18

## Exercise 8.6 Questions from examination papers

1. (a) 11.4 g   (b) $109 \, cm^3$   (c) $1.48 \, dm^3$

2. (b) Multiply by the Avogadro constant
   (c) (i) $0.167 \, mol \, dm^{-3}$   (ii) $1.00 \, dm^3$   (iii) $37.3 \, cm^3$

3. (a) (ii) 54.1%
   (b) (i) $NO_3^-(aq) + 6H_2O(l) + 8e^- \longrightarrow NH_3(g) + 9OH^-(aq)$
   (ii) $[Al(OH)_4]^-$
   (c) (i) $NH_4^+$ and $NH_3$, $H_2O$ and $OH^-$   (ii) Combination of
   ions is exothermic. Position of equilibrium moves
   towards LHS with rise in temperature.
   (d) (i) $NH_4^+$ can donate a proton to $NH_2^-(am)$
   (ii) $Na^+NH_2^-(am) + NH_4^+Cl^-(am) \longrightarrow$
   $$Na^+Cl^-(am) + 2NH_3(l)$$

4. (a) (i) +4, +6   (ii) oxidised, its ox. no. increases
   (b) (i) $1.64 \times 10^{-4} \, mol$   (ii) $1.64 \times 10^{-4} \, mol$
   (iii) $3.28 \times 10^{-3} \, mol \, dm^{-3}$   (iv) $0.210 \, g \, dm^{-3}$
   (c) Sufficient to preserve, just below the maximum.

5. (a) (i) N changes from ox. no. −3 to 0, Cl changes from 0 to −1
   (ii) Cu no change: ox. no. +2 throughout, Cl ox. no. −1
   throughout
   (b) (i) $2MnO_4^- + 16H^+ + 5C_2O_4^{2-} \longrightarrow$
   $$2Mn^{2+} + 10CO_2 + 8H_2O$$
   (ii) 69.3%

6. (e) $[NaClO] = 0.84 \, mol \, dm^{-3}$

7. (b) (i) $2MnO_4^- + 16H^+ + 5C_2O_4^{2-} \longrightarrow$
   $$2Mn^{2+} + 10CO_2 + 8H_2O$$
   (ii) autocatalysis (by $Mn^{2+}$)   (iii) $6.09 \, g \, dm^{-3}$

(c) (i) $3MnO_4^{2-} + 2H_2O \longrightarrow 2MnO_4^- + MnO_2 + 4OH^-$
(ii) oxidation, chlorine

8. (a) (ii) $Cr_2O_7^{2-} + 3SO_3^{2-} + 8H^+ \longrightarrow 2Cr^{3+} + 3SO_4^{2-} + 4H_2O$
(iii) $16.0 \text{ cm}^3$   (iv) $X = 39$

9. (a) (i) $FeSO_4$   (b) (i) $8H^+, 5Fe^{3+}, 4H_2O$   (ii) $5.5 \times 10^{-8}$ mol
(iii) $2.75 \times 10^{-6}$ mol   (iv) 15.4 mg per 100 g

10. (b) (iii) 17.4%

11. (a) (i) $2KMnO_4 + 16HCl \longrightarrow 2MnCl_2 + 2KCl + 5Cl_2 + 8H_2O$
(ii) only $\frac{10}{16}$ of the Cl in HCl forms $Cl_2$
(i
(b) (i) HCl is oxidised by $KMnO_4$   (ii) 0.79 g

12. (a) $C_2O_4^{2-}$, +3   (b) $x = \frac{1}{2}$
(c) $2MnO_4^- + 16H^+ + 5C_2O_4^{2-} \longrightarrow 2Mn^{2+} + 10CO_2 + 8H_2O$

13. (a) starch, blue to colourless
(b) $12 \text{ mol dm}^{-3}$   (c) $1.44 \text{ dm}^3$

14. (b) (ii) The powder has lost water of crystallisation.
(c) (i) $3.54 \times 10^{-3}$ mol   (iii) $7.08 \times 10^{-2} \text{ mol dm}^{-3}$
(iv) $x = 10$

15. (a) $I_2 + 2S_2O_3^{2-} \longrightarrow 2I^- + S_4O_6^{2-}$
(b) $[KIO_3] = 2.80 \times 10^{-3} \text{ mol dm}^{-3}$, [thio] = $0.0186 \text{ mol dm}^{-3}$

16. (c) (i) $0.0206 \text{ mol } H_2$   (ii) NaOH from Na = 0.0412 mol,
NaOH total = 0.0417 mol.% $Na_2O$ = 1.55%, % Na = 94.8%

17. (b) 51%

18. (c) (i) $0.167 \text{ mol dm}^{-3}$
(ii) $1.00 \text{ dm}^3$
(iii) $37.3 \text{ cm}^3$

19. (c) (iv) 85%

## Exercise 9.1 Problems on relative atomic mass

1. 85.6 u

2. 69.8 u

3. 24.3

4. $3 \,^{35}Cl : 1 \,^{37}Cl$; 35.5 u

5. 6.93

6. 1, $^1H$; 2, $^2H$; 3, $^1H^2H$; 4, $^2H_2$; 17 $^{16}O^1H$; 18, $^{16}O^2H$ and $^1H_2^{16}O$;
19, $^1H^2H^{16}O$; 20, $^2H_2^{16}O$

7. 39.1 u

8. 58.70

## Exercise 9.2 Problems on mass spectra of compounds

1. 122, 124

2. (a) Br   (b) $CH_3^+$   (c) $CH_3Br$   (i) $CH_3^{79}Br^+$, $CH_3^{81}Br^+$
(ii) $^{13}CH_3^{79}Br$, $^{13}CH_3^{81}Br$

3. (a) 122 u   (b) $C_6H_5$
(c) 45 peak could be due to $CO_2H^+$, 17 peak could be due to $OH^+$.
(d) 105 peak could be due to $C_6H_5CO^+$

4. (a) 46, 47; The species with a 47 peak contains $^{13}C$.
(b) $CH_3^+$ and $C_2H_5^+$
(c) $46 = C_2H_5OH$, $45 = C_2H_5O^+$, $31 = CH_2OH^+$

5. (a) 4   (b) $C_2H_3^{35}Cl_3$, $C_2H_3^{35}Cl_2^{37}Cl$, $C_2H_3^{35}Cl^{37}Cl_2$, $C_2H_3^{37}Cl_3$
(c) 138

## Exercise 9.3 Problems on nuclear reactions

1. (a) $a = 1, b = 0$   (b) $a = 17, b = 8$   (c) $a = 4, b = 2$
(d) $a = 210, b = 83$   (e) $a = 4, b = 2$   (f) $a = 4, b = 2$
(g) $a = 14, b = 6$   (h) $a = 16, b = 7$   (i) $a = 207, b = 83$
(j) $a = 4, b = 11$   (k) $q = 0, p = 1, r = 35$   (l) $c = 3, d = 1$

## Exercise 9.4 Questions from examination papers

1. (a) electron, lightest
(b) for acceleration by electric field, for deflection by magnetic field
(c) $^{10}B : ^{11}B = 1 : 4$
(d) (i) $B_2H_5$   (ii) $B_4H_{10}$

2. (a) (i) only one isotope $^{31}P_4$   (ii) weak intermolecular forces
(iii) giant molecular structure
(b) (i) $P_2O_3$   (ii) $M_r$   (iii) 7.33 g

3. (a) $^{34}_{16}S$   (b) $1s^2 2s^2 2p^6$
(c) Positive ions are deflected towards the detector.
(d) mass of n = $1.65 \times 10^{-24}$ g, $A_r(^{13}C) = 12.99$, $A_r(C) = 12.01$

4. (d) $A_r = 83.93$, Kr
(e) Empirical formula $C_3H_2NO_2$ molecular formula $C_6H_4N_2O_4$

## Exercise 10.1 Problems on gas molar volume

1. (a) $5.60 \text{ dm}^3$   (b) $6.00 \text{ dm}^3$

2. 2.50 mol

3. $83.8 \text{ g mol}^{-1}$

4. $6.25 \times 10^{-2}$ mol

5. $39.9 \text{ g mol}^{-1}$

## Exercise 10.2 Problems on partial pressure

1. $2.53 \times 10^4 \text{ N m}^{-2}$

2. $O_2$, 30.3 kPa; $H_2$ 70.7 kPa

3. methane $1.20 \times 10^5$ Pa, ethane $3.61 \times 10^5$ Pa,
propane $2.40 \times 10^4$ Pa

4. $CO_2$ 400 kPa, $N_2$ 200 kPa; total = 600 kPa

5. (a) $p(N_2) = 3.00 \times 10^4 \text{ N m}^{-2}$   $p(O_2) = 2.63 \times 10^4 \text{ N m}^{-2}$,
$p(CO_2) = 1.88 \times 10^4 \text{ N m}^{-2}$
(b) $p(N_2) = 3.00 \times 10^4 \text{ N m}^{-2}$   $p(O_2) = 2.63 \times 10^4 \text{ N m}^{-2}$

6. (a) $p(NH_3) = 6.00 \times 10^4 \text{ N m}^{-2}$   $p(H_2) = 3.75 \times 10^4 \text{ N m}^{-2}$
$p(N_2) = 5.25 \times 10^4 \text{ N m}^{-2}$
(b) $p(H_2) = 3.75 \times 10^4 \text{ N m}^{-2}$   $p(N_2) = 5.25 \times 10^4 \text{ N m}^{-2}$

## Exercise 10.3 Problems on the ideal gas equation

1. $64.0 \text{ g mol}^{-1}$

2. $46 \text{ g mol}^{-1}$

3. $86 \text{ g mol}^{-1}$

4. (a) 47   (b) $C_2H_6O$, ethanol

5. $85 \text{ g mol}^{-1}$

6. empirical formula $PF_5$, $M_r = 126$, molecular formula = $PF_5$

7. $91 \text{ g mol}^{-1}$   8. $65 \text{ g mol}^{-1}$

## Exercise 10.4 Problems on partition

1. B

2. 13.6 g

3. 4.76 g

4. 20.0

5. 0.0514 mol

6. (a) 4.74 g   (b) 4.95 g

7. (a) 4.75 g   (b) 4.95 g

8. 25.0

9. $1.18 \times 10^{-2}$

10. (a) 3.0 g   (b) 3.5 g

**Exercise 10.5** Questions from examination papers

1. (b) Mass of 1 mol of $^{12}C$ = 12.000 g
   Mass of 1 atom of $^{12}C$ = $1.993 \times 10^{-23}$ g
   No of atoms mol$^{-1}$ = $6.02 \times 10^{23}$ mol$^{-1}$
   (c) (i) charge (ii) $2Fe^{3+}$, $2Fe^{2+}$
   (d) (i) $2.00 \times 10^7$ Pa (from $PV = nRT$)
       (ii) $2.5 \times 10^7$ Pa (iii) 9.00 mol dm$^{-3}$

2. (b) (i) 30.6 dm$^3$
   (c) (ii) Mass of $I_2$ = 0.0127 g, Mass extracted = 0.0125 g
       (iii) Use in two successive portions.

**Exercise 11.1** Problems on electrolysis

1. 0.265 g

2. 0.403 g (a) Double (b) Double (c) No change

3. 1.244 g Ca; 2.207 g Cl$_2$

4. 0.394 g Au

5. (a) 24 cm$^3$ H$_2$ (b) 0.207 g Pb

6. 1200 minutes

7. 2.48 hours

8. 4.48 dm$^3$ H$_2$; 2.24 dm$^3$ O$_2$

9. $1.84 \times 10^4$ C   10. 1070 s

**Exercise 11.2** Problems on pH

**Section 1**

1.

| | pH | pOH | | pH | pOH | | pH | pOH | | pH | pOH |
|---|---|---|---|---|---|---|---|---|---|---|---|
| (a) | 8 | 6 | (b) | 4 | 10 | (c) | 7 | 7 | (d) | 2.2 | 11.8 |
| (e) | 4.5 | 9.5 | (f) | 1.5 | 12.5 | (g) | 0.60 | 13.4 | (h) | 8.3 | 5.7 |
| (i) | 6.2 | 7.8 | (j) | 1.0 | 13.0 | | | | | | |

2. (a) 12 (b) 11 (c) 6.0 (d) 12.7 (e) 11
   (f) 12.9 (g) 12 (h) 9.7 (i) 6.8 (j) 4.6

3. In mol dm$^{-3}$, the values are:
   (a) 1.00 (b) $5.01 \times 10^{-5}$ (c) $4.47 \times 10^{-3}$ (d) 0.0132
   (e) $7.08 \times 10^{-5}$ (f) $1.45 \times 10^{-8}$ (g) $6.17 \times 10^{-10}$
   (h) $2.00 \times 10^{-14}$ (i) $3.16 \times 10^{-1}$ (j) $2.34 \times 10^{-3}$

4. (a) 0.784 (b) 1.05 (c) 13.3 (d) 12.7 (e) 13.4

5. (a) 2 (b) 12 (c) 2.3

6. (a) 3 (b) 9 cm$^3$ (c) 0.9 cm$^3$

**Section 2**

1. HCl is completely ionised, but CH$_3$CO$_2$H is partially ionised. As H$^+$ from CH$_3$CO$_2$H are neutralised, more molecules of CH$_3$CO$_2$H ionise.

2. pH = 7

3. C

4. (a) 7.0 (b) 6.6

5. (a) $1.74 \times 10^{-5}$ mol dm$^{-3}$ (b) $3.97 \times 10^{-10}$ mol dm$^{-3}$
   (c) $1.38 \times 10^{-5}$ mol dm$^{-3}$ (d) $2.00 \times 10^{-9}$ mol dm$^{-3}$

6. (a) (i) $2.04 \times 10^{-3}$ mol dm$^{-3}$ (ii) $2.78 \times 10^{-5}$ mol dm$^{-3}$
   (b) 15 cm$^3$, (c) $2.78 \times 10^{-5}$ mol dm$^{-3}$ (d) pH = 12.7

7. (a) $1.35 \times 10^{-9}$ mol dm$^{-3}$, $7.41 \times 10^{-6}$ mol dm$^{-3}$
   (b) $7.41 \times 10^{-6}$ mol dm$^{-3}$ (c) $1.82 \times 10^{-6}$ mol dm$^{-3}$

8. (a) 1.30 (b) 12.2

**Exercise 11.3** Problems on buffers

1. B

2. (a) 3.14 (b) 2.84

3. (a) $1.26 \times 10^{-5}$ mol dm$^{-3}$ (b) 4.90

4. 3.34   5. 1.0 mol

**Exercise 11.4** Problems on standard electrode potentials

1. Ag, I$^-$

2. I$_2$, Fe$^{3+}$

3. (a) +0.40 V (b) +0.26 V (c) −0.27 V

4. Yes, reaction **a** goes R to L and **b** goes L to R; then
   $E^{\ominus}$ = +1.36 − 1.19 V = + 0.17 V

5. (a) Cu (b) Zn(s) + Cu$^{2+}$(aq) $\longrightarrow$ Zn$^{2+}$(aq) + Cu(s)

6. $E^{\ominus}$ = +1.70 − 0.54 V = +1.16 V
   2Ce$^{4+}$(aq) + 2I$^-$(aq) $\longrightarrow$ 2Ce$^{3+}$(aq) + I$_2$(aq)

7. (a) +0.94 V (b) +0.44 V (c) −0.67 V
   (d) +0.78 V (e) +0.63 V

**Section 2**

1. Fe(s) + Fe$^{3+}$(aq) $\longrightarrow$ 2Fe$^{2+}$(aq)

2. Cr$_2$O$_7{}^{2-}$(aq) + 14H$^+$(aq) + 6Fe$^{2+}$(aq) $\longrightarrow$
   $\qquad\qquad$ 2Cr$^{3+}$(aq) + 6Fe$^{3+}$(aq) + 14H$_2$O

3. (a) 2Fe$^{3+}$(aq) + 2I$^-$(aq) $\longrightarrow$ 2Fe$^{2+}$(aq) + I$_2$(aq)
   (b) 2Ag$^+$(aq) + Cu(s) $\longrightarrow$ 2Ag(s) + Cu$^{2+}$(aq)
   (c) No reaction (d) No reaction
   (e) Br$_2$(aq) + 2Fe$^{2+}$(aq) $\longrightarrow$ 2Br$^-$(aq) + 2Fe$^{3+}$(aq)

4. (a) Cl$_2$ and Br$_2$ (b) Cl$_2$, Br$_2$ and I$_2$

5. (a) Cd(s) + 4OH$^-$(aq) + Ni(OH)$_2$(s) $\longrightarrow$
   $\qquad\qquad$ Cd(OH)$_2$(s) + NiO$_2$(s) + 2H$_2$O(l)
   (b) +0.81 − 0.49 = +0.32 V

6. (a) Cu(I) $\longrightarrow$ Cu(II) + Cu has positive $E^{\ominus}$.
   Cu$^+$ + e$^-$ $\longrightarrow$ Cu; $E^{\ominus}$ = +0.52 V
   Cu$^+$ $\longrightarrow$ Cu$^{2+}$ + e$^-$; $E^{\ominus}$ = −0.15 V
   Add: 2Cu$^+$ $\longrightarrow$ Cu + Cu$^{2+}$; $E^{\ominus}$ = +0.37 V
   (b) 2I$^-$ $\longrightarrow$ I$_2$ + 2e$^-$; $E^{\ominus}$ = −0.54 V
   combined with either Cu$^{2+}$ + e$^-$ $\longrightarrow$ Cu$^+$; $E^{\ominus}$ = +0.15 V gives
   $E^{\ominus}$ = −0.54 + 0.15 = −0.39 V
   or with Cu$^{2+}$ + 2e$^-$ $\longrightarrow$ Cu; $E^{\ominus}$ = 0.34 V gives
   $E^{\ominus}$ = −0.54 + 0.34 = −0.20 V
   combined with Cu$^{2+}$ + I$^-$ + e$^-$ $\rightleftharpoons$ CuI; $E^{\ominus}$ = + 0.92 V gives
   $E^{\ominus}$ = −0.54 + 0.92 = 0.38 V
   therefore 2Cu$^{2+}$ + 4I$^-$ $\longrightarrow$ 2CuI + I$_2$

**Exercise 11.5** Questions from examination papers

1. (a) (i) $1.53 \times 10^{-5}$ mol dm$^{-3}$
       (ii) $K_a$ increases, [H$^+$] increases, pH decreases
   (b) (i) $\frac{2n}{3}$ (ii) $\frac{2n}{3} \div \frac{n}{3} = 2$
       (iii) $K_a$ = [H$^+$] [X$^-$]/[HX] = $2.1 \times 10^{-4}$ mol dm$^{-3}$

2. (a) 12 cm$^3$ (ii) S, pH = 12.8
   At $\frac{1}{2}$ equivalence pH = 4.7, at 2 × equivalence pH = 12.8
   (c)

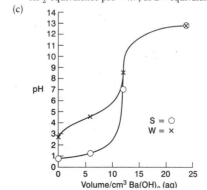

S = ○
W = ×

Volume/cm$^3$ Ba(OH)$_2$ (aq)

**238**  ANSWERS

3. (c) (i)  I. +2 to +3 = +1
II. $Cl_2(aq) + 2Fe^{2+}(aq) \longrightarrow 2Fe^{3+}(aq) + 2Cl^-(aq)$
III. +1.36 − 0.77 = +0.59 V
(b) (ii)  I. $VO_3^-(aq) + 2H^+(aq) \longrightarrow VO_2^+(aq) + H_2O(l)$
II. $2VO_2^+(aq) + 2I^-(aq) + 4H^+(aq) \longrightarrow$
$\qquad 2VO^{2+}(aq) + I_2(aq) + 2H_2O(l)$
III. +1.00 − 0.54 = +0.46 V

4. (a) $OH^-(aq), O_2(g) \mid Pt$
(c) (i)  $Au(s) \mid Au(CN)_2^-(aq), CN^-(aq) \vdots OH^-(aq), O_2(g) \mid Pt$
(ii) +0.28 V  (iii) Since $E^\ominus$ is positive, $O_2$ is acting as the oxidant and the position of equilibrium is towards the RHS.
(d) linear
(e) (i)  +3, −1
(ii)  release of poisonous As compounds into the atmosphere

5. (a) (ii) endothermic; $K_w$ increases with a rise in temperature
(iv) pH = 6.63
(b) (ii) $NH_4^+$ and $NH_3$, $H_2O$ and $OH^-$

6. (a) $H_2CO_3$ donates a proton to $H_2O$
(b) (i) $K_a = [H_3O^+(aq)][HCO_3^-(aq)]/[H_2CO_3(aq)]$  (ii) 2.85
(c) Citric acid; has the higher $K_a$
(d) Buffer: $H^+(aq) + A^-(aq) \rightleftharpoons HA(aq)$
$\qquad OH^-(aq) + HA(aq) \rightleftharpoons H_2O(l) + A^-(aq)$

7. (c) negative  (d) $E^\ominus$ = +1.30 V
(e) $E^\ominus$ must be positive = +0.28 V – 0.234 V
$\qquad Ni^{2+}(aq) + Co(s) \longrightarrow Ni(s) + Co^{2+}(aq)$

8. (a) (i)

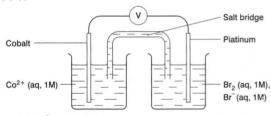

(ii) $E^\ominus$ = +1.35 V
(iii) $Br_2(aq) + Co(s) \longrightarrow 2Br^-(aq) + Co^{2+}(aq)$
(iv) $Br_2$, accepts electrons
(b) Manganate(VII) because acidic dichromate is a less powerful oxidant (lower $E^\ominus$) than chlorine.

9. (b) $H_2$ has the less negative $E^\ominus$
$\qquad 2H^+(aq) + D_2(g) \rightleftharpoons H_2(g) + 2D^+(aq); E^\ominus = +0.004$ V
(c) $AgF(s) + Cl^-(aq) \longrightarrow F^-(aq) + AgCl(s); E^\ominus = +0.56$ V
(d) (i) $Ag/Ag^+$, with $E^\ominus$ = +0.8 V, is a weaker reductant than $H_2(g)$
(ii) HI
$\qquad Ag(s) + I^-(aq) + H^+(aq) \longrightarrow AgI(s) + \frac{1}{2}H_2(g); E^\ominus = +0.15$ V

10. (a) (i) $K_a = [H^+(aq)][CH_2ClCO_2^-(aq)]/[CH_2ClCO_2H(aq)]$
(ii) $CH_2ClCO_2H$ has the highest $K_a$
(b) (i) 3.09  (ii) 12.7
(iii)

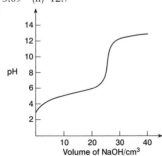

Volume of NaOH/cm³

(iv) phenolphthalein, changes colour at the equivalence point

11. (a) (i) no effect
(ii) Equilibrium moves towards LHS.
(b) +7
(c) (i) The drops of iodine solution lose their colour as they react with thiosulphate:
$\qquad I_2(aq) + 2S_2O_3^{2-}(aq) \longrightarrow 2I^-(aq) + S_4O_6^{2-}(aq)$
When all the thio has reacted the addition of further iodine turns the solution gradually brown.
(ii) +2.5  (iii) +6  (iv) $Cl_2$ is a stronger oxidant than $I_2$

12. (a)(i) $HF(aq) + H_2O(l) \rightleftharpoons H_3O^+(aq) + F^-(aq)$
(ii) The H—F bond is stronger than other H—Halogen bonds.
(b) (i) $O_2(g)$ is formed. $F_2$ oxidises $H_2O$.
(ii) $2Cl_2(aq) + 2H_2O(l) \longrightarrow O_2(g) + 4H^+(aq) + 4Cl^-(aq)$
(iii) little difference in $E^\ominus$ values; $Cl_2$ disproportionates
(c) (i) $3ClO^-(aq) \longrightarrow ClO_3^-(aq) + 2Cl^-(aq)$; Heat
(ii) Part of $ClO^-$ is oxidised from ox. no. +1 to ox. no. +5 in $ClO_3^-$, while part is reduced to ox. no. −1 in $Cl^-$.
(d) (i) $AgBr: Ag^+(aq) + Br^-(aq) \longrightarrow AgBr(s)$
(ii) $2KBrO_3(s) \longrightarrow 3O_2(g) + 2KBr(s)$
(e) (i) Reduction: $H_2O_2(aq) \longrightarrow O_2(g) + 2H^+(aq) + 2e^-$
Oxidation: $2IO_3^-(aq) + 12H^+(aq) + 10e^- \longrightarrow$
$\qquad\qquad\qquad\qquad\qquad I_2(aq) + 6H_2O(l)$
(ii) Bubbles of gas would appear. The solution would turn brown.

13. (a)(i) $H^+(aq) + e^- \rightleftharpoons \frac{1}{2}H_2(g)$
(ii) $KMnO_4$ is a stronger oxidant in acidic conditions
(iii) $IO_3^-(aq) + 5I^-(aq) + 6H^+(aq) \longrightarrow 3I_2(aq) + 3H_2O(l)$
(iv) Reduce $Cr^{3+}(aq)$ by placing a piece of Cr in a solution of $Cr^{3+}(aq)$.
(b)(i) $I_2$ ($Cr_2O_7^{2-}$ has a higher value of $E^\ominus$ than $I_2$)
(ii) 0.762 g

14. (a) $Cr^{3+}(aq) + 4OH^-(aq) \longrightarrow Cr(OH)_4^-(aq)$
$Cr(OH)_4^-(aq) + 2H_2O_2(aq) \longrightarrow CrO_4^{2-}(aq) + 4H_2O(l)$
$2CrO_4^{2-}(aq) + 2H^+(aq) \longrightarrow Cr_2O_7^{2-}(aq) + H_2O(l)$
Destroy $H_2O_2$ to avoid oxidation of $CrO_4^{2-}$ to $CrO_4^-$.
(b) $(NH_4)_2Cr_2O_7(s) \longrightarrow Cr_2O_3(s) + N_2(g) + 4H_2O(g)$
N is oxidised from ox. no. −3 to ox. no. 0; Cr is reduced from ox. no. +6 to ox. no. +3.
(c) In acidic solution the equilibrium between $Cr_2O_7^{2-}$ and $Cr^{3+}$ lies over to the RHS. No.
(d) Lead(II) is more stable than lead(IV), e.g. oxidation of HCl to $Cl_2$

15. (b)(i) $Al^{3+}(l) + 3e^- \longrightarrow Al(l)$  (ii) 269 kg

16. (a) pH = 3.34
(b) pH = 12.7
(c) (i)

(ii) pH 8-10
(d) (i) The acid is diprotic and neutralisation takes place in two steps.

17. (d)(iii) 13.2 g

**Exercise 12.1** Problems on standard enthalpy of reaction

1. $-64$ kJ mol$^{-1}$

2. $-847$ kJ mol$^{-1}$, exothermic

3. $-106$ kJ mol$^{-1}$, Mg reduces $Al_2O_3$

4. (a) $-890$ kJ mol$^{-1}$   (b) $+317$ kJ mol$^{-1}$   (c) $-55$ kJ mol$^{-1}$

**Exercise 12.2** Problems on neutralisation and combustion

1. $-57.3$ kJ mol$^{-1}$   2. $-56.3$ kJ mol$^{-1}$   3. $-49.6$ kJ mol$^{-1}$

4. $-56.4$ kJ mol$^{-1}$   5. $-59.1$ kJ mol$^{-1}$   6. (c) $356 \times 10^3$ kJ

7. $-1360$ kJ mol$^{-1}$   8. $-534.2$ kJ mol$^{-1}$   9. **C**

**Exercise 12.3** Problems on standard enthalpy of formation

1. $-367$ kJ mol$^{-1}$

2. $\Delta H^{\ominus}$(c) $< 3\Delta H^{\ominus}$(a) because bond delocalisation makes benzene more stable than calculated from bond energy terms, and less enthalpy than expected is released when benzene $\longrightarrow$ cyclohexane. Hydrogenation of the first double bond destroys bond delocalisation in benzene, and $\Delta H^{\ominus}$(b) is therefore positive, showing that the enthalpy content of the system has increased

3. $-46$ kJ mol$^{-1}$   4. $-78.2$ kJ mol$^{-1}$   5. $+14$ kJ mol$^{-1}$

6. (a) $-297$ kJ mol$^{-1}$
   (b) $-394$ kJ mol$^{-1}$

7. $-0.3$ kJ mol$^{-1}$

8. $-86$ kJ mol$^{-1}$

9. 184 kJ mol$^{-1}$. First find $\Delta H^{\ominus}_C(CH_2{=}CH_2) = -1456$ kJ mol$^{-1}$; then use this to find $\Delta H^{\ominus}_F(CH_2{=}CH_2)$

10. **B**

11. $-126.5$ kJ mol$^{-1}$

12. $-1370$ kJ mol$^{-1}$

13. (a) $-484$ kJ mol$^{-1}$   (b) 108 kJ mol$^{-1}$   (c) $-75$ and $+33$ kJ mol$^{-1}$
    (d) $-118$ kJ mol$^{-1}$   (e) $-11$ kJ mol$^{-1}$

**Exercise 12.4** Problems on bond energy terms

1. (a) **B**   (b) **D**   (c) **C**   (d) $C_2H_6 - 76$; $C_2H_4 + 48$ kJ mol$^{-1}$
   (e) $-95$ kJ mol$^{-1}$   (f) $-200$ kJ mol$^{-1}$   (g) $-343$ kJ mol$^{-1}$
   (h) $+264$ kJ mol$^{-1}$. The difference is the bond delocalisation energy of benzene.
   (i) $+136$ kJ mol$^{-1}$, 28 kJ mol$^{-1}$ higher than the value from combustion. Butadiene is more stable than it is calculated to be because it is stabilised by bond delocalisation.
   (j) (i) $+74$ kJ mol$^{-1}$   (ii) $-20$ kJ mol$^{-1}$ Reaction (ii) will occur.

**Exercise 12.5** Problems on the Born–Haber cycle

1. $-372$ kJ mol$^{-1}$

2. (b) $-380$ kJ mol$^{-1}$

3. $-775$ kJ mol$^{-1}$

4. (a) $-387$ kJ mol$^{-1}$   (b) $-246$ kJ mol$^{-1}$ $\Delta H^{\ominus}_F$ ($CaCl_2$) has a larger negative value than $\Delta H^{\ominus}_F$(CaCl).

5. **C**   6. $+4$ kJ mol$^{-1}$

7. Calculation of $\Delta H^{\ominus}_{Solution}$ shows that in solubility
   (a) LiCl $>$ NaCl   (b) NaF $<$ NaCl

**Exercise 12.6** Problems on standard entropy change and standard free energy change

1. In J K$^{-1}$ mol$^{-1}$   (a) $+20.0$   (b) $-199$   (c) $-163.5$
   (d) $-121$   (e) $+176$   (f) $-90.1$
   (g) $+285$   (h) $+681$

2. (a) $+$   (b) $+$   (c) $-$   (d) $-$   (e) $+$   (f) $-$

3. (a) $+0.202$ kJ mol$^{-1}$ at 25 °C   $+0.127$ kJ mol$^{-1}$ at 100 °C above 227 °C
   (b) $+2.56$ kJ mol$^{-1}$ at 25 °C; No

4. (a) $-2.91$ kJ mol$^{-1}$   (b) $+2.91$ kJ mol$^{-1}$, trans

**Exercise 12.7** Questions from examination papers

1. (a) By definition
   (d) $-402$ kJ mol$^{-1}$   (e) increase
   (f) $\Delta H^{\ominus}_{Ionisation}$ (Cs) is less endothermic than $\Delta H^{\ominus}_{Ionisation}$ (Li) but $\Delta H^{\ominus}_{Solvation}$ (Cs) is less exothermic than $\Delta H^{\ominus}_{Solvation}$ (Li). The sum $\Delta H^{\ominus}_{Ionisation} + \Delta H^{\ominus}_{Solvation}$ is less exothermic for Cs than for Li.

2. (a) (i) $-126$ kJ mol$^{-1}$   (ii) $-378$ kJ mol$^{-1}$
       (iii) Ethylbenzene; less enthalpy is released when ethylbenzene is hydrogenated.
   (b) (ii) $-364$ kJ mol$^{-1}$

3. (b) (i) $-846$ kJ mol$^{-1}$   (ii) $\Delta H^{\ominus}_{Sublimation}$ (Li)
       (iii) larger for Li because the smaller atoms are closer together.
   (c) (i) sizes and charges of ions   (ii) some degree of covalent character
       (iii) almost 100% ionic bond
   (d) (i) The Li 2s electron is promoted to a higher energy level. When it falls back to the ground state the value of $\Delta E$ corresponds to the wavelength of red light.
       (ii) Rb is dangerously reactive.

4. (b) $-2609$ kJ mol$^{-1}$

5. (b) $-125$ kJ mol$^{-1}$
   (c) Broken: $(C-C) + \frac{1}{2}(O{=}O) + 2(C-H)$;
       Made: $(C{=}C) + 2(H-O)$; $O{=}O = 478$ kJ mol$^{-1}$

6. (b) (i) $\Delta H_2$ ionisation energy of Rb, $\Delta H_4$ electron affinity of Cl, $\Delta H_6$ standard enthalpy of formation of RbCl
       (ii) $\Delta H_5 = -692$ kJ mol$^{-1}$
       (iii) The lattice energy of LiCi is more exothermic because the smaller Li$^+$ ions are closer than Rb$^+$ ions to the Cl$^-$ ions.
   (c) covalent

7. (b) $-145$ kJ mol$^{-1}$   (c) $-1642$ kJ mol$^{-1}$   (d) lack of oxygen
   (e) $+44$ kJ mol. Energy is required to separate the molecules in a liquid against the intermolecular attractions.

8. (b) $-114$ kJ mol$^{-1}$   (c) $-425$ kJ mol$^{-1}$   (d) 2 mol of solid form 2 mol of solid
   (e) $\Delta H_F$ for $MgCl_2$ is 6 times as exothermic as $\Delta H_F$, for MgCl.

9. (b) $\Delta H_{Solution} = \Delta H_{Lattice\ dissociation} + \Delta H_{Hydration}$
       Calculation of values of $\Delta H_{Solution}$ shows that $\Delta H_{Solution}$ decreases down the group, therefore solubility decreases down the group.
   (c) (i) $IO_4^-(aq) + I^-(aq) + 8H^+(aq) \longrightarrow I_2(aq) + 4H_2O(l)$
       (ii) 0.060 mol dm$^{-3}$
       (iii) K$^+$ displaces equilibrium to the LHS, reducing $[IO_4^-(aq)]$. The brown colour of iodine fades.

10. (a) solid dissociates to form a gas, e.g
    $CaCO_3(s) \longrightarrow CaO(s) + CO_2(g)$; $\Delta S$ is positive
    (b) $+6$ J K$^{-1}$ mol$^{-1}$ Graphite has an ordered structure.
    (c) (i) $\Delta G^{\ominus} = \Delta H^{\ominus} - T\Delta S^{\ominus}$ and $\Delta H^{\ominus} = T\Delta S^{\ominus}$
    (d) $\Delta S = 66$ J K$^{-1}$ for 54 g

11. (a) $\Delta G^{\ominus}$ is negative   (b) Energy of activation is high.
    (c) (i) $2NaHCO_3(s) \longrightarrow Na_2CO_3(s) + CO_2(g) + H_2O(g)$
        (ii) $\Delta G^{\ominus} = 14.9$ kJ mol$^{-1}$
        (iii) When $T$ is high enough, ($-T\Delta S^{\ominus}$) becomes sufficiently negative to outweigh $\Delta H^{\ominus}$ and make $\Delta G^{\ominus}$ negative.

12. (c) $\Delta H_R = -416$ kJ mol$^{-1}$

13. (a)(i)  0   (ii)  $\Delta S^\ominus = 216$ J K$^{-1}$ mol$^{-1}$
    (iii) $\Delta H^\ominus = 206$ kJ mol$^{-1}$   (iv)  954 K (from $\Delta G^\ominus = 0$)
    (b)(ii)  $K_p = p(H_2O)(g)/[p(CO(g)) \times p(H_2(g))] = 5.99 \times 10^{-7}$ Pa$^{-1}$

## Exercise 13.1 Problems on finding the order of reaction

1. 2

2. 1 w.r.t. A   2 w.r.t. B

3. D

4. Rate $= k[X]^0$; 0

5. 10.0 mol$^{-1}$ dm$^3$ s$^{-1}$

6. Rate $= k[A][B]$   $2.0 \times 10^{-3}$ dm$^3$ mol$^{-1}$ s$^{-1}$

7. (a) 1   (b) 2   $1.67 \times 10^5$ mol$^{-2}$ dm$^6$ h$^{-1}$ or 46.4 mol$^{-2}$ dm$^6$ s$^{-1}$

8. Rate $= k[N_2O_5]$   (a) 0.150 mol dm$^{-3}$ s$^{-1}$   (b) 1.80 mol dm$^{-3}$ s$^{-1}$

## Exercise 13.2 *Problems on first-order reactions

1. 1.25 hours

2. $1.10 \times 10^{-5}$ s$^{-1}$

3. Gradients/mol dm$^{-3}$ min$^{-1}$:  (a) −0.842   (b) −0.386
   (c) −0.200   (d) 0      0.365 min$^{-1}$ or $6.10 \times 10^{-3}$ s$^{-1}$   1

4. (a) Initial rates/mol dm$^{-3}$ min$^{-1}$:  **1** $6.30 \times 10^{-3}$  **2** $1.27 \times 10^{-2}$
   **3** $1.90 \times 10^{-2}$
   (b) 1   (c) 3.75 min   7.50 min   1
   (d) Rate $= k[A][B]$   (e) $1.06 \times 10^{-2}$ dm$^3$ mol$^{-1}$ s$^{-1}$

## Exercise 13.3 Problems on second-order reactions

1. 50.0 dm$^3$ mol$^{-1}$ s$^{-1}$

2. 1 w.r.t. A   2 w.r.t. B   Rate $= k[A][B]^2$   4.0 dm$^6$ mol$^{-2}$ s$^{-1}$

3. (a) (i) 1   (ii) 1   (iii) 0
   (b) Rate $= k[ClO^-][I^-]$

4. (a) (i) 1   (ii) 1   (b) Rate $= k[A][B]$, dm$^3$ mol$^{-1}$ s$^{-1}$

## Exercise 13.4 *Problems on radioactive decay

1. 170 min

2. 64.5 h

3. $6.25 \times 10^{-6}$ g

4. 14 days

5. (a) 0.03125 ($= 1/2^5$)   (b) $9.3 \times 10^{-10}$ ($= 1/2^{30}$)

## Exercise 13.5 *Problems on activation energy

1. 29.0 kJ mol$^{-1}$

2. 25.5 kJ mol$^{-1}$

3. 13.6 kJ mol$^{-1}$

## Exercise 13.6 Problems on reaction kinetics

1. $2.4 \times 10^{-5}$ mol dm$^{-3}$ s$^{-1}$

2. (a) (i) 0   (ii) 1   (b) Rate $= k[HNO_3]$

3. $2.95 \times 10^{-6}$ mol dm$^{-3}$ s$^{-1}$

4. $1.36 \times 10^{-7}$ mol dm$^{-3}$ s$^{-1}$

5. (a) 1   (b) $7.48 \times 10^{-4}$ s$^{-1}$

6. (a) (i) 1   (ii) 3   (b) Rate $= k[P][Q]^3$
   $1.19 \times 10^{-3}$ M s$^{-1} = k \times 10^{-7} \times M^4 \longrightarrow k = 1.19 \times 10^4$ mol$^{-3}$ dm$^9$ s$^{-1}$

7. Time for $[A]_{Initial}$ to fall to $\frac{1}{2}[A]_{Initial}$ = Time for $\frac{1}{2}[A]_{Initial}$ to fall to $\frac{1}{4}[A]_{Initial}$ = 30 minutes.
   Thus half-life is independent of $[A]$: the reaction is first order.

8. (a) 0   (b) no change   (c) 1st order w.r.t. HCl

9. (b) (i) 1   (ii) 15 h   (iii)  4.8% h$^{-1}$
   (c) (i) 15 h   (ii) 19.2% h$^{-1}$

10. (a) 20 p, 27 n   (b)  $^{47}_{21}$Sc   (c) 0.25, no, too long

## Exercise 13.7 Questions from examination papers

1. (b) (i)  2: A plot of rate against concn$^2$ gives a straight line.
       (ii) Rate $= k[CH_3CHO(g)]^2$
       (iii) $k = 3.74 \times 10^{-2}$ mol$^{-1}$ dm$^3$ s$^{-1}$

2. (a) (i)  so that $[H_2O_2]$ and $[H^+]$ remain approximately constant.
       (ii) to avoid decomposition of $H_2O_2$ before the start of the reaction.
       (iii) to study the effect of change in one variable only, [KI]
   (b) (i)  The graph is linear: $\frac{1}{t} \propto$ [KI]   (ii) $n = 1$
   (c) (i)  $H^+$ takes part in a fast step, not the r.d.s.
       (ii) mol$^{-n}$ dm$^{3n}$ s$^{-1}$

3. (b) (i)  The time for concn to fall from 0.5 M to 0.25 M = time
       for concn to fall from 0.25 M to 0.125 M = $0.6 \times 10^4$ s,
       therefore the reaction is first order.
       (ii) Rate $= k[CH_3CO_2C_2H_5]$   (iii) $2.95 \times 10^{-5}$ mol dm$^{-3}$ s$^{-1}$

4. (a) (i)  Order of reaction is not known.
       (ii) $CH_3COCH_3$ 1, $I_2$ 0, $H^+$ 1
       (iii) $I_2$ takes part in a fast step, which is not rate-determining.
       (iv) Rate $= k[CH_3COCH_3][H^+]$
   (b) (i)  +38 kJ mol$^{-1}$
       (ii)

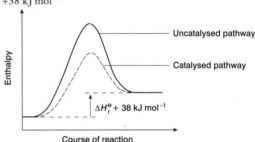

5. (b) (ii)  The graph of 1/time against concn shows that the
        reaction is not zero order (the rate is not independent of
        [HCl]) and not first order (the rate is not $\propto$ [HCl].
        (iii) The graph of 1/time against [HCl]$^2$ is linear therefore the
        reaction is second order.
        (iv) Rate $= k[HCl]^2$
        Rate = Gradient = 1.00 mol dm$^{-3}$/$8.33 \times 10^{-2}$ s$^{-1}$
                     = 12.00 mol dm$^{-3}$ s$^{-1}$
        $k = 12.00$ mol$^{-1}$ dm$^3$ s$^{-1}$

6. (a) 1 w.r.t. A, (from 1 and 3), 1 w.r.t. B (from 1 and 2), 2 w.r.t C
   (b) Rate $= k$ [A][B][C]$^2$
        $k = 0.10$ mol$^{-3}$ dm$^9$ s$^{-1}$, Rate $= 1.0 \times 10^{-5}$ mol dm$^{-3}$ s$^{-1}$
   (c) $k$ increases with a rise in temperature.

7. (a) The ratio (surface area/mass) increases.
   (b) (i)  2: $9.6 \times 10^{-4}$, 3: 0.010, 4: $8.1 \times 10^{-4}$, 5: 0.0346
       (ii) $k = 15.0$ mol$^{-1}$ dm$^3$ s$^{-1}$

8. (a) 1 w.r.t $O_2$ (from 1 and 4), 2 w.r.t. NO (from 1 and 2)
   (b) (i)  Rate $= k$ [$O_2$] [NO]$^2$   (ii) 0.7 mol$^{-1}$ dm$^3$ s$^{-1}$
   (c) $1.4 \times 10^{-4}$ mol dm$^{-3}$ s$^{-1}$

## Exercise 14.1 Problems on equilibria

1. (a) $K_p = \dfrac{p_C \times p_D}{p_A \times p_B}$   (b) $K_c = \dfrac{[C][D]}{[A][B]}$
   (c) $K_{Het} = \dfrac{p_D}{p_B}$   (d) $K_{Het} = \dfrac{[D]}{[A][B]}$

2. 144

3. 4.0

4. 64

5. 2.25

6. 36.5 kPa

7. $3.39 \times 10^{-15} \, N^{-2} \, m^4$

8. C

9. $9.26 \times 10^{-13} \, Pa^{-2}$

10. (a) $9.33 \times 10^{-4} \, N^{-1} \, m^2$   (b) $0.200 \, atm^{-1}$

11. (a) 31.25 atm   (b) 60.00 atm   (c) 3.23%
    (d) (i) $K_p = p(C_2H_5OH)/[p(C_2H_4) \times p(H_2O)]$
    (ii) $2.48 \times 10^{-3} \, atm^{-1}$

## Exercise 14.2 Questions from examination papers

1. (a) 0.523 mol gas, 0.103 mol methanol
   (b) (i) $K_c = 10.3 \, mol^{-2} \, dm^6$
   (c) (ii) $x(CH_3OH) = 0.198, x(CO) = 0.233, x(H_2) = 0.569$
   (iii) $p(H_2) = 0.905 \, MPa, p(CO) = 0.371 \, MPa,$
   $p(CH_3OH) = 0.315 \, MPa$
   $K_p = 1.04 \, MPa^{-2}$

2. (b) (i) $K_p = p(H_2O(g)) \times p(SO_3(g))/p(H_2SO_4(g))$
   (ii) I. $K_p$ at 400 K = 2.06 Pa, $K_p$ at 493 K = 172 Pa

   I. Endothermic: increases with rise in temperature
   II. No, the ratio $p(H_2O(g)) \times p(SO_3(g))/p(H_2SO_4(g))$
   = 0.200 Pa, which is less than $K_p$ = 2.06 Pa at 400 K.

   (c) (ii) tetrahedral   (iii) NaI
   (d) **A** = ethene, **B** = ethanol (by catalytic hydration) **C** = ethane (by catalytic hydrogenation)

3. (b) (i) Equilibrium moves towards the LHS, in the endothermic direction.
   (ii) No effect because there is no change in the number of moles of gas.
   (c) Equilibrium is reached faster. The position of equilibrium is unchanged.
   (d) (ii) HI 0.80 mol, $H_2$ 0.10 mol, $I_2$ 0.10 mol   (iii) 0.156
   (iv) dimensionless
   (v) Inits cancel in the expression for $K_c$.

4. (a) (i) $K_c = \dfrac{[NH_3]^2}{[N_2][H_2]^3}$

   (ii) $K_c = 0.183 \, mol^{-2} \, dm^6$
   (b) $[H^+] = 7.94 \times 10^{-12} \, mol \, dm^{-3}, [OH^-] = 1.26 \times 10^{-3} \, mol \, dm^{-3}$
   (c) (ii) 25.0 $cm^3$ ethanoic acid + 10.6 $cm^3$ NaOH(aq)

5. (a) $PCl_5$ 0.244 mol; $PCl_3$ 0.156 mol; $Cl_2$ 0.156 mol
   (b) (ii) 0.0108 $mol \, dm^{-3}$
   (c) (ii) $Cl_2$ 0.281   (iii) $PCl_5$ 109.7 kPa   (iv) $K_p$ = 44.9 kPa

6. (a) (i) Shorter: collision frequency increases   (ii) reduced: reaction from L to R involves a decrease in the amount of gas
   (b) $K_p = p(H_2(g)) \times p(CO(g))/p(H_2O(g))$
   (c) (i) $K_p$ = 9.92 kPa

7. (a) (i) I. $\frac{1}{2}$ II. The product appears in it.   III. $O_2$ does not appear in the rate equation. No.
   IV. There is a decrease in pressure as 3 volumes of gas form 2 volumes of gas therefore follow the reaction with a manometer attached to the reaction vessel. Initial rate $\propto [SO_2]^n$ where n = order w.r.t. $SO_2$.
   V. Rate decreases to $\frac{1}{3}$ of its valve.
   (ii) I. Rate(b) = $k_b K_c [SO_2][SO_3]^{-1/2}$

   II. No effect: both forward and back reactions are speeded up.

(b) intermolecular
(c) (i) I. reacts with both acids and bases
   II. e.g. $Al(OH)_3$
   III. $[Fe^{3+}]$ increases from pH7 to 2 as Fe reacts with acid. No increase from pH7 to 14.
   (ii) $Al_2Cl_6$ dissociates $\longrightarrow$ $2AlCl_3$

## Exercise 15.1 Problems on organic chemistry

1. $C_2H_5$   $C_4H_{10}$

2. 88 $g \, mol^{-1}$   $C_3H_7CO_2H$

3. $C_{11}H_{10}O$

4. $C_{16}H_{13}O_3N$

5. **A** is $C_6H_{12}$ = $CH_3CH_2CH=CHCH_2CH_3$   **B** is $C_2H_5CO_2H$

6. **A** is $C_6H_5CO_2CH_3$   **B** is $CH_3OH$   **C** is $C_6H_5CO_2Na$
   **D** is $C_6H_5CO_2H$

7. (a) $C_3H_7ON$   (b) propanamide $C_2H_5CONH_2$

8. (d) 3 mol $Br_2$: 1 mol $C_6H_5OH$
   $C_6H_5OH + 3Br_2 \longrightarrow C_6H_2Br_3OH + 3HBr$

9. (a) 55.9 $g \, mol^{-1}$   (b) $C_4H_8$
   but-1-ene $CH_3CH_2CH=CH_2$ and but-2-ene, $CH_3CH=CHCH_3$

10. (a) CHO   (b) $C_4H_4O_4$
    (c) butenedioic acid, *cis* $HCCO_2H$ and *trans* $HO_2CCH$
    $$\overset{\|}{HCCO_2H} \qquad \overset{\|}{HCCO_2H}$$

11. (a) $C_4H_{10}O$   (b) $C_4H_{10}O$   (c) $(CH_3)_3COH$
    $CH_3CH(CH_3)CH_2OH$   $CH_3CH_2CH_2CH_2OH$
    $CH_3CH_2CH(CH_3)OH$

12. **A** is $C_8H_9ON=C_6H_5NHCOCH_3$   **B** is $C_6H_7N = C_6H_5NH_2$

## Exercise 15.2 Questions from examination papers

1. (a) $C_2H_4O$. The peak below 3000 $cm^{-1}$ corresponds to an alkyl group, those at 1725 $cm^{-1}$ and 1250 $cm^{-1}$ to an ester group. The molecular formula $C_4H_8O_2$ corresponds to the esters $HCO_2CH(CH_3)_2$, $HCO_2CH_2CH_2CH_3$, $CH_3CO_2CH_2CH_3$ and $CH_3CH_2CO_2CH_3$.
   (b) **A** = $C_2H_5CO_2H$, **B** = ethane, **C** = $C_2H_5COCl$,
   **D** = $C_2H_5CH_2OH$, **E** = $C_2H_5CO_2CH_2C_2H_5$
   (c) 51%

2. (d) (i) Use a manometer to measure the pressure of nitrogen produced.
   (iii) The graph of ln rate against $1/T$ has gradient = $-12.3 \times 10^3 \, K = -E_A/8.3 \, J \, K^{-1}$
   $E_A = 100 \, kJ \, mol^{-1}$

3. (a) $C_4H_8O$
   (b) (i) $>C=O$   (ii) $CH_3CH_2CH_2CHO$, $CH_3CH_2COCH_3$, $(CH_3)_2CHCHO$
   (c) (i) **Z** = butan-2-one   (ii) $K_2Cr_2O_7$, acid
   (d) **Z** does not reduce Fehling's solution to $Cu_2O$
   (e) 57 = $C_2H_5CO^+$, 43 = $CH_2CO^+$, 29 = $C_2H_5^+$
   (f) intermolecular attractions between $C^{\delta+} = O^{\delta-}$ dipoles
   (g) (i) $CH_3CH_2C^*HOHCH_3$   (ii) optical   (iii) chiral C atom

4. (a) (ii) I. Rate = $1.0 \times 10^2 \, mol \, dm^{-3} \, s^{-1}$
   II. Concentration of $HNO_3$ decreases.
   (b) (i) 1   (ii) ketones
   (iii) **T** = $C_6H_5CO_2H$, **S** = $CH_3COC_2H_5$,
   **P** = $CH_3CHOHC_2H_5$, **Q** = $CH_3CHBrC_2H_5$,
   **R** = $C_2H_5CH(CH_3)OCOC_6H_5$
   (iv) I. KOH(aq)   II. $LiAlH_4$(ether)
   (v) $-COCH_3$ and $-CHOHCH_3$; **S** and **P**
   (vi) mirror images

5. (a) (i) $M_r = 118$   (ii) $C_2H_3O_2$, $C_4H_6O_4$ Amount of acid =
$\frac{1}{2} \times$ amount of NaOH therefore the acid is diprotic:
$HO_2CCH_2CH_2CO_2H$
  (b) two in the ratio 2 : 1 due to 4H in $CH_2$ groups and 2H in $CO_2H$
groups
  (c) (ii) 0.177 g

6. (a) $C_4H_4O$   (b) $C_4H_4O$ has $M_r = 70$
  (c) (i) $4C(s) + 2H_2(g) + \frac{1}{2}O_2(g) \longrightarrow C_4H_4O(l)$
   (ii) 4050 kJ mol$^{-1}$
  (d) +3940 kJ mol$^{-1}$   (e) 111 kJ mol$^{-1}$

7.

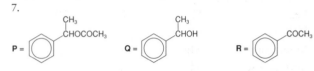

**R**: 120 = molecular ion, 105 = $C_6H_5CO^+$, 77 = $C_6H_5^+$,
43 = $COCH_3^+$, 51 = $C_4H_3^+$

## Exercise 16.1 Synoptic questions from examination papers

1. (a) (i) $K_c = [SO_3]^2/\{[SO_2]^2[O_2]\}$, unit mol$^{-1}$ dm$^3$
   (ii) The reaction is exothermic.   (ii) Increase pressure, use a
catalyst, remove the product as it is formed.
   (iv) pollutants
  (b) (i) decrease because 3 moles of gas form 2 moles of gas
   (ii) $\Delta S$ (surroundings) = $-\Delta H^{\ominus}/T$ = +280 J K$^{-1}$ mol$^{-1}$
   (iii) $\Delta S$(system) = $-280$ J K$^{-1}$ mol$^{-1}$
  (c) (i) pH = 0.82
  (ii)

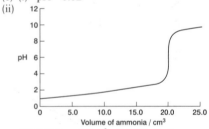

   (iii) 0.075 mol dm$^{-3}$   (iv) The indicator must change on the
acid side of pH 7.

2. (b) 0.20 mol $CO_2$. If 14.78 g $ZCO_3 \longrightarrow$ 0.2 mol $CO_2$,
$A_r(Z) = 13.9$. If 14.78 g $Z_2CO_3 \longrightarrow$ 0.20 mol $CO_2$, $A_r(Z) = 7$,
and Z is Li.
  (c) $Ca^{2+}$ gives a white ppt of $CaSO_4(s)$ but no ppt with NaOH(aq).
$Cu^{2+}$ oxidises $I^-$ to $I_2$. $MX_2$ is $CaI_2$.
$$2Cu^{2+}(aq) + 2I^-(aq) \longrightarrow 2Cu^+(aq) + I_2(aq)$$
  (d) $\Delta H_r = -2.8$ kJ for 0.1 mol

3. (a) (ii) $2MnO_4^-(aq) + 5H_2O_2(aq) + 6H^+(aq) \longrightarrow$
$2Mn^{2+}(aq) + 5O_2(g) + 8H_2O(l)$
   (iii) $[H_2O_2] = 3.80 \times 10^{-2}$ mol dm$^{-3}$   (iv) $-1$
  (b) $-140$ kJ mol$^-$
  (c) (i) $^{226}_{88}Ra \longrightarrow {}^4_2He + {}^{222}_{86}Rn$
   (iii) I. $1.106 \times 10^{-3}$ mol
    II. $3.319 \times 10^{-3}$ mol
    III. 74.3 cm$^3$

4. (a) $\Delta H_r = -45$ kJ mol$^{-1}$
  (b) (i) decreases yield because the reaction is exothermic
   (ii) increases yield because 2 moles of gas form 1 mole.
  (c) $4.21 \times 10^{-8}$ Pa$^{-1}$

(d) (i)

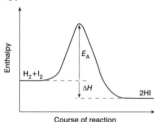

   (ii) phenolphthalein
   (iii) The equilibrium $CH_3CO_2^-(aq) + H^+(aq) \rightleftharpoons$
$CH_3CO_2H(aq)$ removes $H^+$ from the solution.

5. (a) $\Delta H_r^{\ominus} = -91$ kJ mol$^{-1}$ $\Delta S_r^{\ominus} = -89$ J K$^{-1}$ mol$^{-1}$
  (b) $\Delta G^{\ominus}$ is negative, $T = 1022$ K (This is the $T$ at which $\Delta G^{\ominus} = 0$)
  (c) (i) The reaction is exothermic.   (ii) to increase the rate at
which equilibrium is reached   (iii) recycling unreacted
gases

6. (b) (i) $X^{3+}$, $A_r(X) = 140 \Rightarrow$ X is Ce
   (ii) $Fe^{2+}(Cl_{12}H_8N_2)_3 \rightleftharpoons Fe^{3+}(C_{12}H_8N_2)_3 + e^-$
  (c) 1. $Fe^{3+}$ oxidises $I^-$ because $Fe^{3+}/Fe^{2+}$ has a more positive value
of $E^{\ominus}$ than $I_2/I^-$.
   2. $I_2$ oxidises $Fe(CN)_6^{4-}$ because $E^{\ominus}$ for $I_2/I^-$ is more
positive than $E^{\ominus}$ for $Fe(CN)_6^{3-}/Fe(CN)_6^{4-}$. In oxidising
power, $Fe(H_2O)_6^{3+} > I_2 > Fe(CN)_6^{3-}$

7. (a) $\Delta H^{\ominus} = -198$ kJ mol$^{-1}$, $\Delta S^{\ominus} = -186$ J K$^{-1}$ mol$^{-1}$,
$\Delta G^{\ominus} = -143$ kJ mol$^{-1}$
  (b) $\Delta G^{\ominus}$ is negative. $T = 1065$ K
  (c) $K_p = p(SO_3)^2/\{p(SO_2)^2 \times p(O_2)\} = 9.07 \times 10^{-3}$ kPa$^{-1}$

8. (a) (i) Rate = $k [H_2][I_2]$
   (ii) $k = 0.0642$ mol dm$^{-3}$ s$^{-1}$   (iii) $E_A = 164$ kJ mol$^{-1}$
   (iv) Enthalpy

  (b) (i) $\Delta S^{\ominus} = 20$ J K$^{-1}$ mol$^{-1}$
   (ii) $\Delta G^{\ominus} = \Delta H^{\ominus} - T\Delta S^{\ominus}$. $\Delta H^{\ominus}$ and $-T\Delta S^{\ominus}$ are negative at
all temperatures, therefore $\Delta G^{\ominus}$ is negative at all
temperatures and the reacation is feasible at all
temperatures. $\Delta G^{\ominus}$ becomes more negative as the
temperature increases.

9. (c) (i) 1. N 11.38%;
    2. O 26.15%;  C 58.4%;  H 4.07%. $C_6H_5NO_2$
   (ii) $C_6H_5^+$ suggests that **T** is an aromatic compound
$C_6H_5NO_2 \longrightarrow C_6H_5^+ + NO_2^+ + 2e^-$

10. (a)(ii) $-860$ kJ mol$^{-1}$   (iii) $-2088$ kJ mol$^{-1}$
$\Delta H^{\ominus}_{solution} = \Delta H^{\ominus}_{lattice\ dissociation} + \Delta H^{\ominus}_{hydration}$
$= +2018$ kJ mol$^{-1}$ $- 2088$ kJ mol$^{-1}$ = $-70$ kJ mol$^{-1}$;
dissolution is exothermic
  (d) (i) N   (ii) pH = 9
   (iii) **X** shows peaks at N—H (str) and N—H(bend); **Y** shows
a peak at O—H (str).

11. (a)(i) 1   (ii) 5   (iii) 2
  (b)Ethanoic acid is incompletely dissociated.
  (c)(i) $C_3H_6O$   (ii) 1. C=O group,  2. not an aldehyde;
$CH_3COCH_3$

# Table of Relative Atomic Masses

| Element | Symbol | Atomic number | Relative atomic mass | Element | Symbol | Atomic number | Relative atomic mass |
|---------|--------|---------------|----------------------|---------|--------|---------------|----------------------|
| Aluminium | Al | 13 | 27.0 | Lead | Pb | 82 | 207 |
| Antimony | Sb | 51 | 122 | Lithium | Li | 3 | 6.94 |
| Argon | Ar | 18 | 40.0 | Magnesium | Mg | 12 | 24.3 |
| Arsenic | As | 33 | 75.0 | Manganese | Mn | 25 | 54.9 |
| Barium | Ba | 56 | 137 | Mercury | Hg | 80 | 201 |
| Beryllium | Be | 4 | 9.0 | Neon | Ne | 10 | 20.2 |
| Bismuth | Bi | 83 | 209 | Nickel | Ni | 28 | 58.7 |
| Boron | B | 5 | 10.8 | Nitrogen | N | 7 | 14.0 |
| Bromine | Br | 35 | 80.0 | Oxygen | O | 8 | 16.0 |
| Cadmium | Cd | 48 | 112.5 | Phosphorus | P | 15 | 31.0 |
| Calcium | Ca | 20 | 40.1 | Platinum | Pt | 78 | 195 |
| Carbon | C | 6 | 12.0 | Potassium | K | 19 | 39.1 |
| Chlorine | Cl | 17 | 35.5 | Selenium | Se | 34 | 79.0 |
| Chromium | Cr | 24 | 52.0 | Silicon | Si | 14 | 28.1 |
| Cobalt | Co | 27 | 59.0 | Silver | Ag | 47 | 108 |
| Copper | Cu | 29 | 63.5 | Sodium | Na | 11 | 23.0 |
| Fluorine | F | 9 | 19.0 | Strontium | Sr | 38 | 87.6 |
| Germanium | Ge | 32 | 72.5 | Sulphur | S | 16 | 32.1 |
| Gold | Au | 79 | 197 | Tin | Sn | 50 | 119 |
| Helium | He | 2 | 4.00 | Titanium | Ti | 22 | 47.9 |
| Hydrogen | H | 1 | 1.01 | Vanadium | V | 23 | 50.9 |
| Iodine | I | 53 | 127 | Xenon | Xe | 54 | 131 |
| Iron | Fe | 26 | 55.8 | Zinc | Zn | 30 | 65.4 |
| Krypton | Kr | 36 | 83.8 | | | | |

# Periodic Table of the Elements

Atomic number | 11    23.0   Relative atomic mass
**Na**
Sodium

| 1   1.000 **H** Hydrogen | | | | | | | | | | | | | | | | | 2   4.00 **He** Helium |

| 3   6.90 **Li** Lithium | 4   9.00 **Be** Beryllium | | | | | | | | | | | 5   10.8 **B** Boron | 6   12.0 **C** Carbon | 7   14.0 **N** Nitrogen | 8   16.0 **O** Oxygen | 9   19.0 **F** Fluorine | 10   20.2 **Ne** Neon |

| 11   23.0 **Na** Sodium | 12   24.3 **Mg** Magnesium | | | | | | | | | | | 13   27.0 **Al** Aluminium | 14   28.1 **Si** Silicon | 15   31.0 **P** Phosphorus | 16   32.1 **S** Sulphur | 17   35.5 **Cl** Chlorine | 18   39.9 **Ar** Argon |

| 19   39.1 **K** Potassium | 20   40.1 **Ca** Calcium | | | | | | | | | | | 31   69.7 **Ga** Gallium | 32   72.6 **Ge** Germanium | 33   74.9 **As** Arsenic | 34   79.0 **Se** Selenium | 35   79.9 **Br** Bromine | 36   83.8 **Kr** Krypton |

| 37   85.5 **Rb** Rubidium | 38   87.6 **Sr** Strontium | | | | | | | | | | | 49   115 **In** Indium | 50   119 **Sn** Tin | 51   122 **Sb** Antimony | 52   128 **Te** Tellurium | 53   127 **I** Iodine | 54   131 **Xe** Xenon |

| 55   133 **Cs** Caesium | 56   137 **Ba** Barium | | | | | | | | | | | 81   204 **Tl** Thallium | 82   207 **Pb** Lead | 83   209 **Bi** Bismuth | 84   209 **Po** Polonium | 85   210 **At** Astatine | 86   222 **Rn** Radon |

Transition elements:

| 21   45.0 **Sc** Scandium | 22   47.9 **Ti** Titanium | 23   50.9 **V** Vanadium | 24   52.0 **Cr** Chromium | 25   54.9 **Mn** Manganese | 26   55.8 **Fe** Iron | 27   58.9 **Co** Cobalt | 28   58.7 **Ni** Nickel | 29   63.5 **Cu** Copper | 30   65.4 **Zn** Zinc |
|---|---|---|---|---|---|---|---|---|---|
| 39   88.9 **Y** Yttrium | 40   91.2 **Zr** Zirconium | 41   92.9 **Nb** Niobium | 42   95.9 **Mo** Molybdenum | 43   99 **Tc** Technetium | 44   101 **Ru** Ruthenium | 45   103 **Rh** Rhodium | 46   106 **Pd** Palladium | 47   108 **Ag** Silver | 48   112 **Cd** Cadmium |
| 57   139 **La** Lanthanum | 72   178 **Hf** Hafnium | 73   181 **Ta** Tantalum | 74   184 **W** Tungsten | 75   186 **Re** Rhenium | 76   190 **Os** Osmium | 77   192 **Ir** Iridium | 78   195 **Pt** Platinum | 79   197 **Au** Gold | 80   201 **Hg** Mercury |

# Appendix: Units

There are two sets of units currently employed in scientific work. One is the CGS system, based on the centimetre, gram and second. The other is the Système Internationale (SI) which is based on the metre, kilogram, second and ampere. SI units were introduced in 1960. The Association for Science Education booklet *Signs, Symbols and Systematics* (ASE, 2000) gives a full account of this system.

Listed below are the SI units for the seven fundamental physical quantities on which the system is based and also a number of derived quantities and their units.

Chemists are still using some of the CGS units. You will find mass in g; volume in $cm^3$ and $dm^3$; concentrations in mol $dm^{-3}$ or mol $litre^{-1}$. Pressure is sometimes given in mm mercury and temperatures in °C.

## Basic SI Units

| Physical Quantity | Name of Unit | Symbol |
|---|---|---|
| Length | metre | m |
| Mass | kilogram | kg |
| Time | second | s |
| Electric current | ampere | A |
| Temperature | kelvin | K |
| Amount of substance | mole | mol |
| Light intensity | candela | cd |

## Derived SI Units

| Physical Quantity | Name of Unit | Symbol | Definition |
|---|---|---|---|
| Energy | joule | J | $kg\ m^2\ s^{-2}$ |
| Force | newton | N | $J\ m^{-1}$ |
| Electric charge | coulomb | C | A s |
| Electric potential difference | volt | V | $J\ A^{-1}\ s^{-1}$ |
| Electric resistance | ohm | $\Omega$ | $V\ A^{-1}$ |
| Area | square metre | | $m^2$ |
| Volume | cubic metre | | $m^3$ |
| Density | kilogram per cubic metre | | $kg\ m^{-3}$ |
| Pressure | newton per square metre or pascal | | $N\ m^{-2}$ or Pa |
| Molar mass | kilogram per mole | | $kg\ mol^{-1}$ |

With all these units, the following prefixes (and others) may be used:

| Prefix | Symbol | Meaning |
|--------|--------|---------|
| deci | d | $10^{-1}$ |
| centi | c | $10^{-2}$ |
| milli | m | $10^{-3}$ |
| micro | $\mu$ | $10^{-6}$ |
| nano | n | $10^{-9}$ |
| kilo | k | $10^{3}$ |
| mega | M | $10^{6}$ |
| giga | G | $10^{9}$ |
| tera | T | $10^{12}$ |

It is very important when putting values for physical quantities into an equation to be consistent in the use of units. If you are, then the units can be treated as factors in the same way as numbers. Suppose you are asked to calculate the volume occupied by 0.0110 kg of carbon dioxide at 27 °C and a pressure of $9.80 \times 10^4$ N m$^{-2}$. You know that the gas constant is 8.31 J mol$^{-1}$ K$^{-1}$ and that the molar mass of carbon dioxide is 44.0 g mol$^{-1}$. Use the ideal gas equation:

$$PV = nRT$$

The pressure $\quad P = 9.80 \times 10^4 \text{ N m}^{-2}$

The constant $\quad R = 8.31 \text{ J K}^{-1} \text{ mol}^{-1}$

The temperature $\quad T = 27 + 273 = 300 \text{ K}$

The amount in moles

$$n = \text{Mass/Molar mass}$$
$$= 0.0110 \text{ kg}/(44.0 \times 10^{-3} \text{ kg mol}^{-1})$$
$$= 0.250 \text{ mol}$$

Then
$$V = \frac{0.250 \text{ mol} \times 8.31 \text{ J K}^{-1} \text{mol}^{-1} \times 300 \text{ K}}{9.80 \times 10^4 \text{ N m}^{-2}}$$
$$= 6.34 \times 10^{-3} \text{ J N}^{-1} \text{ m}^2$$

Since
$$\text{J} = \text{N m} \quad (1 \text{ joule} = 1 \text{ newton metre})$$
$$V = 6.34 \times 10^{-3} \text{ N m N}^{-1} \text{ m}^2$$
$$= 6.34 \times 10^{-3} \text{ m}^3$$

Volume has the unit of cubic metre. This calculation illustrates what people mean when they say that SI units form a *coherent system of units*. You can convert from one unit to another by multiplication and division, without introducing any numerical factors.

# Index